PLANNING FOR A BETTER URBAN LIVING ENVIRONMENT IN ASIA

Planning for a Better Urban Living Environment in Asia

Edited by

ANTHONY GAR-ON YEH and MEE KAM NG
Centre of Urban Planning and Environmental Management,
University of Hong Kong

Routledge
Taylor & Francis Group

LONDON AND NEW YORK

First published 2000 by Ashgate Publishing

Reissued 2018 by Routledge
2 Park Square, Milton Park, Abingdon, Oxon OX14 4RN
711 Third Avenue, New York, NY 10017, USA

Routledge is an imprint of the Taylor & Francis Group, an informa business

A Library of Congress record exists under LC control number: 00134824

ISBN 13: 978-1-138-72896-7 (hbk)
ISBN 13: 978-1-315-19019-8 (ebk)

Contents

List of Figures

List of Tables

List of Contributors

A.T.M. Nurul **Amin**   Division of Human Settlements Development, Asian Institute of Technology, Bangok, Thailand.

Jamal H. **Ansari**   Professor, Department of Physical Planning, School of Planning and Architecture, New Delhi, India.

Ray W. **Archer**   Professor, Division of Human Settlements Development, Asian Institute of Technology, Bangkok, Thailand.

Che Musa **Che Omar**   Chairman, Urban and Regional Planning Programme, School of Housing, Building and Planning, University Sains Malaysia, Penang, Malaysia.

Hon Shyan **Chong**   Division of Global Environment Engineering, Graduate School of Engineering, Kyoto University, Kyoto, Japan.

Chong-Won **Chu**   Professor, Department of Urban Engineering, College of Engineering, Seoul National University, Republic of Korea.

Takatsune **Fukami**   Professor, Department of Social Engineering, Tokyo Institute of Technology, Tokyo, Japan.

Takahiro **Hisa**   Department of Environmental Engineering, Faculty of Engineering, Osaka University, Osaka, Japan.

Baozhe **Hu**   Professor, Department of Urban Engineering,
                                 University of Tokyo, Tokyo, Japan.

Ikaputra   Department of Environmental Engineering,
                                 Faculty of Engineering, Osaka University,
                                 Osaka, Japan.

Mir Shahidul **Islam**   Professor, Department of Urban and Regional
                                 Planning, Bangladesh University, Bangladesh.

Kunihiro **Narumi**   Professor, Department of Environmental
                                 Engineering, Faculty of Engineering, Osaka
                                 University, Osaka, Japan.

Mee Kam **Ng**   Associate Professor, Centre of Urban Planning
                                 and Environmental Management, University of
                                 Hong Kong, Hong Kong.

Yukio **Nishimura**   Research Centre for Advanced Science and
                                 Technology, University of Tokyo, Tokyo, Japan.

Zhaoyu **Qian**   Professor, Department of Urban Planning, Tongji
                                 University, Shanghai, People's Republic of
                                 China.

Mahbubur **Rahman**   Professor, Department of Architecture,
                                 Engineering University, Bangladesh.

Asteya M. **Santiago**   Professor, School of Urban and Regional
                                 Planning, University of the Philippines, Manila,
                                 Philippines.

Hidehiko **Sazanami**   Ex-director, United Nations Centre for Regional
                                 Development, Nagoya, Japan

Naoki **Tahara**   Department of Geotechnical and Environmental
                                 Engineering, Nagoya University, Nagoya, Japan.

Mamoru **Tohiguchi**   Professor, Division of Global Environment Engineering, Kyoto University, Kyoto, Japan.

Xiangrong **Wang**   Department of Environmental Science, East China Normal University, Shanghai, People's Republic of China.

Sadao **Watanabe**   Professor, Department of Urban Engineering, University of Tokyo, Tokyo, Japan.

Chung-Tong **Wu**   Dean, Faculty of the Built Environment, University of New South Wales.

Anthony Gar-On **Yeh**   Assistant Director and Professor, Centre of Urban Planning and Environmental Management, University of Hong Kong, Hong Kong.

Wahyu Subroto T. **Yoyok** Department of Environmental Engineering, Faculty of Engineering, Osaka University, Osaka, Japan.

1 Introduction: Planning for a Better Urban Living Environment in Asia

ANTHONY GAR-ON YEH AND MEE KAM NG

Rapid economic growth and increasing population in Asia, where over 40 per cent of the world's urban population is living, have led to many problems in different urban sectors: environmental degradation, housing shortages, traffic congestion, vehicular and pedestrian conflicts, problems in land management, etc. Squatters, high density development and urban sprawl are some typical spatial manifestations of these problems. Most Asian countries also face the coexistence of a large informal economy fuelled by rural-urban migration, and a restructuring formal economy as a result of the growing interdependency of the global economy. Urban planners in Asia, therefore, are given the challenging tasks of meeting rapidly evolving economic needs and preserving cultural heritage unique to Asia. However, the institutional setup often fails to meet these challenges. While an effective and efficient planning system is indispensable in dealing with all these issues, coordination of the central and local governments, non-government organisations, private business sectors and the public in the planning process is generally poor.

Not only are these issues at the top of the agenda for policy makers, research and practising planners, they also pose vexing problems to planning educators in Asia.

The central thesis of all the chapters in this volume, outcomes of the Second International Congress of the Asian Planning School, is to investigate urban development issues and various responses to plan for a better urban living environment in Asia. Chapters are grouped together according to various themes: urban and regional planning, environment, housing, redevelopment and conservation, and planning education.

Sazanami in chapter 2 provides an overview of the challenges and prospects of large cities in Asia. He argues that increasing industrialisation and urbanisation have led to many interrelated sectoral and spatial problems which can only be resolved by innovative planning and implementation mechanisms. The problems outlined by Sazanami are substantiated by the papers on urban and regional planning. In chapter 3, the origin and effects of the existence of a huge informal sector in Asian megacities are explored. Amin contends that with the existence of a huge labour surplus in the agricultural sector, rural-urban migration is inevitable which in turn will lead to income disparities and degradation of the urban environment. To better manage the urban economy and environment, it is necessary for the urban planning system to accommodate the informal sector.

Che Omar in chapter 4 further substantiates the general problems discussed in chapter 2 through a case study of urban development and planning practice in Malaysia. The paper argues that planners cannot handle urban problems properly unless the economic restructuring and related processes are thoroughly understood. Similarly, Ng in chapter 5 outlines the economic restructuring processes in the small territory of Hong Kong and its immediate hinterland, the Pearl River Delta in the People's Republic of China. She argues that the economic integration of the two places in the context of their political reunification in 1997 has exposed two root problems of planning in the region: the lack of coordination between the local and cross-boundary authorities in physical and infrastructure planning, and a bias towards the interests of the local bureaucrats and private sectors in the development process. Unless tackled properly, these problems will be stumbling blocks for continuous regional development.

Megacities in Asia are characterised by high population densities. To promote pedestrians' enjoyment of urban life, Hu, Nishimura and Watanabe (chapter 6) examine through simulation modelling how pedestrian movements can be facilitated by manipulating the built environment. While they focus on pedestrian movement, Yeh (chapter 7) discusses how urban planning and management measures can help create a better high density environment. Using the example of Hong Kong, he shows that high density if better planned and managed can be interesting and pleasant.

To improve the urban environment in a modern city, Qian and Wang suggest the adoption of an ecological plan. In chapter 8, they use Huizhou, Guangdong Province, China, as a case study and put forward an ecological plan which focuses on the management of the urban living environment according to ecological principles.

Islam in chapter 9 outlines all the environmental problems in the densely populated areas of Bangladesh. He argues that better planning and management are essential to overcome these environmental problems.

Housing issues also assume great importance amidst rapid economic and urban growth in Asia. Chu (chapter 10) evaluates the Five-Year Two Million Housing Unit Construction Plan of South Korea (1988–92) which was formulated to relieve housing problems. It shows how the plan has affected the housing market and the macroeconomy in South Korea. He also discusses future housing policies in the country. Rahman, instead of reviewing government policies, discusses the concept of attainability to solve the urban housing crisis in Bangladesh. He argues in chapter 11 that there are ways to extend the affordability of housing to lower-middle income households without increasing the level of expenditure. Through a survey in Dhaka, he proves that attainable housing schemes when run by housing agencies and financial institutions, can help solve housing problems in Bangladesh.

Housing is not just a shelter. It also embodies many cultural and socio-economic characteristics unique to a country. Shophouses, as argued by Tohiguchi and Chong in chapter 12, not only facilitate the traditional Asian residential style of work-residence proximity, the architecture is also an excellent historical footprint of various races and culture. The chapter tries to analyse the formation, distribution, composition and conditions of shophouses and their significance in the design of the townscape. Chapter 13 is another sectoral study which illustrates the importance of culture in utilising space internal to a house. Narumi, Tahara, Subroto and Yoyok, in a survey done in 1991 to 1992, found that when people in Java, Indonesia, renovated their houses, they modified them according to a pattern of 'bipolar interior space'. According to this principle, contact between the 'two-worlds', the private and the public realms, takes place in what local people called a *ruang tamu,* which is located at the front portion of the house and is demarcated from the interior. This principle also leads to the existence of the kitchen at the 'back' to protect their private life from the public world. They argue that this internal arrangement of space conforms to the hierarchy of Javanese social intercourse.

To satisfy housing and economic needs, sometimes urban redevelopment is necessary. Archer discusses the land pooling/readjustment technique (LP/R) in urban redevelopment with landowner participation in chapter 14. The LP/R technique is used to manage the unified design, servicing and subdivision of separately-owned land parcels for urban development. Through two case studies in Japan and Taiwan respectively, he shows some varieties of LP/R techniques, problems encountered in the redevelopment process and planning lessons that can be learnt from the experiences.

Through generalising and examining various redevelopment projects in mixed commercial and industrial areas, Fukami outlines Japan's urban renewal system in chapter 15. In Japan, urban renewal generally refers to rebuilding wooden houses into fire-proof structures together with the provision of roads and social amenities through the 'conversion of rights and buildings before and after the project'. However, he argues that this tool may no longer be effective in renewing the high-density urban core as Japan enters into a post-industrial age and a new settlement pattern featured by sub-urbanisation.

Instead of redeveloping the built environment, Hisa, Ikaputra and Narumi (chapter 16) illustrate how the noble's residence in Yogyakarta, Indonesia, has been transformed from a house of an aristocratic family to a residential community meeting dynamic social needs of the evolving economy keeping intact its great traditional Javanese architecture.

The last three chapters of the book focus on planning education in Asia. Ansari in chapter 17 believes that planning education in India should go beyond its engineering, architectural and scientific emphases to meet the challenges imposed by the changing urban situations. She argues that as urban institutions are ill-equipped to tackle the urban problems and the government has adopted urban reforms to enlarge the private sector's role in urban management, planning education needs to be reformed to cope with the development.

Santiago in chapter 18 shares the experience of the School of Urban and Regional Planning at the University of the Philippines, in promoting the country's planning environment and training professional planners for the government and private sectors. She shows how the School has evolved under the radically divergent political and socio-economic environment both within and outside the University. The paper also projects the School's future directions and opportunities.

Wu asks the question 'whither Asian planning education?' in chapter 19. The paper reviews the challenges facing Asian planning educators and raises questions about the type of planning education that might be equal to the tasks ahead. He argues that the economic context and issues confronting Asian planners are unique. It is therefore important for Asian educators to develop planning programmes that are appropriate to the current conditions of Asia.

2 Challenges and Future Prospects of Planning for a Better Living Environment in the Large Cities in Asia

HIDEHIKO SAZANAMI

Introduction

Asia has the majority of the world's population, vast lands that range from tropical to arctic, and numerous large cities. Its political systems, economies and cultures also reveal a great deal of diversity.

As we all know, nowadays the role of major urban regions in Asia is becoming more and more crucial. A country's wealth, human resources, government authorities, educational and cultural institutions, and economic functions are all concentrated within large cities. Furthermore, taxes collected from these regions account for a major part of the national revenue, and the economic activities there contribute significantly to the growth of a nation's GNP. However, these urban areas also have problems related to overpopulation, such as insufficient public institutions, decreasing public services, overcrowded residential areas, and traffic congestion. Despite these problems, people continue to move to large cities in order to leave the poverty of the countryside and to search for job opportunities. We must therefore focus on the coexistence of both positive and negative aspects in large cities. Such dualities are common in Asia: slums and empty lots exist next to modern buildings and historical sites, and automobiles travel next to carts pulled by cattle. The present challenge lies in the removal of these 'paradoxes' from large urban areas and minimising their negative aspects.

Toward the Development of a New Urban Society

Past Regional Development

The NIEs, ASEAN, South Asian, and Pacific island countries as well as other developing countries (including China) have each approached regional development based on their different economic and political structures. After the Second World War, two methods have been used consecutively by the developing countries of Asia for regional development: the regional resource development method and the comprehensive regional development method. The former was carried out by many countries which had won independence from colonialism and hoped to consolidate the national economy. In order to industrialise and modernise, they emphasised resource development to increase productivity and to stimulate import substitution industries. The establishment of productive infrastructure, mainly carried out in the 1960s, was aimed at transforming the social structure from traditional to modern. By this transformation these nations sought to improve the coordination of various branches of the government and the economy. However, there was often a severe shortage of human and material resources, and due to the absence of a developed market economy, these countries had to first rely heavily on national economic development policies made by the central governments.

Internationally, the rush to establish import-substitution industries led to economic dependence; domestically, problems such as the polarisation of traditional and modern industries, the stagnation of farming, and unemployment emerged. The result was the opposite of the goal of consolidation; divisions were created between the industrial/urban and the rural, the modern and the traditional, and the rich and the poor.

Based on the lessons of the 1960s resource development method, a second method, the comprehensive regional planning method, was formulated in the 1970s. This approach was characterised by the emphases on the transformation of organisational structures through market mechanisms, as well as the attainment of equality in the process of increasing productivity. Obtaining modern production materials and technologies, the establishment of various economic systems, and the organisation of local residents were necessary in order to carry out these fundamental changes. This meant that the government had to become involved even more deeply than before. Under this system, the functions of supply and distribution related to development measures, as well as the organisations which received them, were strengthened. However, these rapidly changing strategies stunted the efficiency of the programmes, and the changes seemed to profit only a small percentage of the population, which widened the disparities between the 'haves' and 'have nots'.

Overall, the regional resource development method aimed mainly to increase material production and economic growth, whereas the comprehensive regional development method sought to achieve social equality in addition to growth. Now, the NIEs are faced with problems that have arisen with rapid development, such as the destruction of natural resources, environmental pollution and insufficient services for the aging and the disabled. Accordingly, welfare and environmental strategies are now becoming a major part of their development efforts.

In the future, developing countries will have to choose their development methods based on the experiences of the developed nations, with optimal consideration on social equality and welfare. Such considerations are especially crucial in the case of urban development, which may lead to serious economic and social impacts. For this purpose, development planning must account for the costs of preventing the destruction of natural resources and pollution, and for the costs of protecting the underprivileged in the urban areas. In other words, it must take into account the costs and effects of development and its social benefits.

Urban Planning from a Social Perspective

It is essential for the Asian developing countries to recognise the realities and problems of large cities, as outlined above, and respond to them with flexibility and with concrete approaches. It should be noted that local communities, which usually depend on self-sufficiency and internal cooperation, have all too often been transformed and destroyed by modernisation. Modernisation, which has been regarded as the goal of development, often succeeds in the distribution of all goods through market mechanisms. However, it also implies shifting factors such as land, labour and trust into the market sphere, which had not been a part of this circulation in the past. Modernisation dissipates traditional mechanisms of local communities which regard family as the fundamental unit, causing problems outlined in previous paragraphs and destroying the social structures of local communities.

In order to tackle these problems, we must first analyse how the public, the community, and the private sectors are established and relate to each other in Asian cities, and then formulate ways in which people's abilities and energies can become productive in a new social system. We must also rethink the future of the market mechanism which the development of large cities rely upon. After investigating these issues, we are then able to link urban and social development strategies and make appropriate adjustments.

Economic Growth and Urban Society

As we witness the economic growth of the various Asian countries after the second world war, we can see that those countries that changed their strategies toward export-oriented industrialisation and encouraged agricultural development under the market mechanism have experienced far greater degrees of growth than those countries which fostered import-substitution productions and regulated agricultural products. The successful nations were not heavily affected by the oil crises of the 1970s and the recession. They managed to increase their output, especially in the area of agricultural production. For example, Thailand is now an exporter of rice, and self-sufficiency in food has been generally achieved in other Asian nations as well.

However, if the United States and the European Community lean toward the conservative side in their trade policies and economic strategies in the future, there is a danger that Asian nations can no longer attain past levels of economic growth. Furthermore, due to the limited amount of funds available in the developing countries, an increase in the share of production responsibilities, which had been the task of developed nations, may lead to further structural divisions and disparities. Since the process of economic development involves not only central or local government organisations, the local communities which augment and support these organisations must also be considered. The role of local communities, which provide a forum for smoothing out social discrepancies and securing the role of family in development, is quite significant.

Local communities are not merely composed of traditional communities. They are a multilayered structure of local government organisations, communities, families and individuals, where the public, the community, and the private sectors interact in a complex manner. Each local community will reflect the unique political, economic, social and cultural situation of the area. There is thus a need to build experience and conduct research in each area.

Management and Policies Concerning Large Cities

Issues in Development

The population of Asia is expected to grow continually at an urbanisation rate of 80 per cent, which is comparable to that of the developed nations, and is anticipated between 2020 and 2050. The majority of large cities in the world will be concentrated in Asia. The beginning of the twenty-first century will signal a new era for urban areas, thus it is vital to develop a long-term urban planning policy that will focus on the next 30 years.

The rapid changes that are taking place in the international environment are shaping the problems of large cities in Asia, such as the flow of capital due to overseas direct investment and the flow of labour due to income disparities. The environmental problems of developing nations which are caused by economic development and industrialisation require immediate attention. Policies for the large cities of Asia must be able to foster autonomy at the local level, improve the standard of living, and meet the challenge of dynamic changes in both domestic and international arenas. Such policies must be practical and flexible.

The two goals of development are: 1) comprehensive economic growth of a region and the nurturing of large cities' potential for development; and 2) through this economic growth, to adjust disparities within the region and to alleviate poverty. Comprehensive economic growth comes only from initiatives such as the stimulation of primary, secondary and tertiary production, the upgrading and expansion of infrastructure, and the restructuring of government organisations which are necessary for these changes. It is especially important to upgrade infrastructures, such as communications, roads, railways, ports, and airports. Countries must actively seek to accomplish these tasks by acquiring aid from overseas. Among ASEAN nations, such as Thailand and Malaysia, new ventures such as the construction of distribution centres and industrial estates, are contributing greatly to the creation of new employment.

Measures that directly aid the poor include 'sites-and-services' projects, relief projects for the unemployed, such as cleaning the roads, and medical services are all currently offered. Also, the efforts to increase employment in the informal sector are particularly notable. This move is very important when public aid is insufficient.

Creating Structures for Management

As large-scale urbanisation in the developing countries of Asia progresses rapidly, the lack of adequate scientific, technological, and social capital foundations will inevitably lead to a deepening of problems such as the expansion of the informal sector, traffic congestion, and housing shortages. The driving force behind the transformation of large cities is economic growth; therefore, how to activate the market mechanism which is crucially linked to this growth will become a key factor. From this standpoint, there are three possibilities for urban policy in these countries: 1) a *laissez-faire* approach, with the market mechanism as its central player; 2) a mixed approach, which combines *laissez-faire* and regulated development by public organisations; and 3) an approach based mainly on public organisations. The first approach is being followed by many countries, but it is dangerous for those countries whose private sectors are not fully developed to rely solely on *laissez-faire* policies. As for the second approach, one possibility is a combination of the public sector and the market mechanism through the 'sites-and-services' method, on the basis of a land subdivision system. This would include the provision of infrastructure and development control measures. Examples of policies under the third approach include large-scale construction of housing, rent support and low-interest, long-term financing for housing, securing housing sites, and the removal of substandard housing and unlawful residents.

As outlined above, there is a call for a flexible approach which is based upon the market mechanism and development control by public organisations. As the economy grows, the middle class prospers along with the increase in private capital investment in housing. In addition, as industrialisation progresses, the number of medium-sized and small businesses will gradually rise, and the role of private investment will expand within city institutions and functions. In large cities in Asia, various businesses participate in both the public and private sectors, and their influence will be crucial in the future. For instance, in the case of urban transportation, although public transportation is certainly important, policies must consider systems such as para-transit in the informal sector, in which utilisation should be maximised.

The Role of City Administration

There are many administrative organisations for the management of urban regions in various Asian nations. In order to improve the efficiency of these organisations and their services, it is necessary to re-examine their sizes so that optimum results can be achieved.

Under a democratic system, it is vital that city administration is closely in touch with residents. If the administrative organisation becomes too large, residents remain unaware of its actions and, consequently, participation becomes difficult. If an administration seeks to remain close to the local community and successfully carry out public services that fulfil the needs of the community, then its size must be kept manageable.

The problem of funds is a major issue for urban development projects in developing countries. How to obtain financing for improving the quality and quantity of infrastructure and how to obtain results efficiently from invested capital are pressing questions. The collection of fees from those who directly receive benefits from services and the adjustment of utility fees including gas, water, and electricity would be starting points for solving these problems. Tax collection is also a vital source of income for local administrative organisations, therefore its efficiency should be re-examined.

The Importance of Integrating the Vitality of Private Groups

Since the mid-1970s, developed nations have sought to integrate the contributions of private organisations in order to efficiently carry out city administration and large-scale urban development while solving the chronic problem of insufficient financing. In the 1980s, numerous major urban development projects were the results of joint public-private efforts. Since the late 1980s, developing countries, which suffered from even more serious financial difficulties, have adopted this method.

First, public organisations who seek to incorporate the private sector should take steps, such as deregulation, to facilitate the acquisition of land and development, and should insure that private developers do not reap excessive profits. Since public organisations cannot be expected to provide substantial funding, possible alternatives for funding include low-interest public financing, foreign capital, financing by private individuals, and cooperative financing can be considered. Private housing construction has already relied heavily on small-scale private sources of funding. Therefore, establishing a system of volunteers and aiding the activities of local communities are necessary.

Private construction efforts can be divided into two major types. One emerged as efforts by the users themselves and their local communities, which grew into the small-scale occupational production organisations of today. This type is a local production system which has the traditional methods at its base. The other type includes more formal housing industries which have flourished since the 1970s and consist of modern building corporations and developers. In the past, the first type was the most common, especially in housing, but currently both types exist in the developing nations of Asia and continue to influence each other.

Despite the fact that traditional production systems have many positive aspects, overall, they are struggling against modern production systems after all. The latter, being in the private sector, have connections to other related industries, such as finance and real estate; this can be of great advantage in the activation of urban economies. However, these efforts tend to lean towards a large-scale, yet uniform urban development, and at the level of city planning that requires traditional methods to suit the needs of each area, small-scale systems are more desirable. They are closely related to the traditional and small-unit departments which are ignored in large-scale systems. Attention to these departments can bring about a new dynamic in cities and stimulate the region.

The Role of Local Communities

Aside from public and private sectors, the local community can be considered as a third sector. In Europe and the USA, a cooperative approach in the area of housing has been prominent since earlier times. For the developing countries of Asia which are undergoing modernisation, the role of local communities is particularly significant.

Developing countries suffer from a lack of sufficient administrative organisations to manage local areas, therefore they cannot provide adequate public services. Under these circumstances, those services which are not made available are covered by individual connections within local communities. A feeling of solidarity among those from the same state or village is fostered to the extent that organisations that offer employment assistance and welfare for particular areas have been founded. Even in the field of local management, the role of the local community is quite important; we must recognise this special status when formulating urban development in developing countries. Residents' participation has already produced numerous positive results in the development of large cities in Asia, and should be further encouraged.

For this purpose, an information disclosure system and personnel which can make use of this information are necessary. In order to ensure that residents are able to acquire information about an area or a city, a democratic system of administration which aims to regularly disclose information must be established along with transformations in information technology. The educational level of people in the developing countries of Asia has been rising at a rapid pace, and the number of people who are demanding a democratic system are increasing greatly. We need to observe carefully how this rise in the level of education will affect large cities in Asia and how it will influence urban management in the future.

Large City Planning Issues

Developed countries, and especially developing countries, often face a lack of strong mechanisms for controlling administrative bodies, resulting in poor coordination between planning-related organisations and the activities of various government sectors. Transportation, housing, and water supplies are pressing issues, but they are usually each assigned to different departments and handled independently, and adjustments between the departments are inadequate. This lack of coordination has frequently led to extreme inefficiency. In order to correct this problem, an administrative body that would manage and control planning for each department becomes necessary.

Land Systems

Recently, due to the rapid economic growth among the developing countries of Asia, the problem of land in large cities has intensified. Traditional ideas of land and property coexist with modern legal codes. Often the conceptual distinction between public- and private-owned land remains vague, or development along modern guidelines exists side-by-side with illegally occupied districts. Thus, land systems and urban planning institutions are hardly firmly established. They are, however, incorporated into the market mechanism of a capitalist economy that is progressing quickly. Under these circumstances, a system of land management including land registry should be instituted as soon as possible.

In developing countries, land is usually acquired through speculative investment and capital investment by individuals and businesses. In recent years, excess capital from Japan and developed countries have been flowing into these countries as direct investments, resulting in the rise in value of land within large city regions. This tendency is especially prominent in city centres, which has made it difficult for small businesses and residents who have been dwelling in these areas to stay. As a consequence, they are forced to move elsewhere. To remedy this situation, a special system of land administration that would protect traditional landholders needs to be instituted. A city centre has accumulated investments made toward the infrastructure of a city, thus in order to put these investments work, land use must be carefully planned. Since farming has traditionally been a major industry in Asia, the idea of public ownership of land is widely accepted. For example, in Indonesia, the right to use land has been more important than its ownership. Developing countries, which are seeking to incorporate market mechanisms and achieve modernisation, remain relatively uninfluenced by Western social systems that seek to create a welfare state. Consequently, public-minded ideas about land are not clearly defined. Since land lies at the base of a nation's social and economic activities, the development of a concept of land ownership that is unique according to each nation's traditional definition is a difficult, yet necessary task.

Urban Transportation

Transportation problems have become severe in large cities in Asia. The provision of roads and funding for large-volume transportation systems have not been able to keep up with the sudden rise in population and the increase in the number of automobiles that followed the growth of GNP and the development of secondary and tertiary industries. Developed nations have extended networks of large-volume transportation systems including subways, but developing nations should instead focus on constructing elevated railways and extending and improving existing above-ground railways, which cost less. In order to alleviate traffic congestion in cities, existing roads and streets should be used efficiently. In Singapore, programmes such as designated tolls for various zones and regulations of automobile entry based on car numbers have been shown to be effective. Cooperation between public organisations and the private sector regarding bus operations, and the assignment of bus-only-lanes during peak hours also produce satisfactory results. In addition, bicycles and three-wheel vehicles continue to be used commonly in developing nations and the designation of special lanes for these modes of transportation are necessary to improve the situation. Providing support for transportation systems in the informal sector, such as minibuses and motored tricycles, should be considered as well.

Large cities in Asia should further invest in transportation in order to maintain and expand it. At the same time, it is necessary to formulate a new strategy of urban planning that includes the optimisation of existing urban land use, the halting of the 'sprawling' tendency of cities, and the creation of secondary city centres, new residential complexes, industrial complexes, and large-scale recreational facilities.

Information Networks

The networks of communication and information among various Asian nations have recently been expanding greatly. In the areas of regional and urban planning, information systems and geographical information systems have been integrated by government organisations, universities, and research institutions. Efficient city management requires not only basic data on land, capital, population, households, industry, and employment, but also data on earnings and expenditures of each sector and the environment. Of course, currently, problems exist in data collection and calculation such as statistical inaccuracy, the lack of specially trained personnel, and the differences in the areas controlled by each administration. The rapid changes in circumstances require a system that can keep up with these changes and update data accordingly.

It is also necessary to propagate among the mid-level technological staff the precious specialised knowledge and experiences accumulated by the central government and large city governments. A system should be developed to this end, and an active transfer and distribution of information and technology should be encouraged.

The establishment of information systems also plays a crucial role in the promotion of resident participation in development planning. Local communities possess the wisdom that stem from social customs and practices and can be translated to insights on development. This knowledge should be spread widely and be converted into an accessible database. Such know-how can contribute greatly to the development of new theories in midterm and long-term planning that have both strategic and future perspectives. Encouraging dialogue between various local organisations and non-governmental organisations should have a positive effect on maximising the development potential of each area.

The Urban Environment

Today, large cities in Asia face another problem, that of a grave environmental crisis. Problems at the national level including the destruction of forests, the drying-up and pollution of water resources, and damage to the environment by the expansion of primary production, industries cannot be separated from the problems of the urban environment that large cities occupy. Amidst rapid modernisation and industrialisation, the social structures of developing nations are changing quickly, as is their urban environment. The following four points outline environmental problems which arose from both urbanisation and industrialisation:

1) the inability of public services to keep up with the expanding population has worsened the living environment;

2) rapid industrialisation has caused pollution; the pollution of water is especially serious in that it has a wide effect on the urban region as a whole;

3) air pollution from automobile exhaust poses yet another serious problem. Large cities in Asia have been slow to develop a system of large-volume transportation. Despite the low levels of income, automobile usage is quite common;

4) large cities are faced with the challenge of acquiring public spaces such as parks and greenery areas, and the effective utilisation of empty lots.

Disaster Prevention Systems

Another issue which must not be forgotten when discussing development in large cities in Asia is safety. Most of these cities are located in low-altitude areas by the sea or rivers, which make them vulnerable to floods.

The population increase in large cities has caused a shortage of land and housing. This in turn has forced people with low incomes who have moved from the countryside, to dwell in areas prone to natural disasters. The vulnerability of cities to disasters is therefore becoming more serious day by day.

The problems of disaster prevention in large cities stem more from the economic and social situation than from the lack of technology. Since the threat of poverty and starvation often outweighs the threat of a natural disaster, urban disasters that are caused by a mixture of social and economic factors have manifested on the scene instead. The systems of disaster prevention formulated in developed countries are slow to be accepted in developing countries because they are often not compatible with the local economic and social situation. It is thus important to develop technologies that are feasible within the current economic standards and social systems of these nations.

Regional Planning

Currently, in developing countries, the provision of the infrastructure of major cities has been rushed. In addition, an evenly distributed development method has not been applied adequately. Recently, a new approach that differs from the across-the-board method of the past and one that is more suited to the changes in industrial structures at the global scale has been formulated. It seeks to promote local independence, give authority to local administrations, and strengthen key regional cities.

Today, many regions have topographical or functional problems because the cities in the regions were established during the colonial period. It is therefore necessary to create new industrial bases, and the following efforts are necessary: 1) to establish infrastructure and related institutions to lure industries to the area; 2) to provide favourable conditions for businesses that move to the area; 3) low-cost training of labourers; 4) to foster and manage subcontractors; and 5) to construct a key regional city that is culturally and socially attractive. To realise these goals, active participation by private businesses, residents, and effective administrative organisations are required. If domestic capital is insufficient, the incorporation of overseas capital should also be planned. For this purpose, Asian developing countries have designated special free trade zones to invite foreign businesses. On the other hand, local initiatives should not prioritise big businesses alone; traditional industries that have unique local flavours, products which are small-scale but require scarce special skills, and mid-sized and small businesses should all be encouraged to grow.

Issues Related to Area Development

Changes within large cities in Asia have materialised in ways which are quite different from those seen in the West. Asian cities have two special characteristics. First is the continuing expansion of large urban areas; in the postwar period, cities with populations of over 10 million have been increasing rapidly in the Third World. Second is the fact that inner city areas continue to be highly active relative to the suburbs and the outlying areas. City centres suffer from the pressures of development and restrictions on space, and their traditional roles are in urgent need of re-examination.

Inner City Areas

In inner city areas, it is often the case that the very energies that lie at the root of development and expansion are in fact destroying the social and economic structures. Valuable natural, historical, and cultural surroundings are being sacrificed for the sake of superficial and short-term economic growth. These losses affect not only the region but the local community in general, and require immediate attention.

In the past, two policies have been carried out under the pressures of social and economic powers that have changed the environment of large cities. One is to remove existing structures and build anew, and the other is to preserve existing structures while improving them. The first policy is typical of a development-centred approach, while the second concentrates upon preservation. Emphasising preservation calls for detailed discussions with the local residents, which costs both time and money; for this reason, many cities have chosen the 'scrap and build' approach.

The problem of development and preservation in large cities includes not only the material environment but the natural and social environment as well. The true costs of the destruction of the environment through development must be assessed. For example, tearing down an old district in an inner city area means that the unique traditional ways of life that have been practised will be destroyed. These unique lifestyles were formed slowly, through many generations of complex interpersonal relationships based on local, family, and traditional ties, and had solved various problems that the residents had faced in the past. This system has also contributed to the productive and economic aspects of the community, such as the medium and small businesses that are unique to the region. It was the norm for residents to help each other on a daily basis, and they were tied by a strong sense of community.

However, with the large-scale development that is occurring again in inner cities, such old districts have begun to quickly disappear. We must reconsider the merits of a profit-oriented approach that focuses upon direct profits from land and buildings as a result of redevelopment. The true costs of such projects must not account for material assessment of land or buildings alone, but also the social and cultural effects such as changes in lifestyle, the dismantling of social organisations, economic effects such as changes in land value, rent, taxes, and business activities, and effects upon the environment such as water, air, earth, and noise pollution. A system of evaluation that prioritises the 'quality of life' is in order.

Suburbs and Outlying Areas

Rapid economic growth in large cities in Asia causes the movement of slum areas from city centres to the surrounding areas. Due to the rise in economic activities in inner cities, the prices of land and rent have increased. Redevelopment has progressed for the sake of high-profit land use, and illegally occupied districts and low-income housing have lost their places in inner cities. For example, in the outlying areas of Bangkok and Manila, vast farmlands and fields have been urbanised in order to accommodate migrants from the countryside and those who have fled from city centres. Most of them are poor. This trend can be observed in other large cities in Asia as well. These suburban residential areas which offer poor living conditions are the 'unplanned' results of development planning or infrastructure upgrading projects. If these areas expand further, even more urban-related problems will arise.

The problem here can be outlined in two parts. First, economic growth and population increase has led to a 'sprawling' effect, in which farmlands and fields are devoured into unplanned residential areas, causing inefficient public investments and the deterioration of living conditions. Second, farmlands near cities generally had been vital sources for food, and their rate of productivity was quite high. Urban sprawl ignores this productivity and crucial role of these lands. If we are to hope that agricultural productivity will continue to rise, then the relationship between cities and farmlands must be secured.

Housing Problems in Asia

The Present Housing Conditions in Developing Countries

As in other Third World nations, living conditions in the major urban centres of Asia have deteriorated in the past 20 to 30 years due to the increase in population. In old city centres, existing buildings are falling apart and the population densities have reached unreasonable levels. In some cities, there are over one thousand people per square hectare. Supplying adequate housing under these circumstances poses a great challenge to the Asian countries.

In poorer neighbourhoods, basic facilities such as water supply, education, medical care, and housing are lacking, and the living conditions are quite bad. On the other hand, these areas are characterised by strong social bonds.

Their support makes it possible for those who have moved from elsewhere to survive in a city. The residents of these slums and illegally occupied areas are starting to be driven out of inner cities to the surrounding areas as economic activities increase in city centres, and these areas are often rebuilt as commercial districts or middle or high-income housing estates.

Solving the Housing Problem

It is necessary for comprehensive urban development to include improving the living conditions of poorer districts as an integral part of city plans. Planning should address social, economic, and material concerns such as land problems, their effects on employment, and community activation and participation. Asia's rapid economic growth has produced a shortage of staff who can devote themselves to solving these problems. Training new personnel is an urgent need at this point.

1) 'Sites-and-services' projects are effective in improving the living conditions of slum areas and housing construction for low-income families. Specifically, minimal services such as water, electricity, sewerage, and paved roads are provided by public organisations, while housing construction is left to the initiatives of the residents. It means that private funds may be actively channelled into these projects as well. Residential buildings to be constructed include one to two-storey structures but consist mainly of single family homes, creating a concentration of low-rise buildings. It will take time before housing can become medium-scale or high-rise and inflammable, as they are in Hong Kong and Singapore. Of course, we must also reconsider whether high-rise buildings are most suited to a city's environment. In building housing as well as community facilities, the overall cost balance in the urban development project must be achieved. These projects should also be able to survive the changes that may take place in the overall urban structure in the future.

2) Construction methods. In providing housing for the poor, it is important to keep construction at low cost. By using resources available within the country and materials that can be produced in the area, local residents will help keep the costs down. For instance, in dry areas, adobe housing made of sunbaked bricks are the mainstream, whereas in humid areas houses made of wood, bamboo, or palm trees are common. However, adobe houses are vulnerable to earthquakes, and organic materials are plagued by insects,

wind, and fire. Improving these weaknesses is a project for the immediate future.

In developing countries, materials such as reinforced steel concrete and reinforced concrete blocks are more frequently used, but the technical standard in using these materials has not yet reached the level where it is able to withstand earthquakes and fires. Also, poorer residents cannot afford to live in these modern buildings. While the problems of technology and economy are being addressed, it is necessary to create building materials and methods which the residents can fully utilise.

3) Nurturing the housing industry. Issues of construction methods, design, and land policies have been mentioned above. Developing countries should not relegate these issues to the area of social development alone, but should seek to address them as part of the economic stimulation plan and strive to nurture housing-related industries. Such an effort would maximise the utility of aid to developing nations.

For economic stimulation to occur, there must be cooperation between the existing industries of construction, construction material and equipment, and real estate; also, infrastructure, urban utility services, as well as financial and loan services should be integrated. With their assembled efforts, they can help create stronger and more efficient housing-related industries. Although a great demand for housing lies dormant at present, it is vital that we promote an active housing industry as we look toward the year 2000.

Conclusion: Towards the Twenty-first Century

In the future, large cities in Asia can be expected to transform themselves into balanced yet dynamic societies that take advantage of their diversity. At present, the people who have moved from the countryside or from overseas have been integrated into these cities and have successfully contributed new energies. Such energies have intermingled with the living experiences that have accumulated among the existing communities and have created rich varieties of urban culture. These conditions which lie at the base of large cities should be recognised and re-examined as we move toward the future. In the previous large city planning projects, there was a tendency to implement standardised rules and regulations in the name of efficiency; however, these projects have often resulted in deteriorating economic and social conditions. In order to remedy this situation, future development must respect and prioritise each region's unique traditions, religions, and culture.

Those who live in the poorer districts of large cities in Asia have managed to remain strong. They are extremely active and creative in trying to improve their own living and production conditions. They also participate in the local community and self-help projects. The 'town-making' process for improving the environment would be best proceeded if local communities took the initiative and were aided enthusiastically by public and private organisations. The system of urban planning must be adjusted accordingly to suit this approach.

The planning projects, which are enforced by only government institutions or private developers, do not accurately reflect the grassroots level of reality. Therefore, the planning system must be changed. A new system must be formulated in which residents themselves can pinpoint problems and solve them according to an approach suited to their lifestyles and their real need.

In the past 20 years, the NIEs of Asia have achieved remarkable economic growth and large cities have played a major role in this success. Today, ASEAN and South Asian countries are making progress in economic growing. The developing countries of Asia will certainly undergo many transformations in the future. It is believed that the twenty-first century belongs to Asia; the activation of energies of the people who function in various economic and social structures of large cities will be crucial in order for Asian nations to progress along with the developed nations.

3 The Compulsions of Accommodating the Informal Sector in the Asian Metropolises and Changes Necessary in the Urban Planning Paradigms

A.T.M. NURUL AMIN

Introduction

At present, planning of all kinds seems to be losing its ground. Economic planning started to lose its appeal even before the socialist system crumbled. The failure of centralised planning tarnished the image of planning in general. Besides the arena of planning, other government policy actions, which are essentially intended to correct market distortions or to make markets work better, are not seen very positively in the contemporary world. Thus, with this prevailing attitude towards policy and planning in general, it is unrealistic to expect a heyday for urban planning in today's world. The eminence of market values has indeed produced a strong lobby for the paradigm of urban management, which subtly seeks to replace urban planning totally, or at least to diminish its significance. The view gaining currency is that since economic signals are so powerful and economic growth is so essential, it is better to focus on managing an urban economy for maintaining (or even further increasing) its higher productivity, relative to the national economy as a whole and rural economy in particular. Urban planning is seen as a futile exercise since there appear to be so many instances of growth overtaking planning assumptions or requirements.

It is, however, not this strengthened laissez-faire view that is posing the major challenge to urban planning today. As this chapter argues, the major problem with urban planning is its core paradigms which essentially reflect Western experience of urban growth. Although such a critique of planning paradigms and instances of their irrelevance to Asian cities are neither new nor rare, it cannot be said with certainty that the planning profession has yet adopted an approach that reflects this critique very well. To ensure an understanding of the Asian reality, this chapter, therefore, first presents some basic economic and demographic data of the 11 most populous countries of Asia. With these basic data, it is argued that the huge population base, the existence of a vast rural-agricultural sector, and the growth of mega size cities at a relatively low level of development have given rise to a vast sector within the very heart of these cities, which are now generally labelled as the urban informal sector (UIS).[1] If not for anything else, the huge presence of this sector calls for an appropriate change in the urban planning approach. It then briefly analyses the economic forces that give rise to economic, settlement and environmental imbalances, arising from the particular nature of the relationship that prevails in the interaction between the rural-agricultural (R-A) and the urban-industrial (U-I) sectors. This relationship has resulted, among other things, in cities of mega size with an awkward presence of an informal sector within their very centre. Based on this analysis, the actual manifestation of this growth and its consequences will then be shown in the later part of this chapter. Finally, an agenda for accommodation of the informal sector to urban planning and environmental management is outlined.

Lest We Forget the Asian Setting of Urban Planning

Despite the so called third wave[2] of development that is now raging through a large part of Asia, the fact remains that it is still largely a continent with a huge mass of population and a vast rural-agricultural sector that sends out millions of migrants to urban areas. These migrants have to cope with the pressures of adjusting and adapting to the growth of exchange economy that has now mostly eliminated the self-sufficient village system that had existed in Asia in the past. Although this kind of change is not atypical of the industrially-developed countries of the West, the scale, the time span and the context make the overall outcome of urban growth in Asia, which is primarily rural-to-urban migration-fuelled, profoundly different than previously known experiences. Similarities and dissimilarities in urbanisation experiences between the Western developed and currently developing countries have been well documented by many studies (McGee, 1971; Williamson, 1988). Without going into the details of all the subtle findings (Williamson, 1988, pp. 427–33), it can be safely said that the real difference between these two groups of countries' urbanisation experiences lies in the large number of people involved in the process and their corresponding urban living requirements.

Sir Arthur Lewis (1954) was so optimistic that he expected labour scarcity would emerge as a result of the urban-industrial sector's growth momentum and, even in the labour surplus economies, that rural-urban difference would eventually disappear. If Lewis was a little more familiar with the multitude of people, as is the case in Asia, his optimism would have been restrained. Indeed, Lewis could not be so optimistic if he had not assumed a population growth rate of one per cent per year, which obviously proved to be wide of the mark. The other assumption in his model is that capitalist, profit-seeking entrepreneurs will bend on using cheap labour, which is so abundantly available in these countries. Unfortunately, this assumption again did not work to the extent that Lewis expected. Many other factors of an socioeconomic and political nature also impeded large-scale investment in the U-I sector, despite the boon of cheap labour. It is not that this basic economic factor endowment did not work as stipulated in the Lewis model. If it had not worked at all, then Korea would not be what it is today, or Thailand would not be experiencing what it is currently. The point is that when the population base is so huge and the enclave of urban-industrial growth is surrounded by a huge rural-agricultural sector, there is bound to be massive migration flow. However, no matter how massive this flow is, it cannot ensure a labour shortage in the rural-agricultural sector as expected by Lewis. To make it worse, what Lewis did not anticipate at all is that a 'subsistence sector' (informal sector) gets created in the middle of large urban-industrial agglomeration (mega-size cities) because of excess migration (note that to Lewis, a subsistence sector exists only in the rural-agricultural sector) compared to the absorptive capacity of the U-I sector.

This latter trend in development is well predicted by Todaro (1969). In his model, he convincingly shows that the actual number of migrants is bound to exceed the number of jobs created by the U-I sector since individual migration-decisions take place on the 'probability' of obtaining a job and not necessarily according to its actual getting. Many of these job seekers finally, or sometimes immediately (because of immediate survival needs), resort to the informal sector. Some recent labour market models, however, (e.g. Cole and Sanders 1985), have recreated a Lewisian-type of optimism in labour surplus by likening the contemporary 'urban subsistence sector' (USS) with that of the 'rural subsistence sector' of Lewis. In a way, it is a resurgence of optimism in labour surplus in the sense that the USS is seen to help the U-I sector to save labour, overheads, and social security payment costs. As a result, the U-I sector acquires international competitiveness, particularly in the export market, because of these savings which are made possible through the UIS (or USS as Cole and Sanders (1985) refer to it).

The huge population in Asian economies is the underlying factor for the pattern of development indicated above. To illustrate this, we draw attention to some basic economic and demographic characteristics of the eleven populous Asian countries as presented in Table 3.1. Population mass in each of these cases is indeed huge, ranging between 43 million to well over a billion. It is not going to be far in the future when a second most populous Asian country (namely, India) joins the billion mark. Of these million- or billion-sized populations, on average, only about one-third currently live in the urban area (as of 1990). However, even that one-third, in absolute terms, has accounted for tens to hundreds of millions who live in the urban areas already. In fact, this is only a small indication of the 'real' problem that might ensue in the years to come. As can be seen in column (4) of the same table, most of these countries are still overwhelmingly rural, which in itself should pose no problem. The problem of this overwhelming rural situation arises from the fact that most of those millions are potential migrants sitting on the fence for an opportune moment to make an urban-bound move. No wonder the Todaro model implies that urban problems cannot be solved at the urban end because the solution to those problems (whether it is of urban unemployment or lack of shelter, service, and infrastructure) will trigger a new flow of migration. The prediction in Todaro's model may have limited applicability for countries which are not as populous, with so huge an R-A sector as the Asian ones. Nevertheless, its message for the populous countries like the ones presented in the table is so easy to appreciate.

To make matters worse, the urban-based move is essentially becoming targeted to a specific city of each country. It can be seen from the same table that millions of people are living in the single largest city of all of these countries. In most cases, the largest city accounts for one- to two-thirds of the country's total urban population. Because of a concentration of wealth, income, economic activities, administration, financial services, education and health facilities, the migration flow is usually one-city bound, basically the largest city, which is in most instances, the capital city of each country.

All data mentioned above are, in fact, widely known. What is not so widely known, however, is the fact that the bulk, often (50–65 per cent) of the working people of these largest cities, earn their living from economic activities different

Table 3.1 **Eleven most populous developing countries of Asia with huge rural-agricultural economy, cities of mega-size and vast informal sector**

Country	Total national population (m.)	Total urban population (m.)	Rural population as % of total	Population in the largest city (m.)
(1)	(2)	(3)	(4)	(5)
1 Bangladesh	116.4	18.6	84.0	4.3
2 China	1,170.7	386.3	67.0	9.2
3 India	862.7	232.9	73.0	12.9
4 Iran	59.9	34.1	43.0	8.1
5 Indonesia	187.7	58.2	69.0	9.9
6 Korea	66.0	44.9	32.0	–
7 Myanmar	42.7	10.7	75.0	3.1
8 Pakistan	121.5	38.9	68.0	7.3
9 Philippines	63.8	27.4	57.0	9.2
10 Thailand	55.4	12.7	77.0	9.2
11 Vietnam	68.1	15.0	78.0	3.5

Table 3.1 cont'd

Country	% of GDP from			% of labour force engaged in			Proportion of people engaged in the IS (nationally)	Proportion of people engaged in the IS of the largest city
	Ag. (6)	Ind. (7)	Services (8)	Ag. (9)	Ind. (10)	Services (11)		
1 Bangladesh	38	15	47	56	10	34	70.0	65.0
2 China	27	42	31	73	14	13	–	–
3 India	31	29	40	62	11	27	–	60.0 (Madras)
4 Iran	21	21	58	25	28	47	–	–
5 Indonesia	22	40	38	54	8	38	68.0	65.0
6 Korea	9	45	46	16	34	50	–	–
7 Myanmar	–	–	–	64	9	27	–	–
8 Pakistan	26	25	49	44	25	31	69.0	–
9 Philippines	22	35	43	41	19	40	–	50.0
10 Thailand	12	39	49	70	11	19	62.6	58.0
11 Vietnam	–	–	–	67	12	21	–	48.0

Note

Japan was not included since it is less relevant for the basic theme of the paper although it originally had many of the characteristics of the 11 countries portrayed here.

Sources: UNDP, Human Development Report, 1993 for total population (pp. 180–1), total urban population (calculated from data on proportion of urban population (pp. 178–9); rural population (pp. 154–5); sectoral contribution to GDP (pp. 186–7); sectoral distribution of labour force (pp. 168–9); and the population crisis committee poster (PPC, 1990, p. 16) for the population of largest city of respective country. The IS data are author's estimates from cities statistics reports and studies on the IS.

to those of the urban-industrial or modern sector. Indeed, the industrial sector, in terms of GDP and employment, still accounts for a relatively small proportion despite remarkable growth and development in most Asian countries. It is important to note that, in terms of employment, the agricultural sector is still the predominant sector for most of these populous countries. Many of them are, however, potential job seekers in the U-I sector. On the other hand, the industrial sector absorbs only a small proportion of the national labour force. Job absorption is mostly taking place in the service sector. Many of the service sector jobs are not stable and many entrants in this sector are secondary labour (females, rural migrants, and immigrants). One consequence of this is the growth of informal economic activities and housing, which in turn poses a problem for urban planning since its growth takes place without any care for the assumptions and requirements of urban planning. Instead of bringing forth a solution to the rising problem, rather, planning becomes part of the problem by imposing unrealistic and unaffordable standards, thereby aggravating the situation further.

One basic premises that urban planning has not reconciled with is the simple fact that 'cities are primarily places for earnings'. This has been well stated recently by Malcolm Harper (1992, p. 143):

> Towns and cities are, above all, 'income generating projects'; people may find better health care, education, sanitation or even entertainment when they live in the urban areas, but the basic reason for most people to come to, or stay in a town is their need to earn a living.

To make cities a better place to live, urban planners often ignore this basic function of cities and their attractiveness to the migrants. As a result, the most effective exercise of urban planning ironically results in the eviction of people from their source of income-earning opportunities in the name of zoning, beautification, and all the catchy labels of planning and development needs. This type of event can be tolerated if the number of people affected by the process is limited. Indeed, the situation often comes close to explosion because of the enormous number of people who are affected by such a process.

Market Logic Consequences of Rural-Urban Linkages

This section uses some basic economic concepts to explain what is likely to happen from the intersectoral linkages between the R-A and the U-I sectors, as a consequence of the basic initial conditions or settings of Asian developing countries as presented in the preceding section. A simple descriptive model is used for illustrative purpose.

An economy is seen to be basically divided, regionally or sectorally. Regionally, the divisions are seen as: 1) rural and urban areas; 2) backward and developed regions; and 3) developed and underdeveloped countries (from a global perspective). Sectorally, an economy is seen to be divided as: 1) rural-agricultural sector (R-A); and 2) urban-industrial sector (U-I), as noted already. The former sector is assumed to produce mostly rural-agricultural goods (for brevity, we refer to these as non-urban goods) and the latter sector is assumed to produce urban-industrial goods (for brevity, we refer to these as urban goods).

As an R-A economy becomes commercialised with the expansion of the exchange economy, it becomes closely linked with the U-I sector of the national (even global) economy. Once exchange between the two sectors starts, the nature of the respective sector's products, demand and supply conditions, degree of competition, production conditions (primarily determined by the level of technology and capital accumulation) and the terms of trade, basically determine each sector's relative gains from this exchange. These links generate very powerful economic relationships when the market or the exchange economy is in full play.

In the preceding section, the emphasis was on demographics: the large national population base, the millions of rural population (many of whom are potential migrants) and the millions who move to settle down in the largest cities of the respective countries. This section turns to the economics of the relationship between the two sectors in light of the model stipulated above.

The economic forces that are seen to propel the growth of the U-I sector in general and large urban agglomerations in particular are:

1) relatively higher *income elasticity* of demand (as income increases, proportionately greater spending is made) of urban goods (largely produced by the U-I sector) *vis-à-vis* non-urban goods (largely produced by the R-A sector);

2) relatively low *price inelasticity* of demand of non-urban goods (because of their importance in the consumption basket) compared to urban goods which are only essential for better living and comforts;

3) much greater scope of deriving large *economies of scale* (as size or volume of production increases, per unit cost is reduced) in the production of urban goods;

4) built-in advantages in the production of urban goods due to *agglomeration economies or external economies* (which arise from savings made through reduction of costs in transportation and better use of people in the labour force by matching their diverse abilities against the widest range of things to be done);

5) highly competitive *market structure* in the production of non-urban goods (an outcome of relatively homogenous productions by numerous producers) compared to monopolistic market structure in the production of urban goods (an outcome of product differentiation and a limited number of producers);

6) unfavourable *terms of trade* (the ratio of average export and import prices) of non-urban goods *vis-à-vis* urban goods;

7) higher *capital accumulation* (generation of surpluses and their reinvestment which increases the capital stock) in the U-I sector compared to the R-A sector;

8) greater *innovations and technological progress* (in terms of product as well as process innovations) in the production of urban goods compared to the case of non-urban goods.

All the eight economic phenomena listed above are very powerful in making urban activities more profitable. They create a built-in advantage for an urban economy in general and the bigger urban agglomeration in particular.[3]

Admittedly, urbanisation does not turn into mega-urbanisation or pseudo-urbanisation, with a huge number of people living and working in the informal sector and the operation of the above economic forces or laws. Rather, the observed pattern of urbanisation arises from the simultaneous presence of mutually reinforcing economic forces with cumulative effects and impacts. This would not warrant space configuration of a mega-urban dimension unless it is compounded further by additional factors of a nature noted below:

1) as scale and agglomeration economies get exhausted (as reflected in higher prices of land, rising living and labour costs, etc.), a variety of responses are adopted by the producers and consumers to economise on land and labour. The response to rising land price is reflected in horizontal (spatially along the corridors of development and the road networks in all directions)

and vertical (high-rise buildings) expansions. The response to labour cost also takes two forms: a) increasing reliance on capital and technology in the modern segment of the U-I economy; and b) increasing reliance on the informal sector for a variety of low-cost services and getting work and businesses done through subcontracting and outwork systems. The more frequent the latter is practised, the more the demand side of mega expansion gets firmly established, despite the onset of diseconomies of scale and agglomeration. This would not guarantee an urban agglomeration to assume a true mega form unless, on the supply side, the pool of labour which had been seeking income-earning opportunities in the urban economy was not 'huge'. Here lies crucially the significant role of the absolute size of rural population, most of whom are potential migrants as suggested before.

2) Similarly, when all the benefits from 'natural monopoly' characteristics in the delivery of urban services tend to get exhausted, a 'spatial natural monopoly' condition will be created. The physical manifestation of this is the emergence of satellite towns and suburban communities, in which servicing becomes possible through separate and new public utility service facilities. With interconnections of these settlements, the main city assumes a size of mega-urban dimensions.

3) More importantly, imprecision in the knowledge of exhaustion of scale and agglomeration economies and the political power of urban conglomerates and other interest groups lead to heavy infrastructure invest-ment in the single most important urban region of an economy. This process forestalls the chance of arriving at an economically optimal size of a city which would otherwise emerge from the market mechanism *per se*.

4) The above mix of economic and non-economic factors would not make an urban economy the size of mega dimensions unless there were certain facilitating factors in place. The singularly important facilitating factor of mega-urban expansion is transportation. However, without economic growth (aided by international capital and technology), transportation in itself will not have much effect. This is why transportation is seen as only a facilitator (albeit a crucial one), not a fundamental causal factor in mega-urban expansion.

The implications of the above pattern of relationship between the R-A and the U-I sector can be noted under three major headings.

1) In terms of *human settlements pattern*, this means:

   a) continuous rural-to-urban migration (every rural resident is a potential urbanward migrant);
   b) continuous inter-urban migration (often from the small to the large cities); and
   c) high incentives for international migration (the ultimate of all such migration is to have a foothold in USA, although previously it has been the West European countries. In recent years, Japan, Korea, Malaysia and even Thailand have become destinations of many international migrants from other populous Asian countries and regions, such as China, the Philippines, Myanmar and Indo-China).

2) The second consequence of R-A and U-I intersectoral linkages is observed in the widening of income disparities as reflected in:

   a) rural-urban income disparity;
   b) inter-regional income disparity; and
   c) interpersonal income disparity.

3) Finally, the most obvious consequence of this inter-sectoral relationship is observed in the massive agglomeration that we now call megacities which are beset with *environmental problems* of alarming proportions as observed in:

   a) life-threatening pollution of air and water;
   b) nerve-breaking and life-halting traffic congestion;
   c) unbearable sites of urban wastes and dumping grounds;
   d) life degrading slums and squatter settlements; and
   e) growth of hazardous occupations.

Thus, the economic logic of resource flows between the R-A and the U-I sectors bear the implication of serious consequences for human settlements pattern, income distribution and living environments. These are no more remote possibilities. We have already lived in such settlements and environments as will be further evident from the next two sections.

Metropolisation in Asian Urbanisation

A metropolitan area is defined to include a central city, neighbouring communities linked to the city by continuous urban development, and more distant communities, if they are supported by economic activities in the city or its suburbs (PCC, 1990). A recent report on the 100 largest metropolitan areas of the world shows that 40 of them are in Asia. While some may consider this as more evidence of ushering in an era of prosperity for the Asia-Pacific rim, a careful scrutiny shows that most of these 40 Asian metropolises belong to the bottom of rank orderings in terms of urban living standards. Indeed, with the exception of eight (Osaka-Kobe-Kyoto, Tokyo-Yokohama, Singapore, Nagoya, Kiev, Taipei, Hong Kong, and Tashkent), the living standards in all Asian cities in the list of world's largest 100 metropolitan areas are either 'fair' or 'poor'. As a matter of fact, most cities score 44 and below which denote a 'poor' score in terms of the urban standard of living (Table 3.2).

The 10 assessment indicators used in the PCC ranking of the cities in terms of urban living standard scores are:

1) public safety (murders per 100,000);
2) food costs (percentage of income spent on food);
3) living space (persons per room);
4) housing standards (percentage of houses with water and electricity);
5) communications (telephones per 100 people);
6) education (percentage of children in secondary school);
7) public health (infant deaths per 1,000 live births);
8) peace and quiet levels of ambient noise (1–10);
9) traffic flow (miles per hour in rush hour); and
10) clean air (alternate pollution measures).

Out of these 10 indicators, we have selected three for further scrutiny of the situation in the Asian metropolises. They are: 1) housing standards; 2) education; and 3) public health. Inspection of these data shows (Table 3.3) that many Asian cities, particularly the South Asian ones, fare very badly in terms of crucial urban needs, such as housing, education and health. For example, the housing standards, as illustrated by the proportions of homes without water and electricity, are ranked in the following order: Calcutta (43 per cent), Surabaya (41 per cent), New Delhi (34 per cent), Shenyang (34 per cent), Bangalore (33 per cent), Kanpur (30 per cent), Nanjing (30 per cent), Pusan (30 per cent), Ahmedabad (29 per cent), Dhaka (27 per cent), Karachi

Table 3.2   Population and urban living standards in the metropolitan areas of Asia

Sl. no.	Metropolitan area	Population (m.)	Urban living standards score*
1	Tokyo-Yokohama	28.7	81
2	Osaka-Kobe-Kyoto	16.8	81
3	Seoul, Korea	15.8	58
4	Bombay, India	12.9	35
5	Calcutta, India	12.8	34
6	Jakarta,	9.9	40
7	Delhi-New Delhi	9.8	36
8	Manila	9.2	43
9	Shanghai, China	9.2	56
10	Tehran, Iran	8.1	39
11	Karachi, Pakistan	7.3	36
12	Beijing, China	7.0	55
13	Bangkok, Thailand	7.0	42
14	Taipei, Taiwan	6.1	69
15	Tianjin, China	5.6	51
16	Madras, India	5.6	42
17	Hong Kong	5.2	67
18	Nagoya, Japan	4.9	77
19	Baghdad, Iraq	4.4	54
20	Dhaka, Bangladesh	4.3	32
21	Bangalore, India	4.1	37
22	Shenyang, China	4.0	42
23	Lahore, Pakistan	3.9	34
24	Pusan, South Korea	3.8	56
25	Wuhan, China	3.7	51
26	Hyderabad, India	3.5	39
27	Ho Chi Minh City	3.5	40
28	Guangzhou, China	3.3	42
29	Yangoon (Rangoon), Burma	3.1	–
30	Singapore	3.0	79
31	Ahmedabad, India	3.0	43
32	Kiev, USSR	2.9	74
33	Surabaya, Indonesia	2.8	36
34	Harbin, China	2.7	52
35	Chongqing, China	2.5	48
36	Tashkent, USSR	2.5	60
37	Nanjing, China	2.4	49
38	Bandung, Indonesia	2.4	52
39	Kanpur, India	2.4	33
40	Pune, India		

*   PCC assigned score based on its 10 indicators.

Source: based on data contained in PCC, 1990 and reproduced from Amin, 1992.

Table 3.3 Housing, education and public health situation in metropolitan areas of Asia

Sl. no.	Metropolitan areas of Asia	Housing standards: % homes without water/electricity	Education: % children not in secondary school	Public health: infant deaths per 1,000 live births	Urban living standards score*
1	Tokyo-Okahama, Japan	0	3	5	100.0
2	Osaka-Kobe-Kyoto, Japan	2	3	5	96.6
3	Seoul, South Korea	0	10	12	93.3
4	Bombay, India	15	51	59	46.6
5	Calcutta, India	43	51	46	36.6
6	Jakarta, Indonesia	15	23	45	63.3
7	Delhi-New-Delhi, India	34	51	40	43.3
8	Manila, Philippines	9	33	36	70.0
9	Shanghai, China	5	6	14	90.0
10	Tehran, Iran	16	42	54	53.3
11	Karachi, Pakistan	25	35	97	33.3
12	Beijing, China	11	3	11	86.6
13	Bangkok, Thailand	24	29	27	63.3
14	Taipei, Taiwan	1	19	5	93.3
15	Tianjin, China	18	29	15	70.0
16	Madras, India	24	44	44	50.0
17	Hong Kong	3	14	7	93.3
18	Nagoya, Japan	0	5	5	100.0
19	Baghdad, Iraq	2	15	40	83.3
20	Dhaka, Bangladesh	27	63	108	20.0
21	Bangalore, India	33	40	48	46.6
22	Shenyang, China	34	17	13	66.6
23	Lahore, Pakistan	20	68	83	26.6
24	Pusan, South Korea	30	16	12	66.6

Sl. no.	Metropolitan areas of Asia	Housing standards: % homes without water/electricity	Education: % children not in secondary school	Public health: infant deaths per 1,000 live births	Urban living standards score*
25	Wuhan, China	14	36	18	70.0
26	Hyderabad, India	19	60	83	30.0
27	Ho Chi Minh City,	13	48	48	56.6
28	Guangzhou, China	17	44	8	66.6
29	Yangoon (Rangoon), Burma	–	–	–	–
30	Singapore	0	18	7	93.3
31	Ahmedabad, India	29	45	46	46.6
32	Kiev, USSR	0	4	16	96.6
33	Surabaya, Indonesia	41	24	75	40.0
34	Harbin, China	13	24	18	76.6
35	Chongqing, China	23	42	11	63.3
36	Tashkent, USSR	5	4	28	86.6
37	Nanjing, China	30	21	15	66.6
38	Bandung, Indonesia	9	47	22	66.6
39	Kanpur, India	30	51	157	23.3
40	Pune, India	22	54	57	43.3

* On the basis of our selection of three out of PCC's 10 indicators.

Source: based on data provided in PCC, 1990 and reproduced from Amin, 1992.

(35 per cent), Madras (24 per cent), Bangkok (24 per cent), Chongqing (23 per cent), Pune (22 per cent) and Lahore (20 per cent).

The situation is equally disappointing in terms of the education and health criteria. The proportion of children in the age group 14–17 years who are not enrolling in school is ranked in the following order: Lahore (68 per cent), Dhaka (63 per cent), Hyderabad (60 per cent), Pune (54 per cent), Kanpur (51 per cent), Bombay (51 per cent), Calcutta (51 per cent), New Delhi (51 per cent), Ho Chi Minh City (48 per cent), Bandung (47 per cent), Ahmedabad (45 per cent), Guangzhou (44 per cent), Madras (44 per cent), Tehran (42 per cent) and Bangalore (40 per cent). The grim health situation is reflected in the data on infant death per 1,000 live births: Kanpur (157), Dhaka (108), Karachi (97), Lahore (83), Hyderabad (83) and Surabaya (75), which contrast with five in Tokyo.

The question that naturally arises from the above indications in the indices of quality of life in the Asian metropolises is: what kind of metropolitan expansion or development is this? Metropolitan life or development is supposed to mean modern and higher standards of living, which is not the case in present-day Asian metropolises. There is little doubt that metropolisation, more in the sense of territorial and population increase, is the dominant feature of urbanisation in Asia as reflected in urban primacy, or predominance of the largest city in the urban system. As already suggested by data in Table 3.1, a high proportion of the national urban population of most Asian countries live in the biggest cities of the respective countries. This is the dominant trend in all Asian countries, except India and China. Even in the cases of India and China, the pattern becomes similar to other countries when urban primacy is measured by the ratio of population of the biggest city of a region/state/province to the respective region/state/province's urban population. For example, Calcutta's true primacy position is revealed when the ratio of its population to the urban population of West Bengal, not urban India as a whole, is measured. Indeed, urban primacy of a city in countries like India and China, with distinct regional and federal characters, should be measured by referring to the largest region where it belongs.

The huge primacy of major Asian metropolises, as reflected in the data presented in Table 3.3, lends support to the observation that 'metropolisation' is a major feature of urbanisation in Asia. Two comments are to be made on this characterisation. First, as the quality of living data show, this metropolisation does not denote metropolitan development of a higher standard of living. Second, the metropolisation and huge primacy, admittedly, is not unique to Asian urbanisation. Some industrialised countries, if not all, did also experience a similar urbanisation experience in their early phase of urban-industrial development. London's position in England, or Paris's primacy in France, 150 years ago were not dissimilar to Bangkok's position *vis-à-vis* Thailand today. However, the difference in time and the basic resource base, as argued before, makes urban planning and management of megacities of Asian developing countries extremely difficult.

The next section of the chapter is an attempt to show that one reason for the above unhappy state of affairs of the Asian metropolises is the presence of a vast sector, namely the informal sector, that provides solace and shelter to the poor, the disadvantaged, the unemployed or the underemployed. Their number and quality of living naturally lower the average scores of a metropolis in terms of the standard of living measures.

The Informal Sector in the Asian Metropolises

The UIS is usually defined to include all economic enterprises and employment that are not protected, nurtured or regulated by government laws and a social security system. The criteria that are conventionally used to distinguish the IS enterprises, employment and habitat from their formal sector counterparts are size, nature of technology, organisational and operational characteristics and coverage by social security systems and protection, etc.

As shown previously in Table 3.1, the IS accounts for about 50–65 per cent of total employment of major urban metropolises of Asian countries. More importantly, the sector seems to be still growing, as indicated by comparative data on its size estimate of early 1970s and 1980s (Table 3.4).

Two points, however, should be noted on the definition, size and growth trend of this sector. First, many economic activities which are now bracketed as belonging to the informal sector have existed before urbanisation and modernisation started to receive new impetus from the Western model of urban-industrial development, which all developing countries are currently pursuing. What is happening now is that certain activities which existed previously are often declared as 'informals' because they no longer seem to belong to the new and modern urban environment. The disappearance of *samlors* in Bangkok or the phasing-out of *becaks* in Jakarta and 'rickshaws' in Dhaka illustrate this point. Many retail and service activities and even some small workshops are also in a similar situation. A good proportion of UIS activities thus denote a somewhat paradoxical outcome of urbanisation and modernisation. In other words, urbanisation or modernisation makes certain people and activities out

Table 3.4   City based estimates of the size of the informal sector (percentage of total employment)

City and country	Early 1970s	Early 1980s
Bangladesh		
Dhaka	57.0	64.6
India		
Calcutta	40.0–50.0	54.0
Bombay	49.5	n.a.
Madras	50.0–70.0	60.0
Delhi	53.8	n.a.
Pakistan		
Karachi	69.1*	n.a.
Indonesia		
Jakarta	41.0	65.0
Philippines		
Metro Manila	43.0	50.0
Thailand		
Bangkok	43.4**	49.0

*   Denotes IS size for the country.
**   Denotes 1976 data.

Source: Amin, 1989.

of place or obsolete, which are consequently labelled as 'informals'. To the extent that this trend is operative in the dynamics of UIS growth, the increase in size of the sector is not necessarily a reflection of any failure of the formal sector (FS) or the conventional process of economic growth and development. Rather, it may be a reflection of the introduction of urbanity and modernity into an economy, albeit not yet pervasive. However, it makes a lot of difference to the outcome of this process, namely the growth of UIS, according to 1) its degree or intensity that encompasses a limited number of people compared to 2) its extensiveness or breadth of coverage that encompasses most urban-dwellers. While the former tends to stimulate an 'enclave'- type development, the latter is expected to reduce the duality of the process. The people and their activities affected by the overall process are increasingly marginalised if they fail to change as required by the process of modernisation and urbanisation. The outcome of this process is the 'marginalisation' of certain people and their economic activities that are euphemistically bracketed with the informal sector (IS).

The other dimension of IS growth dynamics can be traced to a more recent trend which is marked by a tendency on the part of a growing number of FS enterprises, and even multinational corporations (MNCs), to reduce the size of their stable workforce by resorting to a variety of the 'putting out' system (e.g. subcontracting). The fold of secondary labour, available from a growing number of women and migrant job-seekers, and UIS enterprises expand as a result of this practice, which is increasingly adopted by many FS enterprises. This may be seen as 'deformalisation' of the formal sector.

The two trends in IS growth discussed above may not necessarily lead to negative consequences. The former may help to identify a target group of people, their economic activities and housing settlements – generally bracketed as 'informals' – which then would be targeted for policy intervention and actions for raising them to the standard of 'formals' (of course, in a gradual and evolutionary fashion). The second process may also be seen positively as a response to the desirability of adopting a production technology according to factor endowment of an economy (i.e. labour and local resource intensive), so that it helps to reduce capital intensity and relax undue standards and rigidities. However, in both the trends of IS growth described above, there are dangers: the informals or the marginalised may never receive the desirable planning and policy attention, whereas many FS stable jobs protected by the social security system (which has been won by long struggles of the working class) would get lost forever. What would be the actual outcome would depend on the degree of success in efficient and imaginative management of the economy by pursuing policy interventions which derive maximum benefits from the ongoing processes? In the context of urban planning, the above growth trends call for accommodation of the sector by the modification of regulations governing land use and building standards.

Accommodation of the Informal Sector

In order to adopt an accommodative approach, a first basic task is to convince the government, planners and policy makers that the UIS contributes significantly to national economic and social development in a variety of ways. The well-documented contributions of the UIS are summarised in Table 3.5 for this purpose. The second level of pursuit is essentially to overcome some well-known doubts and concerns with respect to the sector. In the following sections, we will outline them with indications of ways and means to remove or reduce them. At the third level, it is important to make a change

Table 3.5   The informal sector – its analytical and development contributions/roles

Analytical contributions/role	Development contributions/role
• Has provided an analytical framework for understanding how people outside the formal sector occupations, who sometimes constitute 40–60 per cent of the working population of a city, subsist.	• Facilitates expansion of non-agricultural occupations.
	• Offers an urban foothold to the migrating rural poor, the landless, the unemployed, the underemployed and the victims of natural disasters and social persecutions.
• Provides an useful understanding of the role of a buffer that this sector plays between the rural subsistence sector and the urban industrial and modern service sector.	• Provides income earning opportunities to the urban poor.
	• Keeps the cost of urbanisation low.
• Has provided an analytical framework to study the labour segmentation in the urban labour market.	• Stimulates growth of market economy.
	• Makes available low-cost labour.
• Provides a base for understanding the linkages between the labour market segmentation and land and housing markets.	• Supplies basic goods and services at affordable prices and conveniences.
	• Provides an environment for growth of a dynamic and patriotic group of national entrepreneurs.
• The understanding of these linkages and segmentations in the labour market help to formulate analytical frameworks that help in drawing policy guidelines.	• Stimulates innovations and technological adaptations among the small informal sector workshops.
• In conformity with equity goals of development, the informal sector concept has sharpened understanding and policy interventions for harmonious growth of cities.	• Enriches the cultural life of the city.
	• Facilitates keeping touch with own soil, people and culture (in the absence of this role of the informal sector, a city would be similar to a colonial city or an enclave superimposed from outside for maintaining neo-colonial links).

Source: reproduced from Amin, 1991, p. 6.

in the overall attitudes towards the informal sector so as to ensure an enabling environment for people to exercise their 'right to work' without hindrance from any quarter. This right to work would not involve 'providing' jobs by the government, neither would it require acceptance of working situations that conflict with desirable patterns of urban planning. Such recognition would rather require urban planners to bring necessary changes in the urban planning paradigms, which largely originated from the Western experience of urban-industrial development that had no significant presence of UIS. Success in these three directions would ensure an optimal environment for the operations, expansions and development of the UIS business.

Doubts and Concerns and Their Resolutions

Institutional restrictions and harassment This takes place in situations when the laws and bye-laws are restrictive towards small business initiatives. Simple change in attitudes would alter restrictive practices and harassment which, in turn, would lead to an increase in income and the ability of IS to pay for urban services, infrastructure uses and health care.

Labour exploitation If seen dialectically, labour exploitation is not a big issue. In the case that someone who cannot have a better job than working as a maid, it should not be seen as a mere mechanism of oppression or exploitation since this situation will not last long. Indeed, the trend of change from a maid status to a garment factory worker or the like has become quite widespread in many labour-abundant Asian countries. This is not to suggest that there will be no need for interventionary policies for reducing labour exploitation and improving terms of employment and working conditions.

Self-exploitation This should pose no problem as long as the income earned is commensurate with the hard and long working hours that characterise most self-employment. However, in an ideal case, the awareness of health hazards associated with long hours of work need to be raised.

Conflict with modern urban spatial planning In the context of this chapter, this conflict is an important area to deal with. This conflict would largely disappear if modifications and adaptation are made to the basic paradigms of urban physical planning, that is, zoning, safety and security standards. Land use standards, building codes, housing construction standards, separation of working and living areas need to be scrutinised in order to make them realistic for accommodating the IS people and their habitats within the context of modern urban spatial planning.

Conflict with creation of optimal investment environment Providing incentives for national or multinational investment requires neither the eviction of hawkers nor the clearance of slums and squatters. City beautification programmes will be a misplaced priority and counterproductive if they would entail a loss of jobs for those who need them most.

Possible social unrest from the presence of an informal sector An apprehension has been that the people of this sector would initiate social revolution. The fear is that the rage against those who have amassed wealth through unfair and corrupt means may even engulf the honest and healthy elite who tend to have a conspicuous urban affluent living. The fact of the matter is that petty-bourgeois aspirations[4] are the dominant feature of the informal sector participants, that is, owners seek to expand their businesses while workers like to own a business or become self-employed. Thus, the informal sector participants cannot be seen to be waiting for initiating revolution. If such a threat exists at all, this can be contained by opening-up opportunities for social, occupational and upward mobility and thereby integrating the UIS physically, socially and economically.

Changes Necessary for Accommodation

1) The need to change attitudes of the dominant group of society (from hostility to accommodation).

2) The legality/illegality question should be shelved as long as one earns a living through honest and time-honoured means. There is no good reason why informal sector activity should not be seen as a legitimate economic activity.

3) Planners and policy-makers must realise that *cities are primarily places for earning*, that is, places to do business for income by exchanging goods and services. The mounting problem of shelter, urban services and transportation will be substantially reduced if cities are seen as centres for earnings and *not* for permanent settlements by total withdrawal from one's respective rural roots.

The supportive legal, institutional and work environment could move in phase, or simultaneously, according to the prevailing environment and resource availability in a country. On the basis of points made above, four major sequential steps can be identified in adopting an appropriate accommodating approach to the sector. The steps are noted as follows:

1) changing the attitude towards the formal sector from hostility to acceptance;

2) moving away from a restrictive to an accommodating approach;

3) viewing accommodation as an enabling strategy; and

4) utilisation of the enabling condition to raise resources for service-rendering regulations.

Right to Work of the Informal Sector

The basic task here is to bring a fundamental change in attitude towards the informal sector by accepting an individual's right to earn a living from an honest occupation. This should be followed by steps and actions which abolish harassment and extortion by police, local touts, 'musclemen' of influential people, and those who represent various law-enforcing agencies.

Accommodating the UIS in Physical Planning Layout

The following examples[5] of physical accommodations of IS, adopted by several Asian cities, illustrate that the idea of accommodation is practicable.

1) The city of Yogyakarta has accommodated the city hawkers in planning by keeping provisions for their operations along the main road side: the walkway has been designed of 2.5 m width to make it possible for the IS to operate.

2) The integrated hawkers management programme of Metro Manila has made it possible for fleamarket hawkers to be organised and assigned spaces for business operations.

3) The footpath typists of Dhaka have been provided with protective sheds

along some of the roads and in front of the Government Secretariat.

4)   In Bangkok, the soi motorbike service has been a remarkable modal complement to the formal modes of transportation as their operations are limited mainly to the inner sois.

An Enabling Environment for UIS

For this purpose, policies should move a little beyond mere accommodation to creating facilities and the adoption of actions which would *enable* the informal sector operators to pursue their operations and businesses, as well as to consolidate their position in society. Enabling policies could, for example, involve the provision of government information for the purpose of mutually beneficial formal-informal sector linkages.

Service Rendering as a Regulatory Approach to UIS

Instead of adopting the traditional regulatory approach, a service-rendering one comprising the following ingredients is considered essential.

1)   Providing access to credit and raw materials for increasing productivity and income.

2)   Services for meeting shelter needs. Here again, there are a number of examples and experiences which shed light for adopting future directions. Yogyakarta Street Vendors Union Co-operative in Indonesia has been successful to construct shelters for the members through their own savings and self-managed construction. With regard to shelter needs, it is important to keep in mind the following:

   a)   interdependence between the informal sector activities and shelter; and

   b)   variability in shelter needs of the informal sector workers.

Depending on the affordability and phase of migratory and life (job) cycle, the shelter needs of the IS workers are bound to be different. This is illustrated in Table 3.6. In light of the illustration contained in this table, target-specific shelter provision can be devised for the UIS workers.[6]

Table 3.6   Variability in shelter needs according to migratory/life (job) cycle phase

Migratory/life (job) cycle phase	Type of shelter need
1 For footholders, adapters, and competitors	1 Rental housing
2 For network extenders and consolidators	2 Settlement improvement with land tenure
3 For successors and identity seekers	3 New housing with amenities

Source: summarised from Sastrosasmita and Amin, 1990.

3) Health and family care is an area of service-rendering which would not require much new capital investment. If committed and patriotic NGO workers or trained social/community volunteers can be organised to provide health and family care services, substantial benefits would accrue to the IS workers in terms of improvement on the quality of life.

4) Education and training needs should constitute a top priority since they cut costs for: i) enterprise development; ii) human resource development; and iii) the improvement of the quality of life. In this area, mere literacy and numeracy raising programmes will do a lot to change the informal sector for the better.

5) Raising of consciousness and awareness is another area of action. Although the benefit of this action may not be so tangible to start with, its effects will be of a profound nature with the passage of time. Reduction of exploitation, harassment and the improvement on the quality of life will not be possible if such services are absent. Trade union workers and leaders, social and community workers, and NGO workers could all contribute to this task.

Desirable Legal Environment

Desirable legal environment can be ensured with:

1) clarity in laws, bye-laws and legal procedures should be placed as the top

priority so that ambiguity or unclarity in laws will not create a situation for harassment, extortion of bribery and evictions, or unnecessary relocation;

2) acceptance of informal sector activities as legitimate economic activities; and

3) application or implementation of labour laws, enterprise establishment acts, labour protection laws. With the understanding that the current situation is under *transition*, the implementation of laws must be done in a gradual and an evolutionary fashion.

Desirable Institutional Environment

A conducive institutional environment for the informal sector can be ensured by simply:

1) removing restrictive practices, such as the requirement of having permanent resident registration for sending children to public school, and the collateral requirement (only formal sector employment!) for government subsidised housing or similar government provided services/ supports. It must be noted that, in a market economy, all access to goods and services should be according to purchasing power but *not* by virtue of a formal sector job or any other *status*; and

2) expanding access of informal sector workers to public institution-created services (shelter, health, education, water, electricity, etc.), on the basis of *equity* and *willingness to pay*.

Desirable Work Environment

A desirable work environment can be ensured by:

1) identification of hazardous occupations (such as scavenging amongst health injurious items and similar recycling work) and ensuring the scavengers, waste-pickers and recyclers know about the hazards involved;

2) provision of toilet and washing facilities in the workplace;

3) provision of an eating facility in or near the workplace;

4) promoting employer-employee relations to improve working conditions; and

5) careful examination of the prevailing interchangeable use of shelter and workplace.[7] Such interchangeable uses are sometimes beneficial but there exist unacceptable situations. While a single room can sometimes be used both for work and rest/sleep, too many people working and sleeping together could be injurious to health. Indeed, among other advantages, shelter-workplace combination facilitates child-rearing.

Summary and Conclusions

The chapter's basic goal is to explain the compulsions to accommodate the informal sector into urban planning and environmental management. The analysis shows that the urbanisation pattern in Asia is marked by a paradoxical trend: growth of mega-size cities with a simultaneous presence of the huge informal sector. It appears that this trend is bound to continue until a high level of economic development is reached by the Asian economies which still have a huge labour pool and a vast rural-agricultural sector. The size of the informal sector may not necessarily get reduced substantially even with a high degree of economic growth since a huge number of people are waiting to enter the U-I sector. Their only alternative is to resort to the income earning opportunities in the informal sector. This chapter explains the fundamental market forces and economic logic that are in operation in the relationship between the R-A and U-I sectors of a typical Asian economy that give rise to: 1) a particular pattern of human settlements (with continuous flow of migration from the rural to the urban area, from the small urban centres and cities to the major metropolises, and the less-developed to the more developed countries); 2) huge inequality in income distribution between the rural and urban areas, between backward and developed regions of a country, and between the urban-industrially less-developed and the urban-industrially more developed countries; and 3) environmental degradation and complications (pollution, congestion and other life-threatening environmental consequences). From this perspective, the chapter has made a general case of an overriding importance for effective policy and planning to reverse these trends. The supremacy of market logic and increasing globalisation in the contemporary world has, instead of reducing, raised the significance of conscious and conscientious policy and public actions, within the context of urban planning.

Available evidence suggests that metropolisation in Asia, instead of reducing the size of the informal sector, is providing new impetus to its growth. The huge population base of major Asian countries, relative to capital and other material resources, and vast qualitative changes in the production structure of metropolitan economies have made the presence of the informal sector particularly relevant for the efficient functioning of the urban economy.

It is ironic that despite living in a 'metropolitan area' and performing a functionally crucial role in the urban economy, 50–65 per cent of the urban working people of a typical Asian metropolis do not have access to basic urban services. It is evident that the traditional approach of urban planning is not very functional in the Asian metropolises, as supported by the investigation of 100 largest metropolises of the world on the basis of 10 different aspects of urban living. Why is it so?

This paper has shown that metropolisation in Asia is taking place in the middle of the huge presence and growth of the informal sector. The existing paradigms of urban planning can neither cope with the presence of such a huge sector nor facilitate the living of those who create their own jobs, shelters and services. The quality of these jobs, standard of these shelters and adequacy of these services are of course substandard. In dealing with this inadequacy, it calls for setting realistic housing or service standards with regard to time, place, context, income, resource-base, and culture. However, in some cases, standards and long term plans may render shelter provision and basic urban services costlier or totally inaccessible. For example, an adoption of a master plan can immediately raise land prices of a particular area and consequently, it is impossible for many households to have access to the land and housing in that area.

This suggests that physical planning is ill-equipped in dealing with economic and social forces. Urban planning seems to work best in spatial terms in the industrially developed rich countries with a relatively large number of people having good income and stable employment, backed up by strong financial and institutional support, and with only a very small proportion of real 'dregs'. However, the Asian metropolises represent a totally different configuration with millions of people having limited or no regular income at all. Planning skills largely learnt from the developed countries thus have only limited application in the Asian metropolises with a limited number of affluent people in the vast sea of low-income or poor people.

This chapter, therefore, makes out a case for the accommodation of the informal sector in the urban economy by bringing changes in attitude and urban planning paradigms which have mostly been defined by western values and technological and development experiences. An accommodating posture and design towards the information sector bear the potential of better management of the urban economy and environment. It may result in higher productivity of the urban poor, and hence, higher income and affordability to pay for urban services and infrastructure. Physical accommodation of the informal sector in the urban void can be an effective tool in urban planning, as illustrated by instances from several Asian cities where such an accommodating strategy has already been adopted.

Notes

1   Of the many alternative nomenclatures that have been proposed for labelling this sector, an interesting one is that of 'non-Westernised sector', which very aptly denotes the economic activities belonging to the informal sector as the ones which are not yet fully Westernised (Hackenberg, 1980). This is interesting to us since this nomenclature indirectly vindicates the view of this chapter that western paradigms of urban planning cannot be applied to the Asian cities precisely because there exist so many 'non-Westernised' entities. This alternative name to the informal sector, as suggested by Hackenberg, has been defined and propagated by ILO and is now widely accepted by the mainstream development literature.

2   The first wave of development in Asia saw the rise of Japan. The second wave brought the four 'little tigers': Hong Kong, Singapore, Taiwan and Korea. The third wave has now encompassed Malaysia, Thailand, Indonesia and Southern China.

3   Interested readers may turn to Amin, 1994 for elaboration of these economic explanations and their consequences for human settlements, income distribution and living environment.

4   Amin, 1981, contains data that show the dominance of such aspirations and forward-looking attitude. Marginalisation or proletarianisation was not found as a major feature that could give rise to revolutionary fervour among the UIS participants.

5   A recently completed PhD thesis (Perera, 1994), which demonstrates the potential and outlines the scope of physical accommodation of UIS in the urban-built environment through imaginative and effective utilisation of urban voids, contains a wealth of information in much details on these and other similar examples.

6   For a comprehensive and implementable target-specific housing scheme for UIS workers, see Sastrosasmita and Amin, 1990.

7   Lipton, 1980 calls this 'fungibility' in resource use. However, he observes that urban informal enterprises bear less potential than their rural counterparts in economising on resources in this way.

References

Amin, A.T.M.N. (1981), 'Marginalisation vs. Dynamism: A Study of the Informal Sector in Dhaka City', *Bangladesh Development Studies*, Vol. IX, No. IV, pp. 72–112.

Amin, A.T.M.N. (1989), 'Macroperspectives on the Informal Sector in Selected Asian Countries', a report prepared for ILO-ARTEP.

Amin, A.T.M.N. (1991), 'A Policy Agenda for the Informal Sector in Thailand', Discussion Paper THA/20, produced for ILO-ARTEP and Department of Labour, Ministry of Interior, Bangkok.

Amin, A.T.M.N. (1992), 'Urban Planning in Metropolitan Area of Asia: The Challenges of Accommodating the Informal Sector', *The Southeast Asia Journal of Tropical Medicine and Public Health*, Vol. 23, Supplement 3 (Asian Workshop on Nutrition in the Metropolitan Area), pp. 119–29.

Amin, A.T.M.N. (1994), 'Economic Logic of Resource Flows between the Rural-Agricultural and the Urban-Industrial Sector: Consequences for Human Settlements, Income Distribution and Living Environment' in Salokhe, V.M. and Chongrak Polprasert (eds), *Proceedings of the 1ST SERD Seminar on Selected Aspects of Environment*, Bangkok: Asian Institute of Technology, pp. 41–56.

Cole, W.E. and Sanders, R.D. (1985), 'Internal Migration and Urban Employment in the Third World', *American Economic Review*, Vol. 75, No. 3, pp. 481–94.

Hackenberg, R.A. (1980), 'New Patterns of Urbanisation in Southeast Asia: An Assessment', *Population and Development Review*, Vol. 6, pp. 391–420.

Harper, M. (1992), 'The City as a Home for Enterprise: Has Anything Changed for the Informal Sector', *Habitat International*, Vol. 16, No. 2, pp. 143–8.

Lewis, W.A. (1954), 'Economic Development with Unlimited Supplies of Labour', *The Manchester School of Economic and Social Studies*, Vol. 22, No. 2, pp. 139–91.

Lipton, M. (1980), 'Family, Fungibility and Formality: Rural Advantages of Informal Non-Farm Enterprises versus the Urban-formal State' in S. Amin (ed.), *Human Resources, Employment and Development – Volume 5: Developing Countries, Mexico City*: Proceedings of the Sixth World Congress of the International Economic Association.

McGee, T.G. (1971), *The Urbanisation Process in the Third World*, London: Bell.

PCC (1990), *Cities: Life in the World's 100 Largest Metropolitan Areas*, Washington, DC: Population Crisis Committee, pp. 16.

Perera, L.A.S.R. (1994), 'Urban Void as a Spatial Planning Tool for Accommodating Informal Sector Enterprises in the Urban Built Environment: An Explanatory Study in Colombo, Sri Lanka', unpublished PhD thesis, submitted to Human Settlements Development Programme, Asian Institute of Technology, Bangkok.

Sastrosasmita, S. and Amin, A.T.M.N. (1990), 'Housing Needs of Informal Sector Workers: The Case of Yogyakarta, Indonesia', *Habitat International*, Vol. 14, No. 4, September, pp. 75–88.

Todaro, M.P. (1969), 'A Model of Labour Migration and Urban Unemployment in Less Developed Countries', *The American Economic Review*, Vol. 59, pp. 138–48.

UNDP (1993), *Human Development Report 1993*, Delhi: United Nations Development Programme.

Williamson, J.G. (1988), 'Migration and Urbanisation' in H.B. Chenery and T.N. Srinivasan (eds), *Handbook of Development Economics*, Vol. 1, Amsterdam: Elsevier Publishers B.V.

4 Urbanisation Versus Urban Planning Practice in Malaysia: Considerations, Prospects and Possibilities

CHE MUSA CHE OMAR

Introduction

Urbanisation and urban growth are of central concerns to urban planners. A large part of planners' time and energy is spent on regulating these phenomena to improve the quality of urban living. As such, one important condition for effective planning an understanding of the dynamics of urbanisation and urban growth. This paper tries to locate these urban phenomena in a larger socio-economic framework of capitalism.

The Process of Urbanisation

Many criteria can be used to measure the urbanization process. However, currently, most writers prefer using demographic statistics. Urbanisation can be defined as a process of growing population concentration whereby the proportion of the total population which is classified as urban is increasing. Thus, the degree of urbanisation can be measured quite easily:

$$U\% = P_u/P_t \times 100 \text{ , where}$$
$$P_u = \text{urban population}$$
$$Pt = \text{total population.}$$

It is a crude way to measure urbanisation as it only takes into consideration the concentration of people to measure a concept which commonly means not only the presence of a relatively large number of people in a settlement but also a way of living which is markedly different from rural living. In fact, urbanization results from far-reaching economic transformations in the national and international plane (Robert, 1978).

Another drawback of the above definition is that the definitions of urban settlement, towns or cities are very different from country to country (Ishak, 1990). In some countries, a town is any gazetted one with a local government, irrespective of the number of people. In the United States of America, for instance, a town may consist of only a few households, with a duly elected mayor. In Malaysia, such a small settlement would not even merit the status of a village. Even within a country, official town boundaries may be outdated or arbitrarily drawn. That data, based on these boundaries, are not good indicators of urbanisation.

Despite its weaknesses, the demographic statistical definition of urbanisation is very simple and allows for some measure of comparison across countries and over time. For instance, according to the United Nations Population Division, between 1950 and 1975, the number of people living in the urban areas of the developing world increased from 275 million to 793 million. By the year 2000, the number of urban dwellers in the developing countries is predicted to increase to 2.12 billion, or almost two-thirds of the world's projected 3.1 billion urban population (Ghosh, 1984). These numbers, though necessarily crude, certainly help urban planners to appreciate the intensity of the problems they are facing or are likely to face in the near future.

In fact, forecasts and studies on the population growth trends of Asian Metropolitan cities in the next decade share a common consensus that the urban population will continue to grow as a result of natural increase and migration to the metropolis. According to the United Nation's forecast, by the year 2010, there will be 204 Asian cities with more than one million people and about half the number of the largest cities in the world will be in Asia (United Nations, 1980). Moreover, by the end of this century, most of the world's mega cities, that is, urban areas with more than 10 million people, will be located in the Asia-Pacific region. The imminent pressure on metropolitan cities is apparent and regional and local governments will be hard-pressed to manage the situation particularly in the areas of metropolitan housing, infrastructure development and environmental management.

Urban Growth in Malaysia

Generally speaking, rapid urban growth in Malaysia is a phenomenon which often coincides with rapid industrialisation. With the new thrust towards becoming an industrialised nation by 2020, the pace of rural-urban migration has been increased. This has resulted in the continual growth of many Malaysian towns (Table 4.1). For instance, the population of Kuala Lumpur grew from 316,200 in 1957 to 648,300 in 1970, 937,800 in 1980 and 1,156,000 in 1990. In other words, the population of metropolitan Kuala Lumpur grew by about 266 per cent between 1957 and 1990, whilst in other major towns, such as Georgetown, Ipoh, Kuala Trengganu and Kuantan, there were substantial population increases after readjustment of town boundaries that were made in the 1980s.

Table 4.1   West Malaysia: population growth of major metropolitan areas (in '000s)

Metropolitan	1957	1970	1980	1990
Kuala Lumpur	316.2	648.3	937.8	1,156.1*
Georgetown	234.9	332.1	250.0	4,661.0
Ipoh	125.8	257.3	283.3	300.3
Johor Bahru	74.9	144.6	188.9	249.9
Klang	69.8	113.6	146.2	196.2
Malacca	75.6	99.8	88.1	105.6
Petaling Jaya	16.6	93.4	218.3	
Seremban	52.1	90.1	136.3	136.3
Kota Bharu	38.1	69.8	108.3	170.6
K. Trengganu	29.4	59.5	96.7	186.6
Taiping	48.2	54.6	55.3	149.3
Kuantan	23.0	43.4	75.6	136.6

Note

*   Kuala Lumpur-Petaling Jaya metropolitan area.

Source: Department of Statistics, 1992.

Kuala Lumpur grew from 316,200 in 1957 to 648,300 in 1970 to 937,800 in 1980 and 1,156,000 in 1990. In other words, the population of metropolitan Kuala Lumpur grew by about 266 per cent between 1957 and 1990. Other major towns, such as Georgetown, Ipoh, Kuala Trengganu and Kuantan also experienced substantial population increases after readjustment of the town boundaries in the 1980s.

Although urban population growth in Malaysia has not reached alarming proportions, the problems associated with rapid urban growth have already been evident. These indicate that the rate of urban growth will soon outpace the ability to provide adequate urban housing, infrastructure and waste disposal systems.

Extensive use of private vehicles in urban areas has also led to congestion and an increase in air-borne pollution. The rapid rural-urban drift has resulted in the negligence of the rural areas, as illustrated by the presence of idle land, shortage of labour, deterioration of rural services and inhabitation of the rural areas by only the very young and very old.

Urbanisation in Malaysia has been gaining momentum, particularly since the declaration of political independence in 1957. Prior to that, the vast majority of the population in the country, amounting to some 85 per cent, were located in the rural areas and only 15 per cent were found in urban centres, each of which has a population of more than 1,000 people (this definition of an urban centre is adopted by the Census of Population). However, over the last 35 years, a tremendous change has occurred in Malaysia largely through the efforts of the government which has implemented a number of five-year development programmes. One of these programmes strongly encourages the development of industries in the major urban centres of Peninsular Malaysia, accompanied by the opening-up of million of acres of virgin forest land for cultivation of rubber, oil palm and other commercial crops. As a result, the percentage of urban population has increased and is now roughly 46 per cent of the population of Peninsular Malaysia. The 46 per cent represents an enormous increase because the latest census of population carried out in 1990 redefined an "urban centre" as a place having a population of over 10,000 persons.

Problems Related to Urbanisation

The process of urbanisation in Malaysia has been accompanied by rural-urban migration, but the scale of such movement is nowhere near those which have been reported for places such as Jakarta in Indonesia (World Bank, 1990). For the City of Kuala Lumpur (Capital of Malaysia), population increased from less than a million in 1980 to 1.1561 million in 1990 and is projected to reach 1.8 million by 2000. The other major urban centres such as Georgetown (Penang), Ipoh, Johore Bahru and Klang have also experienced increases of population.

Owing to the prevailing urban growth, approximately 6.5 million or 36 per cent of the total population of Peninsular Malaysia are found in large metropolitan areas (refer to urban settlements of 75,000 and more persons) or major towns (refer to urban settlement of 10,000 to 74,999 persons). Another 10 per cent are located within smaller towns of 5,000 to 9,999 persons. It has been found that the level of urbanisation in Peninsular Malaysia has been clearly and positively associated with the level of development. In other words, periods of rapid economic growth have been concurrent with periods of rapid urban growth. To date, Malaysia has enjoyed a sustained period of rapid economic development which has brought considerable improvements to the well-being of its people. The urban sector should play a more important role in facilitating the future development of the country by having much stronger urban planning policies and programmes such as:

1)   directions for investment programmes according to priority projects;

2)   priorities for institutional developments which direct and motivate urban planning policies and programmes for development;

3)   to identify and prioritise key policy issues and formulate feasible policies; and

4)   to have a good information system and reliable data for urban planners and policy makers.

Urban Development Policies

The National Development Policy (NDP) has emphasized the restructuring of the service and industrial sectors, formal large-scale, natural resource-based enterprises as well as the promotion of more ethnically balanced residential patterns in housing areas (FMP, 1986). A growing urban sector has provided opportunities for restructuring, as illustrated in the 1990 census in which the population of Kuala Lumpur is beginning to look more balanced between the Malays and non-Malays. The restructuring of urban areas was first made possible by the promulgation of the New Economic Policy (NEP) by the government, which was replaced by NDP in 1990. The NEP was enunciated after the racial riots in May 1969, and the targets were eradicating rural poverty, eliminating the identification of race with occupation, and the enhancement of the share of Malays in urban wealth and properties.

The NEP was implemented by the Second Malaysia Plan (1970) and now the NDP continues with the target of letting the Malays own 30 per cent of urban land and wealth. This is to be achieved through a variety of fiscal, administrative as well as political measures, backed up by vigorous programmes of education and training for the Malays to adjust to urban life. The political leaders have promised that the target will be achieved without sacrificing the rights and social status of the non-Malays. On the other hand, the non-Malays are requested to support the policy by assisting and training the Malays to participate in commerce and industry either as partners or on their own.

The Sixth Malaysia Plan, which was launched in 1990, is a great forward thrust made by both the private and public sectors to eradicate urban and rural poverty. One of the areas in which considerable efforts and attention have been devoted is in the field of housing (SMP, 1990). The number of housing units seems to be adequate for all the people within the country if only measured in terms of the number of shelters. However, the 1980 population census, which included a census of housing, had shown that there was a considerable disparity in the quality of housing available. The largest area of concern is among the urban poor who occupy squatter huts which consist of flimsy structures and are generally put up through the self-help principle.

The Sixth Malaysia Plan also emphasises the need to build more low-cost housing, defined as houses which could be built and sold at a market price of below RM35,000 Malaysian per unit (US$13,000). The public sector housing has set for itself a target which is even lower than that, namely houses to be built and sold at RM25,000 Malaysian per unit (US$9,600). This conclusion has been reached as a result of the fact that approximately 50 per cent of the urban population are earning below RM500 Malaysian per month (US$190). If 15 per cent of their wages is to be devoted to housing, these people could only afford between RM70 to RM80 per month to pay the hire-purchase payments as well as electricity, water and other public utilities services. With this limited amount of money, they could only afford houses of around RM25,000 in which payments are to be made over 20 to 25 years at an interest of 8.5 per cent per annum.

The initial concern for the need to build low cost housing has now been gradually replaced by the realisation of the need for an integrated approach to housing issues. Ideally, people should be housed in areas with opportunities for employment in the proximity, education facilities for the children, marketing and shopping venues for the housewives, and light industrial establishments where part-time jobs are available to women to supplement their family income.

As a result, professional institutions have taken an active part in promoting and fostering the idea of a balanced approach to housing. Although the planning machinery needs to be improved and the number of qualified planners is still rather limited within the country, the progress, to date has been encouraging. It must be noted that although politics does play a crucial part in planning decisions, in general, technical considerations have not been completely ignored.

Issues in Urban Planning Practice

Problems encountered in Malaysian cities have been partly related to the need for restructuring. Efforts to solve the problems are being made through administrative, political and fiscal measures. Problems of squatters or marginal settlements are also present in the Malaysian urban centres but the proportion of urban population involved is much less than those in the major cities of our neighbouring countries. Approximately 25 per cent of the population in Kuala Lumpur are estimated to be accommodated in marginal settlements built, through self-help principles, on land belonging to the state or on private land where the owners are absent or have no immediate use of it. Some of these squatter huts are occupying valuable central locations and plans have been drawn up to gradually replace them with modern developments. At the same time, other places are made available to accommodate the squatters so as to absorb them into society. The process of replacing the squatters is necessarily a lengthy and protracted one. This is largely due to the need to negotiate with the squatters on the questions of compensation and provision of alternative accommodation. The government has adopted a humanitarian policy and in almost every case the squatters are compensated even though they are occupying land illegally (USM, 1993). This increases the cost of redevelopment and, to a considerable extent, slows down the replacement process.

Another urban problem confronting cities and towns in Malaysia is the increasing volume of motorised traffic. Like most cities in the world, motor vehicles has overtaken many towns where roads and bridges are never intended to cater for such a large volume of traffic. Even the capital city of Kuala Lumpur suffers from massive traffic jams particularly during the peak periods and whenever there is a rain storm causing floods and havoc. The problem in Malaysian towns is aggravated by the fact that almost the entire urban population is dependent on the motor car for transportation. There are no alternative forms of transportation, such as railways or waterways, and almost all transportation within the towns is facilitated by private motorcars and public service vehicles.

Over the years, city administrators and municipal authorities have attempted in a piecemeal fashion to deal with the problems. Kuala Lumpur, in particular, has commissioned the services of traffic and transportation consultants to prepare transportation master plans for the city as a whole. However, the recommendations made by the consultants have not been implemented in total, but only in part. This has aggravated the problems and has led to a clamour, especially among the professionals, for a proper development strategy for Kuala Lumpur and other major cities in Malaysia. It must be realised that the construction of new residential accommodation, new office and commercial accommodation, shops, cinemas, markets, and places of entertainment is going on all the time and these have further added to traffic congestion. The classical approaches such as widening existing roads, constructing multi-level junctions, installing traffic lights in urban areas, and constructing flyovers at busy junctions have been carried out throughout the city but the overall problem remains largely unsolved.

In the meantime, cities are sprawling outwards. In the case of Kuala Lumpur, it is rapidly developing into a mega city and has the potential to form a mega urban region spreading from Kuala Lumpur to Kelang with a distance of around 40 km. The whole of the Kelang Valley, in fact, will become a continuous urban region and the Kelang Valley Development Proposals are being planned to cater for such an eventuality. It has also been suggested that alternative development corridors should be created from Kuala Lumpur to Seremban, and from Kuala Lumpur to Rawang and beyond (Che Musa, 1992).

The Need for Solutions

There are indeed many problems and issues that must be resolved by the government and national resources have to be distributed among various sectors. These pressing issues will require the government's attention and the cooperation from the private sector to improve the environmental conditions and to achieve the avowed objective of upgrading the quality of life for all Malaysians.

The major issues confronting city administrators and acting as sources of frustration to city dwellers are:

1) persistent and continuing proliferation of squatters and unauthorised dwellings within, and on the fringes of urban areas;

2) rising tide of vehicular traffic, leading to mounting casualties on roads and massive traffic jams which are now a daily occurrence in major towns and cities;

3) urban sprawl whereby, in the absence of properly drawn development plans, haphazard development of housing estates is permitted;

4) general deterioration of the physical and social environment, particularly in the urban residential areas; and

5) industrial development being permitted without regard to the principals of separation of conflicting purposes.

The presence of squatters in the urban areas of our country can only be interpreted as a reflection of the inability of both the public and private sectors to produce suitable housing and to bring such housing within the reach of the poor. Some squatter areas occupy the central area locations although most of them are concentrated on the urban fringes. At one time squatter areas were considered to be an eyesore and their presence constitutes an uneconomical use of the land. This is still true today but the authorities and the public, in general, are sympathetic to the squatters who need help rather than condemnation.

Rehousing squatters in planned urban settlements, such as multistorey flats and low-rise high density terrace houses, has been attempted by various city and municipal authorities and statutory bodies, such as the State Economic Development Corporation (SEDC) and Penang Development Corporation (PDC) in Selangor and Pulau Pinang respectively. However, there are more than 60,000 squatter families in the Kuala Lumpur-Petaling Jaya area alone (USM, 1992) whilst the rate of construction of low-cost housing by the various authorities amounts to less than 10,000 units per annum (DBKL, 1985). As a consequence, it is unlikely that the full replacement of the squatter huts by planned low-cost housing can be achieved within the next generation.

As an alternative, the government and various municipal and statutory authorities should attempt to upgrade the squatter areas by a process known as kampong improvement. This has been successfully implemented in Jakarta and various cities in Indonesia. It should also be attempted in this country to ensure that the unfortunate inhabitants of our squatter areas are provided, at least, with basic amenities such as piped water, water-borne sanitation, electricity supply and good all-weather footpaths. If necessary, these amenities could be supplied on a communal basis to ensure that the physical environment is upgraded and improved. Children in the squatter areas would then have the opportunity to grow up healthily and be educated so that they can contribute to society in the future.

Planning System and Practice in Malaysia

The need for environmental planning and an integrated approach of planning as a whole is fully realised by the authorities. Indeed, the department of the environment has been set up within the Ministry of Technology, Science and Environment and an environmental protection act has been promulgated to protect our environment against bad land use practices. The necessary legislation and instrument for environmental protection is already in place but the implementation of the policies will depend on the manpower resources available. It is hoped that, in the years to come, their efforts will bear fruit in the form of a well balanced environment where the interests of all groups and sectors are adequately catered for.

Unfortunately, during all these developments, the agency responsible for physical planning, namely the Department of Town and Country Planning, has been relegated to a secondary role. It must be admitted that the economists have dominated the scene and physical planners have been called in to assist only when all major decisions have already been made. This sad situation has arisen partly due to the lack of sufficient manpower resources within the planning departments.

The Government passed the Local Government Act and the Town and Country Planning Act in 1976 which together brought about the reorganisation of local government authorities, involving the integration of several contiguous authorities into larger integrated authorities. The Town and Country Planning Act is basically an adaptive act in which local authorities are encouraged to prepare regional plans covering their areas to guide future developments. For specific areas coming under development, the integrated local authorities could direct the planning agencies to prepare development plans covering the action areas. However, to date, progress has been slow in reorganising the local authorities and the planning agencies have not commenced actions to prepare development proposals for action areas.

The Department of Town and Country Planning has been unable to prepare structure plans and local plans for smaller urban settlements. Indeed, at present, only a few urban settlements within the country has gazetted or published structure plans to guide future development of the urban areas. Some planning consultants have been commissioned to prepare structure and local plans for urban centres although few have been completed to date (Table 4.2).

Table 4.2   Status of structure plan preparations in peninsular Malaysia (as of June, 1993)

Local authority	Inception report	Report of survey	Draft structure plan	Completion date
Kuala Lumpur	1978	1980	1982	1984
Seremban	1980	1982	1986	1987
Johor Bahru	1982	1984	1985	1985
Ipoh	1982	1984	1986	1987
Kota Bahru	1980	1984	1987	1989
Kota Star	1980	1988	1990	1991
K. Terengganu	1982	1984	1988	1991
Penang Island	1983	1985	1987	1991
Klang	1984	1985	1986	1988
Bangi	1984	1985	1987	1990
Kangar	1985	1987	1991	1992
Seb. Perai	1985	1987	1992	1993 *
Kemaman	1985	1987	1991	1992
Taiping	1987	1990	1992	1993 *

Note

*   Near completion.

Source: Town and Country Planning, Federal, 1993, various structure plans.

The need to control the size of city and urban centres has long been realised. However, we still have a long way to go for urban centres to be under the control of properly prepared development plans, setting the limits of growth and determining the pattern into which the city will be allowed to develop.

To be effective, a planning system must be able to command an efficient use of resources, be responsive to planning needs, have a vertical integration of various planning stages, have inter-agency co-ordination, and co-operation among all parties involved in the process of implementation, monitoring and evaluation (Figure 4.1). The Malaysian Planning System needs to be modified to achieve these ideals. Yet, any proposal must also consider constraints in Malaysia and resources available.

The following steps should be taken for an effective and efficient development control through the structure and local plan concepts, which integrate economic and financial planning with physical planning:

1) all parts of the Town and Country Planning Acts (Act 172) should be adopted for all local planning authorities;

2) all local planning authorities should complete their structure plans along the proposed lines;

3) all local planning authorities should prepare local plans to guide development control based on the draft structure plan;

4) All local planning authorities should ensure that all planning applications for development must comply with the requirements by the TCP Act; and

5) layout plan approval procedures should be simplified and a fixed time limit should be applied.

Anyhow, to most urban planners or town planners, their main duty is to prepare development plans to promote, guide, regulate and control the use and development of land in the urban areas. Since development plans have important implications on the political economy of the city and the overall society, urban planners, in performing their jobs, play an important role in society, as suggested by the first Malaysian Director General of Town and Country Planning in Malaysia '... there's so much need (as I see) to be done. Do something good today as if you are going to die tomorrow!' (Norimah, 1984).

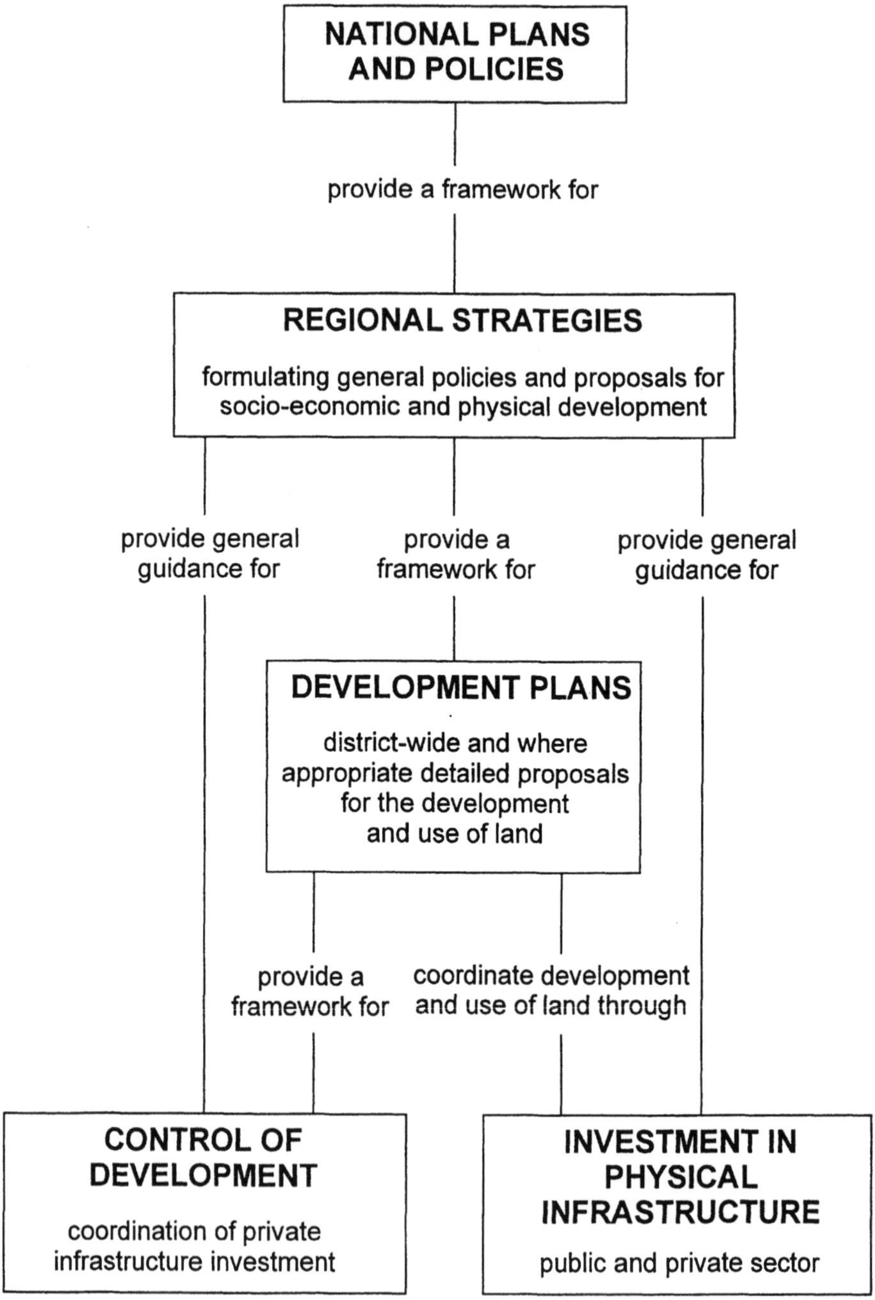

Figure 4.1   The inter-relationship between planning policies at various levels

Conclusion

The more discerning readers may have read between the lines that major urban problems in Malaysia are related to rapid urbanisation. These problems, if left unchecked, could eventually lead to crises. Why are these problems demanding urgent attention? Most of the reasons have been touched upon earlier in this paper. To recapitulate, they include the following:

1) overcrowding squatter settlements and related housing shortage;

2) increasing traffic congestion, breakdown of road systems, and ineffective piecemeal efforts to cope with the situation;

3) urban sprawl and uncontrolled growth of conurbations.

The problems of urbanisation in Malaysia have not rapidly degenerated into crises because our population is relatively small. With approximately 18 million people, about 65 per cent are located in the rural areas. Therefore, the situation can be contained and kept largely within tolerable limits in the short term. However, it is not likely to improve and may get worse if no counter-measures are taken by planners in Malaysia. The efforts of the government represent part of this effort but more could be done, especially by the private sector if they are encouraged and given the opportunity to participate and contribute. Indeed, there is no doubt that most planners want to contribute to the development of the Malaysian society. Looking at the Malaysian towns, it is true that without urban planning and development control, urban living as appeared today would be impossible.

References

Che Musa (1992), 'Vision 2020: The Strategic Planning for Kuala Lumpur/Klang Valley as a Mega Urban Region of Malaysia', paper presented at the International Conference on Managing the Mega-Urban Regions of ASEANS Countries: Policy Challenges and Responses Asian Institute of Technology, Bangkok, Thailand 30 November–3 December.

Department of Statistics (1992), *Population and Housing Census*, Kuala Lumpur: Department of Statistics.

Ghosh, P.K. (ed.) (1984), *Urban Development in the Third World*, Westpoint, Conn: Greenwood Press.

Ishak, M. (1990), 'Urban Growth and Rural Planning for Peninsular Malaysia Priorities and Directions', paper presented at the 12th Congress of the Eastern Regional Organisation for Planning and Housing (EAROPH), *Major Planning Issues in the 1990s: Decade of Turning Point*, South Korea, 5–7 September.

Norimah, M.D. (1984), 'So Much Need To Be Done', interview with Rt Hon. Datuk Haji Mohd. Rosli Hj. Buyong, *Journal Jabatan Perancang Bandar dan Desa; 30* (in Malaysian).

Robert, B.R. (1978), *Cities of Peasants: The Political Economy of Urbanisation in the Third World*, London: Edward Arnold.

United Nations (1980), *Pattern of Urban and Rural Population Growth*, UN Publications Sale No. E79XIII9, New York.

Universiti Sains Malaysia (1993), *Kajian Bancian Setinggan, Dewan Bandaraya Kuala Lumpur*, School Housing, Building and Planning, USM: Kuala Lumpur (in Malaysian).

World Bank (1990), *World Development Report*, New York.

5   An Extended Metropolis? A Growth Triangle? Towards Better Planning for the Hong Kong and Pearl River Delta Region[1]

MEE KAM NG

Introduction

Before China adopted the open door policy in 1978, regional planning was not directly relevant to Hong Kong, which was an international city and a small open economy.[2] Hong Kong and China were supposed to be two distinct economic and political entities.[3] There was not much coordination in terms of physical planning and development between the two places. Spatial planning in the Territory was Hong Kong-centred. Also, Hong Kong's immediate hinterland, the Pearl River Delta, was then primarily an agricultural region with few dynamic economic activities.

However, China's open door policy coupled with the restructuring of the economy in Hong Kong have changed the situation dramatically. Since the early 1980s, southern China has been transformed from an agricultural backwater into special economic zones and open economic regions to lure foreign investments. In Hong Kong, the escalating land prices and labour costs have pushed the labour-intensive industries to seek lower-cost production sites. As a result, the two economies integrate rapidly, leading to various development problems, such as transportation bottlenecks, pollution and incompatible land uses.

This chapter aims at gaining a better understanding of these planning problems. In order to have a fuller appreciation of the issues involved, recent discussion on regional development in Asia will be summarised in section two. Section three discusses the restructuring of Hong Kong's economy and

China's open door policy which have led to growing economic integration between the two places. It will be argued that the lack of coordination between the local and transborder authorities in the region in infrastructure planning and provision, and a bias towards the interests of private businessmen and powerful bureaucrats in the development process have led to numerous inter- and intra-regional development problems. Sections four and five suggest some planning responses and concludes the paper respectively.

The Asian Regional Division of Labour (ARDL) and Regional Development in Asia

In the age of colonialism and imperialism, most Asian countries could be classified as the 'South' in the old international division of labour and were responsible for exporting primary products. In return, they had to pay much higher prices for importing manufactured goods from the 'North', the developed Western countries.

However, since the 1960s, a new international division of labour (NIDL) emerged. A limited number of countries in Latin America and Asia (the 'South') have started to export industrial products as well. This was partly related to the global sourcing strategy adopted by the multinational corporations (MNCs) in core countries to search for lower-cost production sites in the periphery. Governmental effort in the less developed countries to provide a favourable environment for this foreign direct investment was also an important factor. Furthermore, the decentralisation of the productive activities to the periphery by MNCs was made possible because of innovations in production, transportation and communication technologies (transactional revolution).

After the second world war, foreign investments in Asian countries were mainly confined to coastal cities, especially those which had served central functions in the colonial era. Rapid industrialisation in these cities had led to various urban problems, serious rural-urban migration and hence rural poverty. In order to tackle these problems, in the 1960s and 1970s, governments in the more developed Asian countries such as Japan, Korea and Taiwan had encouraged industrial decentralisation through various policies (Ng, 1994). However, the decentralisation process did not last long. The restructuring of these economies and keen competition between themselves amidst rising protectionism in the developed countries have forced the labour-intensive and low value-added industrial sectors within these economies to search for yet cheaper production sites outside their own countries.

As a result of the NIDL and the subsequent transnationalisation of economic activities from Japan and Asian NIEs since the 1980s, an Asian regional division of labour has emerged (ARDL). Rapid urbanisation and industrialisation can be seen in various parts of Asia. However, unlike the previous *city-based urbanisation* processes, there is a rapid development of a *region-based urbanisation* process (McGee, 1995). McGee calls these regions *desakota* zones. Others call them 'mega-urban regions' (MUR), 'dispersed metropolis' or 'metropolitan interlocking regions'. Derived from Bahasa Indonesia, *kota* means town or city and *desa* means village. Usually, *desakota* refers to sub-regional clusters of villages, towns and rural-based enterprises (McGee, 1995, p. 7). When MUR spans national boundaries, it is called a growth triangle (GT).[4]

The unprecedented pace of development in MURs or GTs is largely driven by the investors' profit-maximisation motive and the local authorities' eagerness to attract foreign investments. These regions receive most of the foreign direct investment. However, many common problems can also be found in these rapidly urbanising and industrialising regions.

The mentality which equates development with economic and industrial growth has resulted in overt competition between local authorities for global investment through increasingly generous subsidies and guarantees, and the regulation of the labour force and the civil society (Douglass, 1995). However, such generous provision for foreign investments has led to financial difficulties for local governments in providing basic economic and social infrastructure.

Also, as 'urban' planning regulations may not apply in the 'rural' parts of MURs/GTs, it is difficult for the state to control development in the non-urban areas. As a result, the originally rich agricultural land may be encroached upon and transformed into other uses, such as polluting industrial uses. With the proliferation of industries and other illegal or incompatible land uses, environmental degradation is inevitable. The situation is worsened when MURs or GTs start to attract immigrants from the less developed regions. Increased population and demand for various social amenities have boosted land prices, worsening, among others the problem of urban poverty.

While all these development issues are not new to urban planners, the spatial extent of the manifested problems is unprecedented. Hence, most countries have difficulties in establishing an appropriate institutional infrastructure to cope with the new development. Management of urban growth in MURs and GTs, therefore, is very often fragmented and chaotic. It is a formidable task to achieve genuine cooperation among the competitive local or cross-border governments.

Therefore, McGee (1991, pp. 21–2; 1992, pp. 18–20) argues that it is important for governments to decide on the priority and urgency of developing MURs. If the government decides that development in MURs should be given top priority in the national development strategy, a complementary, equitable and acceptable internal division of labour within the MUR should be formulated. Communication systems should be improved to facilitate economic interaction. Development control should be extended to non-urban areas in order to avoid the existence of incompatible land uses and environmental degradation. This is especially important for transborder regions where a unified response to matters such as pollution, coordination of transport, communication systems and shared infrastructure is required (Parsonage, 1992, p. 21). As argued by Parsonage (1992, p. 19), the need to develop an integrated and institutionalised coordination mechanism is more important in GTs because of the more complex and varied interests involved.

For the implementation of a metropolitan-based management strategy, a redistribution of decision-making power is necessary. In order to enhance the ability of local control, there is need to develop further a capacity to generate income, particularly from land and property tax; and to develop, among other things, human resources at all levels to manage MURs/GTs. These in turn would require genuine coordination, understanding and careful planning among various local or cross-border authorities. Furthermore, governments need to address the roles of various actors such as the private sector, community groups and non-governmental organisations in MURs or GTs' governance.

With this background knowledge, let us now trace the genesis and major problems in the development of the Hong Kong-Pearl River Delta region.

Development in the Hong Kong and Pearl River Delta Region

Genesis

Unlike a typical MUR, the political economies and institutional set-up in the Hong Kong-Pearl River Delta region are more complicated: there are two Special Administrative Regions (Hong Kong in 1997 and Macau in 1999), together with the long established Special Economic Zones (Shenzhen and Zhuhai), the Pearl River Delta Open Economic Zone, and Guangzhou which is the provincial capital and one of the coastal open cities.

Similar to other MURs/GTs, the origin of this region has much to do with the ARDL, that is, the restructuring of a core economy (Hong Kong) and its need to search for cheaper production sites and the adoption of more flexible economic policies in a less developed economy (China) in order to attract foreign investment. Ever since the 1970s, manufacturing industries in Hong Kong have been facing serious problems both domestically and internationally. The escalating production costs within Hong Kong have made the once internationally competitive labour-intensive industries less viable. Also, rising protectionism in developed countries has posed increasing difficulties for the export-oriented industries. Coupled with keen competitions from other Asian economies, the labour-intensive manufacturing industries in the Territory have to find lower-cost production sites.

The open door policy adopted by China since 1978 has provided a breathing space for Hong Kong's labour-intensive and low value-added industries. In 1980, Special Economic Zones (SEZs), not unlike the outward processing zones established in Taiwan in the 1960s, were set up in China to attract foreign investment. Three of these SEZs are located in Guangdong Province. In 1985, part of Guangdong Province was designated as the Pearl River Delta Open Economic Zone, including four cities (Foshan, Jiangmen, Zhongshan and Dongguan) and 12 counties (Shunde, Nanhai, Xinhui, Panyu, Taishan, Kaiping, Baoan, Zhengcheng, Doumen, Enping, Heshan and Gaomaing), covering an area of 22,700 km². In December 1987, this small Pearl River Delta Economic Open Zone was expanded to include seven cities and 21 counties, constituting an area of 45,005 km².[5] Excluding Hong Kong and Macau, the existing population in the region is more than 20 million.

Investment from Hong Kong has proved to be very important for Guangdong's economic takeoff. Economic cooperations between Hong Kong and the Pearl River Delta since 1978 are usually in the form of *sanzi qiye* ('equity joint ventures' with Chinese and foreign capital, Chinese and foreign 'cooperative joint ventures', and 'wholly foreign-owned ventures') and *sanlai yibu* enterprises (enterprises engaging in export processing, assembling and manufacturing according to foreign materials, designs and parts, and compensation trade). By the end of June 1991, there were about 20,000 Hong Kong enterprises conducting 'sanlai yibu' investment in Guangdong employing more than two million workers (HKTDC, 1991, p. 3).

Hong Kong's productive investment in the Pearl River Delta has changed Guangdong's economic structure. In 1978, 73.7 per cent of the total labour force in Guangdong were engaged in the agricultural sector. However, in 1997, the traditional agricultural sector employed only 41 per cent of the labour force, whilst the remaining 59 per cent were employed in the secondary and tertiary sectors (GSB, 1998, p. 121). In other words, workers in the non-farming sectors had increased from about six million in 1978 to about 22 million in 1997 (op. cit.). Between 1986 and 1990, more than 9.2 million rural labourers transferred from farming to non-farming occupations, of which the majority switched to working in rural enterprises (Wen and Zheng, 1992, p. 64). At the same time, the commercialisation rate of the agricultural sector in Guangdong had also increased from 43.7 per cent to 75 per cent in 1997 (GSB, 1998, p. 271).

Similar to other MURs, local authorities in the Pearl River Delta compete fiercely with one another to attract foreign investment. This, together with the administrative complexity in the region, have prevented genuine cooperation between local and transborder authorities to plan for the region as a whole. Therefore, development plans made by local authorities are self-centred, and planning responses to development issues are ad hoc and reactive. Let us examine these problems further.

Lack of Coordination Between Local/Cross-border Authorities

Similar to other MURs, development planning in the Pearl River Delta has been poorly managed.[6] Coordination between Hong Kong and southern China in terms of planning and developing transport and other infrastructure has been slow, ad hoc and reactive. For instance, a cross-border Infrastructure Coordinating Committee (ICC) between Hong Kong and China was not formed until January 1995, more than a decade after the integration of the two economies.

The lack of coordination and unnecessary competition among local authorities is manifested in the development of airports and ports. For instance, three international airports have either been under construction or completed in Hong Kong, Macau and Shenzhen, within an area of less than 1,500 km². The respective designed capacities of these airports are 87 million, 12 million and 4.5 million passengers per year. Also, Zhuhai is building another local airport of international standards while Guangzhou is planning to have a new international airport to replace the existing Baiyun Airport. The ultimate capacity of the new airport in Guangzhou would be 65 million passengers per year.

Competition is fierce in terms of port development in the Hong Kong-Pearl River Delta region as well. The Chinese ports of Chiwan, Shekou, Yantian, Ma Wan, Zhuhai and Dongjiaotou, and Hong Kong all lie within a 35-nautical mile circle. In fact, the Shenzhen city alone has eight harbour zones (including ports and/or terminal facilities): Shekou, Mawan, Chiwan,

Dongjiaotou, Yantian, Huangtian, Shayuyong, and the Shenzhen inland river area. All eight ports are privately owned and they have a high degree of autonomy.

According to the Shenzhen Ports and Harbour Administration Bureau, the city will increase its cargo handling capacity to 60 million tonnes with 122 berths by the year 2000, and 120 million tonnes with 162 berths by 2010. With a handling capacity of 23 million tonnes, Shenzhen harbours recorded a throughput of only 33.57 million tonnes in 1997. In fact, many of the ports are under-utilised. For instance, the planned capacity and actual throughput in 1997 in Shekou and Yantian were 50 million and 9.8 million tonnes, and 40 million and 3.6 million tonnes respectively.

In Hong Kong, under the Port and Airport Development Strategy (PADS), the first four berths of Container Terminal 8 came to operation in July 1993. Container Terminal 9 will be built at southeast Tsing Yi Island while Container Terminals 10 and 11 will be built at Lantau, the location of Hong Kong's new international airport.

Serious duplication and a lack of coordination also exist in the development of roads, railways, water supply, and energy provision, etc. To deal with these problems, it was reported that in October 1994, Guangdong Province finally initiated a planning exercise for the Pearl River Delta Economic Zone (Lam, 1995, p. B3). A Planning and Coordinating Group, comprising mayors of cities and representatives of relevant Ministries and Departments in the Pearl River Delta, was formed under the leadership of the Deputy Governor of the Guangdong Province. Coordinations in terms of networking infrastructure development, spatial distribution of primary, secondary and tertiary economic activities, building a hierarchy of cities, environmental planning to combat pollution problems and ecological degradation, and human resources planning, etc. are emphasised. As all these will involve a redistribution of power and resources among various local authorities, it is doubtful if a solid and feasible plan can be worked out. In fact, it was reported that while Guangzhou, the provincial capital, was very enthusiastic about the plan, other cities in the Delta were not very positive about the proposals (Yun, 1995, p. B3). Although the Pearl River Delta Region Plan was finalised in 1995, the lack of financial resources and enforcement power present difficulties for the Provincial Government in ensuring its implementation (Ng and Tang, 199).

In terms of transborder planning and coordination between Hong Kong and the Delta, it was only in 1995 that a proper channel was established (the ICC) to discuss various bold ideas generated by different local authorities in the region. It was reported that Shenzhen has proposed the development of a multi-billion dollar mass transit railway to link up with the light rail network in Tuen Mun and Yuen Long in Hong Kong, and a road and rail bridge across Deep Bay; and the Zhuhai Special Economic Zone has suggested the construction of a Zhuhai-Hong Kong Bridge to link Zhuhai with Lingding Island and eventually to Tuen Mun in northwest Hong Kong (Yeung, 1994).

Biased Development towards the Interests of Private Sectors and Local Authorities[7]

While the central government in China provides the open door policy, localities are expected to fund the construction of their own infrastructure in order to attract foreign investment. With limited resources, the largely autonomous local authorities tend to emphasise the provision of those infrastructures which can facilitate capital accumulation. Without proper checks and balances in the land use planning process (Ng and Wu, 1995), cadres in the local authorities may also use their public power to steer spatial development for their own benefits.

When local authorities lack the capital to provide proper infrastructure, private investment is solicited. However, as private developers are only interested in profit maximising projects, investments have been unevenly distributed in different sectors. Transportation facilities, which are not well developed in Guangdong and yet are vital to the circulation of commodities and hence capital, are popular channels for foreign investments. The Guangzhou-Shenzhen-Zhuhai 123-kilometre super-highway, the first phase of which was completed in 1994, is one of the show cases. It was estimated by the Jardine Fleming Broking Limited that the super-highway would bring about HK$20 billion net profit to the Hopewell Construction Company within 13 years after its operation in 1994. In fact, other property tycoons have also signed contracts with Chinese authorities to build super-highways within Guangdong. For instance, Guangdong Enterprises (Holdings) and Bechtel Group had established an equal joint venture to finance and build a US$400 million toll-high way in Shenzhen (Wong, 1994).

Tycoons from Hong Kong have also invested in Guangdong's energy sector. Again, Gordon Wu's Hopewell Holdings had built big power plants in Guangdong Province. The Kadoorie family's China Light and Power (CLP), the larger of the colony's two electric utilities, has a 25 per cent stake in the Daya Bay nuclear project, a US$3.7 billion nuclear power station 50 km from Hong Kong.[8]

Port development in the Pearl River Delta has also attracted container terminal operators from Hong Kong, mainly the Hong Kong International Terminals Limited which controls 50 per cent of the Territory's container port capacities. The Hong Kong International Terminals Limited has a 70 per cent stake in the development of the deep-water Yantian Port (SCMP, 8 July 1994) and the Jiuzhou and Gaolan Ports in the Zhuhai Special Economic Zone.

In recent years, real estate development, a popular speculative activity within Hong Kong, has developed rapidly in the Pearl River Delta. This situation is not dissimilar to the GT at Singapore-Johore. The escalating property prices and mushrooming of construction sites have brought various social and environmental problems to the dwellers in the Pearl River Delta.

The bias toward building productive facilities or infrastructure has led to a neglect of providing physical and social infrastructure which is not profitable to provide and yet vital to the prosperity of the booming economy. The need for a proper flood control mechanism is a case in point. Because of the pursuit of economic growth, valuable agricultural land has been turned to barren open development zones waiting for development that may never come. Also, with the end of the commune systems, little cooperation has been made among local authorities to plan and develop mitigation measures for the flood-prone Pearl River Delta area.[9] It was reported that in July 1994, heavy rainfall in northern and eastern Guangdong had affected more than 4.5 million people in 248 towns and villages and some 12,000 people had been made homeless (Zheng and Liang, 1994). Economic losses amounted to $257 million.

In addition, as regulations on industrial safety have not been vigorously implemented, industrial accidents have been increasing. It is only in 1994 that the Labour Law was adopted specifying workers' rights regarding salary, working hours, vacation time, social insurance, and labour disputes. However, it remains to be seen how effective its implementation is and how much the workers who toil in the sweatshops know about their legal rights.

Furthermore, given the stress on economic growth, for self benefits or otherwise, local authorities seem to have little idea about the need to control development and plan the cities as better living and working places for the people. Incompatible land uses are tolerated in the race for more development. For instance, it was discovered after the disastrous explosion of a dangerous chemical storage site at Qungshuihe, Shenzhen, that the hazardous site was surrounded by different kinds of potentially explosive and toxic material, oil and petroleum depots, timber factories, and food and vegetable stores and markets. Since planning and development control mechanisms in China are 'black-box' operations (Ng and Wu, 1995), the lives of the ordinary people become increasingly vulnerable in the course of 'development'.

The stress on industrial development has led to serious encroachment of agricultural land. Almost all farmland along the major roads in the Pearl River Delta have been converted into industrial use. Cultivated land in Guangdong had decreased nine per cent from 41.68 million mou in 1978 to 34.48 million mou in 1991 (GSB, 1998, p. 271).

Encroachment of agricultural land and rapid industrial development, coupled with weak and ineffective development control mechanisms have resulted in many problems. Environmental degradation is one. Also, land prices escalate as a result of brisk economic development. The rising land prices have further attracted speculative activities which contribute to inflation problems.

Similar to other GTs, the Pearl River Delta has attracted many migrant workers from other provinces. Foreign investors are happy to employ these migrant workers as they are cheaper. Some immigrants from provinces outside Guangdong engage in agricultural activities as local farmers turn to non-agricultural pursuits. However, the existence of these outside workers cause a lot of social management problems. Movements of these workers (over a million) to and from their native places have caused great strain to the transportation, especially the railway, network. Many unlucky ones who cannot find proper jobs in factories have ended up engaging in illegal activities and begging, etc. Theft, drug problems, gambling, prostitution, and labour disputes can also be found.

The eagerness to attract labour-intensive and low value-added industries and a neglect to invest and upgrade its own research and development (R&D) capability mean that the region is not equipped to upgrade its industrial structure. In 1989, 90 per cent of enterprises in Guangdong failed to reach the level of comprehensive technological expertise, which is set as an international standard for 1980s enterprises (Zheng, 1991, p. 47). A survey conducted in 1989 showed that engineering staff and technicians employed in Guangdong's rural enterprises amounted to 41,000, only 1.2 per cent of the 3.32 million labour (of which one-third were illiterate or semi-illiterate) (op. cit., p. 48).

Since Hong Kong, the major source of industrial investment in Guangdong, is not particularly strong in industrial development and hence not being able to help in upgrading industrial capacity in Guangdong, the prospect of Guangdong becoming a strong industrial base is uncertain.

Some Suggested Planning Responses

As shown in the above case study, two fundamental problems for regional development planning can be found: the lack of a political will among local authorities to coordinate in terms of physical planning and infrastructural development; and the development tends to bias towards the interests of the private sectors and the cadres themselves. As a result, unnecessary competition to attract investment have resulted in duplication of efforts and wasting of resources. The economically-biased development strategies adopted by various local authorities have created many environmental and social problems.

How can we improve the situation? A two-pronged strategy is suggested. Regional planning should be effectuated and the overall goal of development in the region should be carefully reconsidered. Recent discussion on Asian urbanisation (section two) has shown that the lack of an integrated national, regional and local development strategy and appropriate institutional arrangements to effectuate the strategy has contributed to most of the problems in MURs and GTs. Hence, it is a high time for the Chinese government to review its coastal-oriented and urban-biased development strategy which has contributed to the growth of the Hong Kong and Pearl River Delta region.

To overcome unnecessary competition among local authorities, the Planning and Coordinating Group in the Pearl River Delta and the Infrastructural Coordinating Committee between Hong Kong and China, while going in the right directions, may still not be effective measures. Perhaps, one of the most urgent tasks is to set up a powerful regional-level organisation.[10] This organisation should be given the mandate to devise a widely consulted and commonly accepted development strategy. More importantly, the organisation should have the power to implement the strategy. Of course, mutually suspicious local and cross-border authorities may not welcome another layer of bureaucracy to rule over them. And the governance issue of balancing top-down planning and bottom-up initiative is always a tough one. However, like other MURs and GTs, the Hong Kong-Pearl River Delta region should start exploring ways of managing a MUR which involves 'one country, two systems' and numerous local authorities.

One of the strategic issues meriting top priority is the regional division of labour between Hong Kong and southern China, and within southern China itself. In the past decade, Hong Kong, like Japan, Taiwan and Singapore, has exported its labour-intensive, low value-added traditional industries elsewhere and focuses on higher value-added economic activities. However, unlike Japan and other NIEs, Hong Kong has been slow in fostering high-tech industrial development. This may be detrimental to industrial development in the region in the long run. It is because the prospect of further developing labour-intensive and low value-added industries is not optimistic in the Pearl River Delta, as its comparative advantage is changing rapidly and the region is facing extremely keen competition from other MURs and GTs in Asia.

Nevertheless, there are formidable difficulties in developing high-tech industries in the region. Within Hong Kong, the small-scale industries have been spending very little in research and development (R&D). As industrial development in the territory dwindles, people have looked towards other speculative projects for making quick money. Also, until recently the government has been very slow in fostering high-tech industries or 'directing' industrial development in the territory. In Guangdong, industries also lack the capacity or resources for upgrading.

The emphasis on profitable economic activities has incurred heavy environmental and human costs in Guangdong's development process. This suggests that planners in the region should reconsider the fundamental goals and objectives of development and the following questions should be asked: What is development for? Who should get (lose) what, where, when and how in the development process?

Of course, this complex problem cannot be resolved by planners alone. Planning problems are always wicked problems. However, it is believed that if top-down efforts and local initiatives are mobilised to devise a regional development strategy which takes into consideration the national economic development strategy, local comparative advantages and aspirations, appropriate policy measures can then be formulated to tackle problems in various aspects.

Conclusion

We have come a long way since the 'Decade of Development' in the 1960s and the 'Second Decade of Development' in the 1970s to realise that development is not just about economic growth. In fact, many of the characteristics and issues that we have witnessed in MURs and GTs are not generically different from the century-old problems found in urban areas. While we visualise a region-based urbanisation process, can we afford not to ask whether this kind of 'development' is what we want, after the failing paths to modernity through industrialisation and urbanisation? Can we turn a blind eye to the huge human costs that have been incurred in these so-called booming regions? After all, what sorts of development are we witnessing? Who should make a judgment? And why?

If we agree that development is more than just economic growth, many of the typical questions that we ask in the planning field need to be addressed seriously: if planning is to arrest market failures in the public interest, then whose interests should we adopt? What should be the ultimate goals and objectives in developing MURs or GTs, if we do not just confine our vision to profit-maximisation? What rules should be adopted in the course of negotiations and compromises to counter a skewed distribution of power among various political economies within a region? How can we maximise the utilisation of private investment for the public interests, including environmental conservation?

If the political dimension of regional development should not be neglected in the theorisation process, perhaps more research should be done on this aspect which seems to be little understood and analysed.[11]

Notes

1   This research was supported by the Urban and Environmental Studies Trust Fund, the Centre of Urban Planning and Environmental Management, the University of Hong Kong.

2   Economic linkages between Hong Kong and China have existed for decades. For instance, historically, the Pearl River Delta has been a source of fresh agricultural produce and water supply to Hong Kong. In addition, legal and illegal immigrants from China had been an important source of cheap labour for the manufacturing industries (see Lethbridge, 1980; Schiffer, 1991).

3   Hong Kong was acquired by Britain from China in three stages: Hong Kong Island was ceded to Britain by the Treaty of Nanking in 1843 after the First Opium War; Kowloon Peninsula and Stonecutter's Island were ceded to Britain by the First Convention of Peking in 1886; and the New Territories by a 99-year lease under the Second Convention of Peking in 1898. Consequently, the New Territories is scheduled to revert to China in 1997, while Hong Kong Island and Kowloon Peninsula theoretically are British 'in perpetuity'. Of course, after the Sino-British Joint Declaration of 1984, the whole of Hong Kong is scheduled to revert to China in 1997 and becomes a Special Administrative Region.

4   The term 'growth triangle' was first coined by the then Deputy Prime Minister of Singapore, Mr Goh Chok Thong, in December 1989 to mean a new form of sub-regional economic cooperation in ASEAN involving Singapore, the southern Johor state of Peninsular Malaysia and the Riau Province of Indonesia.

5   The enlarged Pearl River Delta Open Economic Zone occupies 25 per cent, 26.5 per cent and 43.5 per cent of Guangdong Province's area, population and gross industrial output in 1991 (GSB, 1992, pp. 67, 68 and 99).

6   In 1986, there was a planning office for the Pearl River Delta region. However, the office was disbanded in 1987.

7   As argued elsewhere (Ng, 1995), Hong Kong's development strategy before the 1980s had been biased towards economic growth. Yet because of the democratisation process, there has been increasing pressure exerted by the young political community on the government to take environmental and social concerns seriously in the planning and development process. However, this section will focus on the biased development strategy in the Pearl River Delta. For details on Hong Kong's situations, please see Bristow, 1984; Cuthbert, 1991; Castells, 1986; Pun, 1989; Yeung, 1987; Taylor, 1987; Taylor and Kwok, 1989; Yeh, 1993.

8   It should be noted that despite local efforts and foreign investments, Guangdong's power deficiency is worsening. Power output rose by more than 15 per cent per year from 1986 to 1991. But gross industrial and agricultural output value rose by 21 per cent a year. It was estimated that the province's peak-hour shortfall was at least 30 per cent in the early 1990s (Goldstein, 1992). The structure of Guangdong's electricity consumption has been changing rapidly as people purchase more electrical appliances. In the late 1980s, households only accounted for two per cent of Guangdong's electricity consumption. Since the 1990s, household consumption has accounted for at least 10 per cent of Guangdong's total power consumption.

9   This phenomenon is not restricted to the Pearl River Delta. According to Liang (1994), in 1994, for China as a whole, more than 1,400 people were killed and another 85 million affected in the flooding season. An estimated $6.16 billion had been lost through water damage.

10 According to Sun, 1994, regional planning in the Pearl River Delta can probably be formulated by a central-government endorsed department at the provincial level.

11 However, please see the exceptional work of Sklair, 1991, Smart and Smart, 1991, and Leung, 1993.

References

Bristow, R. (1984), *Land-use Planning in Hong Kong: History, Policies and Procedures*, Hong Kong: Oxford University Press.

Castells, M. (1986), 'The Shek Kip Mei Syndrome: Public Housing and Economic Development in Hong Kong', Centre of Urban Studies and Urban Planning Working Paper No. 15, University of Hong Kong: Centre of Urban Studies and Urban Planning.

Cuthbert, A.R. (1991), 'For a Few Dollars More: Urban Planning and the Legitimation Process in Hong Kong', *International Journal of Urban and Regional Research*, 15, 4 December, pp. 575–93.

Douglass, M. (1995), 'Global Interdependence and Urbanisation: Planning for Bangkok Mega-Urban Region', in McGee, T.G. and Robinson, I.M (eds), *The Mega-Urban Regions of Southeast Asia*, Vancouver: UBC Press, pp. 45–77.

Goldstein, C. (1992), 'China's generation gap: massive power programme fails to match soaring demand', *Far Eastern Economic Review*, 11 June, pp. 45–51.

Guangdong Statistics Bureau (GSB) (1998), *Statistical Yearbook of Guangdong 1998*, China Statistics Press (*in Chinese*).

Hong Kong Trade Development Council (HKTDC) (1991), *Survey on Hong Kong Domestic Export, Re-export and Triangular Trade, Nov. 1991*, Hong Kong.

Lam, C.C. (1995), 'Co-prosperity' in a special page on the Pearl River Delta Economic Zone, *Ming Bao*, 16 January, pp. B3 (*in Chinese*).

Lethbridge, H. (1980), *Hong Kong: the Business Environment*, Oxford: Oxford University Press.

Leung, C.K. (1993), 'Personal Contacts, Subcontracting Linkages, and Development in the Hong Kong-Zhujiang Delta Region', *Annals of the Association of American Geographers*, 83(2), pp. 272–302.

Liang, C. (1994), 'Neglecting Flood Control Will Prove Costly in Long Run', *China Daily*, 10 August, p. 4.

McGee, T.G. (1991), 'The Emergence of Desakota Regions in Asia', in N. Ginsburg, B. Koppel and T.G. McGee (eds), *The Extended Metropolis: Settlement Transition in Asia*, Honolulu: University of Hawaii Press.

McGee, T.G. (1995), 'Metrofitting the Emerging Mega-Urban Regions of ASEAN: An Overview', in McGee, T.G. and Robinson, I.M (eds), *The Mega-Urban Regions of Southeast Asia*, Vancouver: UBC Press, pp. 3–26.

Ng, M.K. (1994), 'Economic Transitions, Regional Disparities and Development Policies: a Comparative Study of Japan, Korea and Taiwan', paper presented at The Fourth Asian Urbanisation Conference, Taipei, Taiwan, 1–5 January.

Ng, M.K. (1995), 'Urban and Regional Planning', in J.Y.S. Cheng (ed.), *From Colony to SAR: Hong Kong's Challenges Ahead*, Hong Kong: Chinese University Press, pp. 227–60.

Ng, M.K. and W.S. Tang (1999), 'Urban System Planning in China: A case study of the Pearl River Delta', *Urban Geography*, 20, 7, pp. 591–616.

Ng, M.K. and Wu, F. (1995), 'A Critique of the 1989 City Planning Act of the People's Republic of China', *Third World Planning Review*, 17(3), pp. 279–93.

Parsonage, J. (1992), 'Southeast Asia's 'Growth Triangle' – an Extended Metropolitan Region? Problems and Prospects', paper presented at the International Conference on Managing the Mega-Urban Regions of ASEAN Countries: Policy Challenges and Responses, 30 November–3 December, Thailand, Bangkok: Asian Institute of Technology.

Pun, K.S. (1989), 'Past and Future Development of Urban Planning in Hong Kong', *Planning and Development*, Vol. 5, No. 1, pp. 7–13.

Schiffer, J.R. (1991), 'State Policy and Economic Growth: a Note on the Hong Kong Model', *International Journal of Urban and Regional Research*, June, Vol. 15, No. 2, pp. 180–96.

Sklair, L. (1991), 'Problems of Socialist Development: the Significance of Shenzhen Special Economic Zone for China's Open Door Development Strategy', *International Journal of Urban and Regional Research*, Vol. 15, pp. 197–215.

Smart, J. and A. (1991), 'Personal Relations and Divergent Economics: a Case Study of Hong Kong Investment in South China', *International Journal of Urban and Regional Research*, Vol. 15, pp. 216–33.

South China Morning Post (SCMP) (1994), 'Maersk Leads the way as YICT Tries to Attract More Lines to New Port', *South China Morning Post*, 8 July.

Sun, H. (1994), 'Institutions Responsible for Formulation and Implementation of Regional Plans for PRD', paper presented at a conference Hong Kong as a Regional Hub, jointly organised by the Hong Kong Institute of Engineers and Hong Kong Institute of Planners, CGO Conference Hall, Hong Kong, 30 September.

Taylor, B. (1987), 'Rethinking the Territorial Development Strategy Planning Process in Hong Kong', *Asian Journal of Public Administration*, Vol. 9, No. 1, June, pp. 25–55.

Taylor, B. and Kwok, R.Y.W. (1989), 'From Export Centre to World City: Planning for the Transformation of Hong Kong', *Journal of the American Planning Institute*, Summer, pp. 309–22.

Wen, S. and Zheng, Y.H. (1992), 'Rural Economic Development and Social Changes in Guangdong Province', in T. Maruya (ed.), *Guangdong: 'Open Door' Economic Development Strategy*, Hong Kong and Tokyo: Centre of Asia Studies, University of Hong Kong and Institute of Developing Economics, pp. 49–78.

Wong, J.S. (1994), 'Venture to Build Shenzhen Tollway', *South China Morning Post*, 20 May.

Yeh, A.G.O. (1993), 'Urban Development of Hong Kong in the 21st Century: Opportunities and Challenges', in Y. Yeung (ed.), *Pacific Asia in the 21st Century: Geographical and Development Perspectives*, Hong Kong: The Chinese University Press, pp. 69–103.

Yeung, C. (1994), 'Meeting Stays on Track Despite Rail Link Plan', *South China Morning Post*, 27 December, p. 2.

Yeung, Y.M. (1987), 'Cities that Work: Hong Kong and Singapore', in R.J. Fuchs, G.W. Jones and E.M. Pernia (eds), *Urbanisation and Urban Policies in Pacific Asia*, Boulder and London: Westview Press, pp. 257–74.

Yun, H.Z. (1995), 'Power Politics', in a special page on the Pearl River Delta Economic Zone, *Ming Bao*, 16 January, p. B3 (*in Chinese*).

Zhen, T.-X. (1991), *A Pearl River Delta Economic Geographical Network centring on Guangzhou, Hong Kong and Macau*, Guangzhou: Zhongshan University Press (*in Chinese*).

Zheng, C. and Liang, C. (1994), '65 Killed as Flooding Batters Guangdong', *China Daily*, 26 July.

6 Toward an Environment Facilitating Human Behaviour and Movement: An Approach to Evaluation and Manipulation of Pedestrian Space by Simulation Modelling

BAOZHE HU, YUKIO NISHIMURA AND SADAO WATANABE

Introduction

Along with rapid urbanisation and motorisation, both in Tokyo and Beijing, many problems have appeared in the central areas in terms of pedestrian-vehicle conflict and overcrowding. The physical shapes of central areas as they evolved over the past century are in many ways inhospitable to pedestrians. Many places are no longer places for people to enjoy on foot.

Main Issue

Through the pedestrian flow surveys on shopping streets in Beijing and Tokyo, it was found that many of these streets are overcrowded. For example, in Qianmen commercial area in Beijing, the unit flow of crowded places, such as the northern part of Qianmen Street and the eastern section of Dashala Street, has reached the situation from being a crowded level of 37.6 to a congested level of 47.2 persons/metre/minute (ps/m/min). In Ginza and Ueno commercial areas in Tokyo, the service level of crowded places, such as 4-Chome intersection in Ginza and the northern section of Ameyoko in Ueno, has come up to a crowded and constrained level with unit flow from 22.1 to 39.0 ps/m/min.

After a close-up study, it was found that overcrowding did not actually happen everywhere, but rather, at certain places in a commercial area. Like the case of Ginza commercial area, the number of pedestrians varies considerably over space at the same time.

Obviously, the main issue of overcrowding is the *imbalance in the use of space*. In more detail, the issue can be broken down into several questions:

1) what are the main physical factors which significantly influence the spatial distribution of pedestrians?;

2) how different is the impact of various built settings on human behaviour and movement in terms of walking speed, flow and standing behaviour?; and

3) how to manipulate pedestrian space in commercial areas where land is fairly precious.

Methodology of the Study

The methodologies used in the present study were *direct observation* and *counting*. *Photograph-taking* and *behaviour mapping* were also used. In fact, all data on the topic are obtained mainly from video, photographs and behaviour maps. Statistical analyses, such as correlation analysis, multiple correlation method and statistical indices analysis, were used to examine the data.

Review and Critique of Previous Related Studies

Although there have been many papers investigating the characteristics of pedestrian movement (Navin and Wheeler, 1969; Fruin, 1971; O'Flaherty, 1973; Pushkarev and Zupan, 1975; Takeuchi and Iwamoto, 1976; Yoshioka, 1978; Hensen, 1988; Hillier, Hanson and Xu, 1993), very few of them specifically indicated the influence of the surrounding built environment on pedestrians. It has been noticed that through our surveys, the feature of pedestrian movement/behaviour and the number of people vary considerably from place to place, even on the same street. These phenomena are considered to be caused mainly by the differences in the surrounding built environment — where it is located, how far it is from station(s) of mass transit, what kind of buildings there are, what the spatial configuration of streets is, etc. It is believed that the understanding of quantitative differences of the impact on pedestrian distribution, movement and behaviour under various built environments is certainly important and meaningful to planning and design. In order to estimate and ensure compatible pedestrian space with the surrounding built environment, it is necessary to clarify why overcrowding frequently appears at certain places. As a result, further studies are considered necessary on:

1) the influential factors on spatial distribution of pedestrians;

2) the impact of different built settings on the features of pedestrian movement, namely, speed, density, flow, capacity of space and standing behaviour; and

3) a simulation modelling to estimate, evaluate and manipulate pedestrian space.

The Influential Factors on the Spatial Distribution of Pedestrians

To understand why the imbalance in spatial distribution of pedestrians exists and to establish a practical method to manipulate pedestrian space of commercial areas where land price is extremely high, it is necessary to first explore the nature of pedestrian distribution and its relations to the built environment.

Analysis Approach and Case Study Area

Based on the 'attraction theory' (Pushkarev, 1975) and the 'configuration theory' (Hillier et al., 1993) of pedestrian movement, it is supposed that the number of pedestrians in a given site might have relations with: 1) the site accessibility, namely the distances to mass transit stations such as the subway, railway and bus, and the distances to parking lots and its sidewalk areas; 2) the attractiveness, namely, the use and floor space of buildings surrounding the site; and 3) the layout of whole area including the location of trip-generative facilities and the spatial configuration of streets (Figure 6.1).

The study area used to testify above the hypothesis is 1–8 Chomes along Chuo street in Ginza which is one of the most important commercial areas in

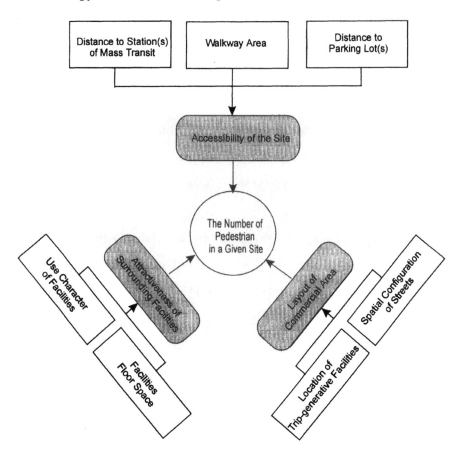

**Figure 6.1   The possible influential factors on the number of
pedestrians in a given site**

Tokyo (Figure 6.2). There are 16 blocks in the study area with a total floor
space of about 450,021 m². including retail facilities, food service, culture
and amusement facilities. Almost 85 per cent of department stores and 76.4
per cent of total commercial floor space in Ginza are located in the study
area.

Data Collection

Current facilities which exist in the areas of study are classified into six
categories. They are retail, food service, daily life-oriented, cultural and
informational, amusement and PR (public relations), and financial and business

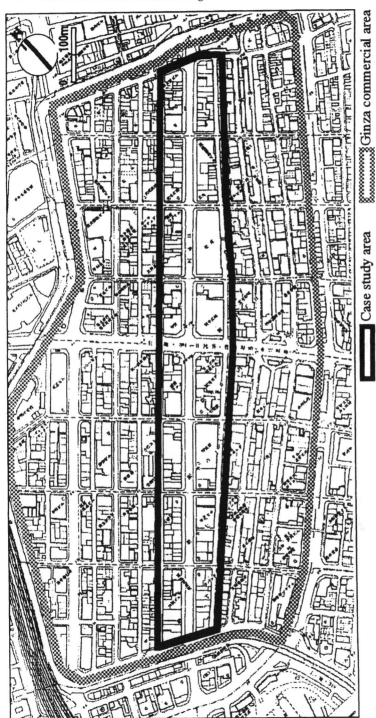

Figure 6.2 Case study area in Ginza of Tokyo

facilities. The floor space of each block by each category of facilities is calculated. The category which takes up the most total floor space is retail facilities of the department stores and speciality shops, especially in 3- to 6-Chomes. The second most is offices, especially in 1-, 2-, 7- and 8-Chomes.

Subway and railway are the dominant means of passenger transport in the Ginza commercial area. The number of passengers using subway and railway stations in Ginza is around 1,934,461 persons a day (Transport Census of Metropolises, 1985), whereas those using buses are comparatively few. Parking-lot capacity in Ginza is also very limited, to only around 4,000 cars (Geographic Description of Japan, 1982).

In the study area, there are two subway stations, one of which is Ginza Station with 880,940 passengers a day, and the other is Ginza 1-Chome Station with 34,502 passengers daily. In addition, there are two railway stations adjacent to the study area which are supposed to have certain influences on pedestrian movement in study area as well. One of the stations is Shinbashi Station located southwest of Ginza with 670,969 passengers daily, and the other is Yurakucho Station northwest of Ginza with 348,050 passengers a day. The distances from the centres of sidewalks on each block to the four stations are measured. The whole width of sidewalk cannot be fully used by pedestrians for walking because of the blockage by street furniture. The effective width of each sidewalk is determined, from which the area of each sidewalk is calculated.

The spatial configuration of streets refers to the way in which streets, alleys and open spaces are linked together to form some kind of 'global' pattern. Through the application of new techniques known as 'space syntax' to the analysis (Hillier and Hanson, 1984; Hillier et al., 1993), the spatial configuration of the study area is analysed in terms of connectivity, control value and integration measure. The integration value of each section of the streets are calculated.

The number of pedestrians is counted from pictures taken from the roofs of 16 buildings in the study area on 13 (Saturday) and 25 (Thursday) June 1992, five times a day at 10:00, 12:00, 14:00, 16:00 and 18:00.

The Analysis of Influence

The impact of building floor space and use character Since there are six categories in the building use, the independence among them is first checked. All correlation coefficients among them are found quite low in value which are under 0.40 except that between retail and office (0.68). Therefore, food service, daily life-oriented facilities, cultural and informational facilities, amusement and PR facilities are considered as independent variables, whereas retail and office are dependent in this case. The correlation between retail and offices is:

$$X_o = 14033 - 0.32597\ X_r \qquad (2\text{-}1)$$

X_o – the floor space of office facilities (m^2)
X_r – the floor space of retail facilities (m^2)

The equation to examine the correlation between pedestrians and use/ floor space of facilities is:

$$P = a+bX_r+cX_f+dX_d+eX_c+fX_a+gX_o \qquad (2\text{-}2)$$

If the mutual dependence of retail facilities and offices is considered, then the equation (2-2) becomes:

$$P = a+bX_r+cX_f+dX_d+eX_c+fX_a+g(14033-0.32597X_r) \qquad (2\text{-}3)$$

P – pedestrians on sidewalks of each block at an instant time
X_f – floor space of food service facilities (m^2)
X_d – loor space of daily life-oriented facilities (m^2)
X_c – floor space of cultural and recreational facilities (m^2)
X_a – floor space of amusement and PR facilities (m^2)
a, b, c, d. e. f. g – the coefficients

According to equation (2-3), the correlation coefficients at each survey time are examined. It is found that the multiple correlation coefficients are fairly high, ranging from 0.76 to 0.97 which indicates that there is a definite correlation between the number of pedestrians and the floor space of facilities.

A further question is which category plays an important role in correlation. The equation for such analysis is:

$$P = a+bX_{1\text{-}6} \qquad (2\text{-}4)$$

X_1 – floor space of retail facilities (m^2)
X_2 – floor space of food service (m^2)
X_3 – floor space of daily life-oriented service (m^2)

X_4 – floor space of cultural and information establishments (m^2)
X_5 – floor space of amusement and PR facilities (m^2)
X_6 – floor space of business and financial facilities (m^2)

The correlation coefficients between pedestrians and floor space of each category of facilities at each survey time are examined. It is found that the most influential category is the retail facilities, including department stores and various speciality shops at all survey times. On both Thursday and Saturday, the highest range of correlation coefficients spreads between 0.59 and 0.84. The second most influential category is the business and financial facilities, comprising offices and agencies, with coefficients ranging from 0.49 to 0.68. Although the correlation coefficients of other categories of facilities are relatively low, the changing feature of those of food service (mainly restaurants) is considerable. The coefficients at 10:00 are '-' (meaning negative impact) on Thursday morning and very low value (0.07) on Saturday morning. However, at 12:00 and 2:00 on Thursday afternoon, they become '+' (meaning positive influence) and even go up to as high as 0.41 on Saturday afternoon.

The influence of distance to mass transit station The influence of distances to stations is analysed by the following equation:

$$P = a+b(r_1/X_1)+c(r_2/X_2)+d(r_3/X_3)+e(r_4/X_4) \tag{2-5}$$

X_1 – distance to Ginza Station (m)
X_2 – distance to Ginza 1-Chome Station (m)
X_3 – distance to Shinbashi Station (m)
X_4 – distance to Yurakucho Station (m)
r_1 – H_1/H r_2-H_2/H r_3-H_3/H r_4-H_4/H
H_1 – daily passengers of Ginza Station (p/day)
H_2 – daily passengers of Ginza 1-Chome Station (p/day)
H_3 – daily passengers of Shinbashi Station (p/day)
H_4 – daily passengers of Yurakucho Station (p/day)
H – $H_1+H_2+H_3+H_4$ (p/day)

Since there are four subway and railway stations in consideration, the independence among the four variables is first checked. The distance to Ginza Station is considered as an independent variable because of the lower correlation coefficients to other variables (0.1-0.35). On the other hand, the distances to Ginza Chome 1 Station, to Shinbashi Station and to Yurakucho Station are three dependent variables with relatively high correlation coefficients ranging from 0.65 to 0.98. The correlation equations between each one are:

$$X_3 = 3621 - 0.0567X_2 \tag{2-6}$$
$$X_4 = 2201.2 + 0.078X_2 \tag{2-7}$$
$$X_4 = 23.749 + 0.2903X_3 \tag{2-8}$$

If the correlations above are taken into account, the equation (2-5) becomes:

$$P = a+b(r_1/X_1)+c(r_2/X_2)+d(r_3/(3621-0.0567X_2))$$
$$+e(r_4/(2201.2+0.078X_2)) \tag{2-9}$$

It is found that, through the analysis of equation (2-9), there is a close correlation between the number of pedestrians and the distances to stations. This is illustrated by the high correlation coefficients ranging from 0.72 to 0.98 at all survey times both on Thursday and Saturday.

The correlation coefficients between pedestrians and the distance to each station are then examined one by one to see the relative significance of each station. It is discovered that the most influential one at all survey times is the distance to Ginza Station. The distance to Yurakucho Station is also quite influential at three survey times on Thursday except at 14:00 and 16:00, whereas the third one is the distance to Ginza Chome 1 Station at 12:00 on Thursday.

The influence of sidewalk area Boris Pushkarev (1975) shows that the sidewalk area is an influential factor on pedestrian movement. It means that if a sidewalk is bigger, more pedestrians will be there than on a smaller one. It seems logical, but through the analysis of the case in Ginza, it is found that all correlation coefficients between sidewalk area and the number of pedestrians are relatively low, ranging from 0.20 to 0.58. This implies that the sidewalk area is, at least, not as influential as the other factors on pedestrian movement. One can easily verify this conclusion by visiting the study area. For instance, on the sidewalks of 4-Chome along Chuo Street, there are more pedestrians than on those of 1-Chome at almost every time of everyday, but the areas of sidewalks of 4-Chome (914 m^2) are comparatively smaller than those of 1-Chome (1,140 m^2).

The impact of facilities distribution in commercial areas From the section above, it is understood that the most influential category on shopping streets is the retail facilities. The impact of the distribution of retail facilities on pedestrian movement in the study area is examined by equation (2-10):

$$P = a+bX_{rg}+cX_{ry}+dX_{rc}+eX_{rs}$$ (2-10)

X_{rg} – retail facilities farther than the block in question from Ginza Station (m^2)

X_{ry} – retail facilities farther than the block in question from Yurakucho Station (m^2)

X_{rc} – retail facilities farther than the block in question from Ginza 1-Chome Station (m^2)

X_{rs} – retail facilities farther than the block in question from Shinbashi Station (m^2)

It is found that the correlation coefficients between the number of pedestrians and the distribution of retail facilities in the study area are relatively high ranging from 0.69 to 0.96. It indicates that there are considerable impacts from the layout of retail facilities on the spatial distribution of pedestrians on shopping streets.

The effect of the spatial configuration of streets The impact of spatial configuration of streets in the study area on the distribution of pedestrians is examined by equation (2-11). It was discovered that there is a remarkable correlation between the spatial configuration of streets and the number of pedestrians, as reflected by relatively high correlation coefficients ranging from 0.54 to 0.95.

$$P = a+bX_{iv}$$ (2-11)

X_{iv} – the integration value of each section of streets

Equations Relating Pedestrians to Built Environment

The tabulation of the aerial pedestrian count by sidewalks in the study area, matching the inventory of buildings and streets, makes it possible to statistically relate pedestrians to the built environment in terms of influential factors such as retail facilities, offices, distances to stations, distribution of trip-generative facilities, and the spatial configuration of streets. The preliminary equations that link the number of pedestrians on a given site to its influential factors at each survey time are:

On weekday:

$$P_{10} = a + bX_r + cX_o + dX_{dg} + eX_{dy} + fX_{rg} + gX_{ry} + hX_{iv} \qquad (2\text{-}12)$$
$$P_{12} = a + bX_r + cX_f + dX_o + eX_{dg} + fX_{dc} + gX_{dy} + hX_{rg} + iX_{ry} + jX_i \qquad (2\text{-}13)$$
$$P_{14} = a + bX_r + cX_f + dX_o + eX_{dg} + fX_{rg} + gX_{iv} \qquad (2\text{-}14)$$
$$P_{16} = a + bX_r + cX_o + dX_{dg} + eX_{rg} + fX_{iv} \qquad (2\text{-}15)$$
$$P_{18} = a + bX_r + cX_f + dX_o + eX_{dg} + fX_{dy} + gX_{rg} + hX_{ry} + iX_{iv} \qquad (2\text{-}16)$$

On weekend:

$$P_{10} = a + bX_r + cX_o + dX_{dg} + eX_{rg} + fX_{iv} \qquad (2\text{-}17)$$
$$P_{12} = a + bX_r + cX_f + dX_o + eX_{dg} + fX_{rg} + gX_{iv} \qquad (2\text{-}18)$$
$$P_{14} = a + bX_r + cX_f + dX_o + eX_{dg} + fX_{rg} + gX_{iv} \qquad (2\text{-}19)$$

P – number of pedestrians at an instant in time on sidewalks (p)
X_r – floor space of retailing facilities (m^2)
X_f – floor space of food service(restaurant) facilities (m^2)
X_o – floor space of business (offices) facilities (m^2)
X_{dg} – distance from the centre of sidewalk to Ginza Station (m)
X_{dc} – distance from centre of sidewalk to Ginza 1-Chome Station (m)
X_{dy} – distance from centre of sidewalk to Yurakucho Station (m)
X_{rg} – retail facilities farther than the block in question from Ginza Station (m^2)
X_{ry} – retail facilities farther than the block in question from Yurakucho Station (m^2)
X_{iv} – integration value
a,b,c,d,e,f,g,h,i,j – coefficients

The independence of variables in preliminary equations, (2-12) to (2-19), is verified. In addition to the correlations between retail facilities and offices, and between distances to Yurakucho Station and to Ginza 1-Chome Station, as mentioned above, there is a close correlation between the distance to Ginza Station and the spatial configuration of streets. The relationship is:

$$X_{iv} = 2.5061 - 0.0012X_{dg} \qquad (2\text{-}20)$$

If the correlations above are considered, in other words, if the equations (2-1), (2-7) and (2-20) are combined, the equations (2-12) to (2-19) become those as shown in Table 6.1.

Table 6.1   Equations relating pedestrians to built environment

(1) 10:00, 25 June 1992 (Thursday)
$$P = 0.0005Xr - 0.0108Xdg - 0.0082Xdy + 0.0002Xrg + 107.357 \tag{2-21}$$

(2) 12:00, 25 June 1992 (Thursday)
$$P = 0.0013Xr + 0.0034Xf + 0.0016Xdg - 0.0132Xdy - 0.0001Xrg + 128.029 \tag{2-22}$$

(3) 14:00, 25 June 1992 (Thursday)
$$P = 0.0012Xr + 0.0054Xf - 0.0667Xdg - 0.0003Xrg + 139.33 \tag{2-23}$$

(4) 16:00, 25 June 1992 (Thursday)
$$P = 0.0004Xr - 0.0671Xdg + 0.0005Xrg + 145.852 \tag{2-24}$$

(5) 18:00, 25 June 1992 (Thursday)
$$P = 0.0012Xr + 0.0005Xf - 0.0894Xdg - 0.014Xdy + 0.0002Xry + 253.673 \tag{2-25}$$

(6) 10:00, 13 June 1992 (Saturday)
$$P = 0.0005Xr + 0.0056Xdg + 0.0005Xrg + 5.969 \tag{2-26}$$

(7) 12:00, 13 June 1992 (Saturday)
$$P = 0.0326Xr + 0.0004Xf - 0.0645Xdg + 0.0006Xrg + 127.237 \tag{2-27}$$

(8) 14:00, 13 June 1992 (Saturday)
$$P = 0.0011Xr + 0.0036Xf - 0.1469Xdg + 0.0007Xrg + 256.871 \tag{2-28}$$

The significance of these equations has been evaluated by more rigorous statistical measures in terms of multiple correlation coefficient, standard error, t-value and _-value (Table 6.2). It is clear that all correlation coefficients of equations (2-21) to (2-28) are fairly high from 0.87 to 0.95 and all standard errors range from 9.84 to 33.15. Based on t-value and _-value, the confidence of each coefficient is listed in column 7, which mostly ranges from 85 per cent to 99 per cent.

Table 6.2   Statistic measures of equations in Table 6.1

Equation	Variable	Coefficient	Standard error of coefficient	t-value	_-value	Coefficient confidence (%)
Equation (2-21)	Retail (Office)	0.0005	0.0003	1.67	0.05	95
R = 0.91	D to G Station (St configuration)	-0.0108	0.0106	1.02	0.15	85
Se = 10.86						
N = 20	D to Y Station	-0.0082	0.0032	2.56	0.01	99
	Distribution of facilities	0.0002	0.0001	2.0	0.025	97.5
Equation (2-22)	Retail (Office)	0.0013	0.0006	2.17	0.025	97.5
R = 0.95	Food service	0.0034	0.0017	2.0	0.025	97.5
Se = 9.84	D to G Station (St configuration)	0.0016	0.002	0.82	0.20	80
N = 20						
	D to Y St (to 1-C. St)	-0.0132	0.0118	1.12	0.15	85
	Distribution of facilities	-0.0001	0.0002	0.5	0.25	75
Equation (2-23)	Retail (Office)	0.0012	0.0011	1.09	0.15	85
R = 0.87	Food service	0.0054	0.0028	1.93	0.05	95
Se = 18.78	D to G station (St configuration)	-0.0667	0.0239	2.79	0.01	99
N = 20						
	Distribution of facilities	-0.0003	0.0003	1.0	0.15	85
Equation (2-24)	Retail (Office)	0.0004	0.0002	2.0	0.025	97.5
R = 0.92	D to G station (St configuration)	-0.0671	0.0221	3.04	0.005	99.5
Se = 18.07						
N = 20	Distribution of facilities	0.0005	0.0003	1.67	0.05	95
Equation (2-25)	Retail (Office)	0.0012	0.0011	1.09	0.15	85
R = 0.95	Restaurant	0.0005	0.0003	1.67	0.05	95
Se = 30.59	D to G station (St configuration)	-0.0894	0.0424	2.11	0.025	97.5
N = 20						
	D to Y station	-0.014	0.0092	1.53	0.10	90
	Distribution of facilities	0.0002	0.0001	2.0	0.025	97.5
Equation (2-26)	Retail (Office)	0.0005	0.0003	1.67	0.05	95
R = 0.94	D to G station (St configuration)	0.0056	0.0051	1.1	0.15	85
Se = 10.75						
N=20	Distribution of facilities	0.0007	0.0002	3.5	0.005	99.5

Table 6.2 cont'd

Equation	Variable	Coefficient	Standard error of coefficient	t-value	_-value	Coefficient confidence (%)
Equation (2-27)	Retail (Office)	0.0326	0.0322	1.01	0.15	85
R = 0.91	Restaurant	0.0004	0.0002	2.0	0.025	97.5
Se = 20.59	D to G station	-0.0645	0.0263	2.46	0.01	99
N=20	(St configuration)					
	Distribution of facilities	0.0006	0.0004	1.5	0.10	90
						99
						85
Equation (2-28)	Retail (Office)	0.0011	0.0005	2.2	0.01	99.5
R = 0.94	Restaurant	0.0036	0.0037	1.0	0.15	
Se = 33.15	D to G station	-0.1469	0.0423	3.47	0.005	99
N=20	(St configuration)					
	Distribution of facilities	0.0007	0.0003	2.33	0.01	

The Impact of the Built Environment on Pedestrians' Behaviour and Movement

To clarify the impact of different built settings on human behaviour and movement, an investigation has been conducted at three different sites in Ginza commercial area in June 1992. The selected three sites are all on the sidewalks of Chuo street in Ginza. One site is in front of Ginza Kintaro (selling various electric toys) at 4-Chome (named hereafter as Site 1) where overcrowding frequently occurs. Another one is located in front of Yamatoya (selling clothes and shoes) at 5-Chome (Site 2), and the third site is in front of Iwasaki Spectacles Shop at 6-Chome (Site 3). Site 2 and 3 are mainly for comparison to examine the effect of different built environments.

The Features of Speed and Flow in Different Settings

A photographic method is used to record the movement of pedestrians at each site. A video camera is placed at a suitable position on the balcony of a top floor of a building beside a site to enable a bird's eye view of the observation site. According to the feature of the hourly flow of Ginza, the investigation has been done at the peak time of 5:30 p.m. and the recording period is 20 minutes at each site. Although the whole width of the sidewalk is 6 metres, the physically available width or so called 'effective width' of each site is around four metres due to the presence of planters and other street furniture such as lighting, trash receptacles and information boards. The observation lengths of three sites are all five metres which are marked by yellow tape (Table 6.3).

Table 6.3 The basic information on survey

Site location	Shops beside	Available width of sidewalk	Observation length of site	Observation time
Chome 4, Chuo Street (Site 1)	Electric toy shop with a vendor	3.8 m	5 m	5:30–5:50 12 June 1992
Chome 5, Chuo Street (Site 2)	Clothes and shoe shop, no vendor	4 m	5 m	5:30–5:50 19 June 1992
Chome 6, Chuo Street (Site 3)	Spectacles shop, no vendor	4 m	5 m	5:30–5:50 26 June 1992

The 20-minute video of each site is then divided into 120 observations, with each observation lasting 10 seconds. The speed of an observation is derived by measuring the speed of two samples of pedestrians. The density of an observation is obtained, when the moving pedestrian under study is at the centre of the observation site, by counting the number of pedestrians within the observed limits where stationary pedestrians are also included, and then dividing this number by the area of the site. The flow of an observation is easily determined from the video by counting the number of pedestrians passing a boundary line in each direction separately. These numbers are then added together as the total flow.

Although the three sites on Chuo street are all located in the central part of the commercial area, they are quite different in terms of their surroundings. Beside Site 1, there is a very popular shop selling electric toys with a small vendor in front. The shop is so totally open, colourful, and well decorated with electric toys that it looks quite attractive. Site 3, in contrast, is in front of a spectacles shop with a small door and an ordinary show window, with nothing special in it. Site 2 is just in between the two sites with a clothes shop next door to it (Table 6.4).

By data analyses, it is found that, firstly, the speed at each site with different built surroundings, varies considerably under each density. For example, the

Table 6.4   The features of surroundings of three observation sites

Site	Name of shop beside	Content of goods	Features	Others
Site 1	Kintaro	Electric toys, doll gaming facilities	Very open, colourful, Excessive decoration	Small vendor, Displaying electric toys
Site 2	Yamatoya	Clothes, shoes	Fairly open, well designed window	–
Site 3	Iwasaki	Spectacles and the goods related	Small door, General showing window	–

mean speed at Site 1, 2 and 3 with the density of 0.5 p/m^2 are 64.446, 67.722 and 75.08 m/min respectively, of which there is 3.276 m/min difference between Site 1 and Site 2, and 10.634 m/min difference between Site 1 and Site 3. Secondly, the respective mean speed of each site is around 67.7, 71.3 and 80.5 m/min for Site 1, 2 and 3, of which 3.6 m/min difference exists between Site 1 and Site 2, and 12.8 m/min difference between Site 1 and Site 3. 80 per cent of the walking speed is under 75 m/min at Site 1, near 80 m/min at Site 2, and around 90 m/min at Site 3.

By the linear regression analysis, the equations describing the relationship between speed and density at each site are developed and presented in Table 6.5.

Table 6.5   Equations of speed and density relationship in different sites

Sites	S=a+bD a (m/min)	S=a+bD b (coefficient)	Correlation coefficient (r)	No. of observations	Equation number
1	73.128	-17.364	0.61652	120	(3-1)
2	76.860	-18.277	0.58553	120	(3-2)
3	86.512	-22.864	0.58709	120	(3-3)

Compared with the results obtained by other scholars, it is clear that the features of Site 1 and Site 2 very much agree to the cases studied by Older (1968) and Yoshioka (1978) about crowd shopping streets in London and in Tokyo respectively. The result of Site 3 is quite close to the equations derived by Oeding (1963) at a shopping street in Germany and those by Fruin (1971) at a shopping street of New York. In contrast, the results are far different from the features of general streets studied by Oeding in Germany and Yoshioka about Nishishinbashi in Japan.

It is found that the unit flow values are also different from site to site at each density. The unit flow of Site 3 at each given density is about 4-9 p/m/min higher than that of Site 1 and 2–5 p/m/min higher than that of Site 2, and the value of unit flow of Site 2 is around 2–3 p/m/min higher than Site 1. This is equivalent to a reduction in width of the sidewalk at Site 1 and Site 2. In other words, more efficient use has been made by pedestrians on the site with less attractive shops, which will be studied in detail later.

Standing in Different Settings

It is found that a significant feature of pedestrian behaviour at Site 1 is that many pedestrians slow down when they move into the site and even stop and stand there for a while to look at the toys displayed, no matter how crowded the sidewalk is. From the recorded video, it is found that around 60.8 per cent of the observations of Site 1 have pedestrians standing inside the site, 36.7 per cent of Site 2 and 15.8 per cent of Site 3.

From the observations with standing pedestrians, it is further found that the standing people usually draw the attention of other people to slow down their pace or even to stop and look in the same direction, such as toys displayed. People seldom try to move away from a crowd, contrary to common belief. They cluster on the sidewalk and partly obstruct the walkway, forming an impediment to pedestrian movement. In other words, they are self-congested. Therefore, the effective width of a sidewalk determined in Table 6.3 for each survey site cannot actually be fully used for through movement because of the presence of standing pedestrians.

The loss in sidewalk width and area due to standing pedestrians, sorted by the number of people, are checked from the video (Table 6.6). It is clear that the mean area and width lost at Site 1, in each case with a different number of standing pedestrians, are higher than those at Site 2 and Site 3, and the loss at Site 2 are higher than those at Site 3.

Because of the loss in sidewalk width caused by the presence of standing pedestrians, the 'real density' for pedestrians to walk through becomes higher than the one previously calculated on the former effective width shown in Table 6.3. Take one of the observations as an example. There are 18 pedestrians including two standing ones. If the effect of standing pedestrians is not considered, the density is 0.947 p/m^2 according to equation (3-1) in Table 6.5. However, if the effect of standing pedestrians is considered, the loss of

Table 6.6   Pedestrians standing and mean losses in walkway area and width

No. of standing pedes- trians	No. of obser- vations	Site 1 Mean area loss	Mean width loss	No. of obser- vations	Site 2 Mean area loss	Mean width loss	No. of obser- vations	Site 3 Mean area loss	Mean width loss
1	23	2.50	0.50	12	2.15	0.43	6	1.85	0.37
2	22	4.40	0.87	15	3.95	0.78	5	2.15	0.43
3	11	4.90	0.98	8	4.45	0.89	5	3.05	0.61
4	10	5.10	1.02	6	4.60	0.92	2	3.20	0.64
5	3	5.50	1.10	2	4.95	0.99	–	–	–
6	2	5.95	1.19	1	4.95	0.99	1	3.60	0.72
7	2	6.45	1.29	–	–	–	–	–	–
Total	73			44			19		
%	60.8			36.7			15.8		

sidewalk in width, due to the presence of two standing pedestrians, is 0.87 meters (Table 6). Therefore the real density for 16 moving pedestrians becomes $D_r = (18-2)/(3.8-0.87)_*5 = 1.1$ p/m^2. There is 0.153 p/m^2 difference between the two figures.

It is obvious that the relations between density, speed and flow will change if the impact of standing pedestrians is considered as shown in the above example. Figures 6.3 and 6.4 show the difference in the relationship between speed and density at Site 1 and Site 2 respectively. From the two figures, it can be deduced that, firstly, the slower mean speed of pedestrians at Site 1 is partly caused by the slowing down behaviour of pedestrians who are attracted by the shops nearby and partly led by the higher 'real density' owing to the presence of standing pedestrians. Secondly, the impact of standing pedestrians at Site 2 is less significant than that at Site 1 because standing pedestrians are seldom present. In conclusion, the high density in the section where popular shops or vendors are located nearby is a result of the high percentage of pedestrians who unintentionally spend more time there. This density is certainly higher than that at sites where shops are less attractive.

The Effect of Different Built Settings

From the analyses above, it is understood that some pedestrians slow down and stand in front of popular shops, vendors and attractive show windows. This behaviour increases the densities for 1.2–18.3 per cent, reduces the

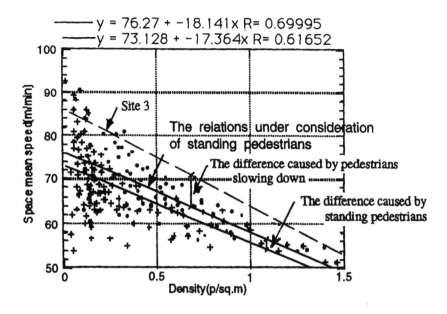

Figure 6.3 **The impact of standing pedestrians on density/speed relation at Site 1**

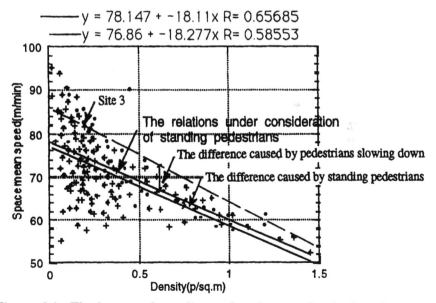

Figure 6.4 **The impact of standing pedestrians on density/speed relation at Site 2**

sidewalk widths for 0.13–0.57 metres compared with those in front of general shops, reduces the efficiencies of sidewalks, namely the pedestrian speed, 5.2–18.8 per cent, and reduces the flow for 13.1–26.2 per cent (Table 6.7).

Table 6.7   The effect of built settings on pedestrian movement and loss of sidewalk

	Site 1	Site 2	Site 3
Mean density (p/m²)	0.3101	0.3066	0.2622
Mean speed (m/min)	67.744	71.256	80.518
Mean flow (p/min.m)	18.756	21.221	23.669
% of observations with standing pedestrians	60.8%	36.7%	15.8%
Loss of sidewalk by standing pedestrians	0.5-1.29 m	0.43-0.99 m	0.37-0.72 m

An Approach to the Issue

At present, there have been several conventional ways to ease 'overcrowding' by expanding pedestrian space, such as narrowing vehicular carriage ways, increasing building heights but reducing site coverage, and building multi-level walkways below and above ground. Nevertheless, it is still essential to estimate compatible walkway space to certain surrounding built environments and to conduct reasonable arrangements and manipulation of pedestrian space by simulation modelling so as to avoid improper allocation of space in valuable commercial areas.

Simulation Modelling

It has been testified that the overcrowded situation of a site is not only determined by the number of pedestrians resulting from the distribution feature of pedestrians in a certain area. Another major determinant of overcrowding is the surrounding built environment of the site which leads to the loss of walkway space due to the standing behaviour and the slow walking speed of pedestrians (Figure 6.5). As a consequence, the simulation process consists of two parts.

In the right part of Figure 6.6, based on the study of influential factors on spatial distribution of pedestrians, the equations relating pedestrians to each unit site within certain built environments can be established as described in Section 2 above. If any changes happen at the sites, the change in pedestrian numbers of each unit site can be immediately estimated from the equations,

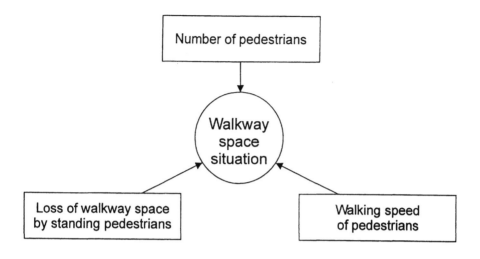

Figure 6.5 The determinants of walkway space situation

by inputting the changed data of those sites according to the redevelopment plan. The left part of Figure 6.6 summarises the study on the impact of different built settings on human behaviour and movement as described in Section 3. It has been understood that the pedestrian speed (S) is influenced by different built settings in terms of popular shops, attractive show windows and vendors. The pedestrian speed (S) and the real density (D) of each unit site can be calculated by selecting an acceptable value of unit flow (F) (Pushkarev, 1975). The equation of this calculation is:

$$D = \quad F/S \qquad\qquad (4\text{-}1)$$

D – the density of an unit site (p/m^2)
F – acceptable value of unit flow (p/min.m)
S – speed of pedestrians in an unit site

Taking the two aspects mentioned above into account, and adding the loss of effective walkway space caused by standing pedestrians (S_1), the total required walkway space (A) can be obtained according to the following equation:

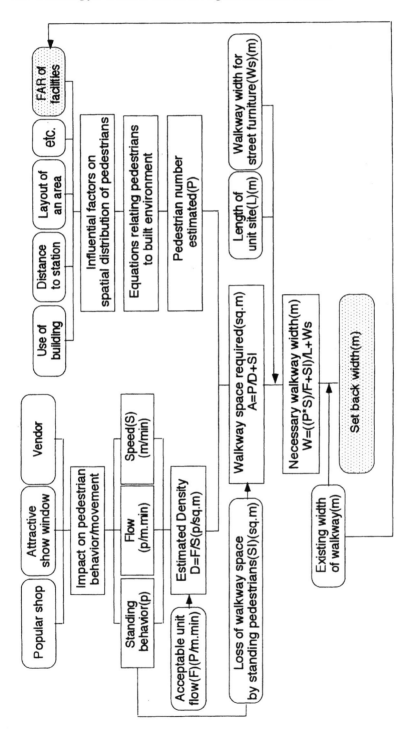

Figure 6.6   The illustration of the simulation process

$$A = P/D + S_1 \qquad\qquad (4\text{-}2)$$

A – the total required walkway space (m²)
P – evaluated pedestrians of a site (p)
S_1 – loss of effective walkway space caused by standing pedestrians (m²)

Further, dividing the total required walkway area calculated above by the sum of the length of unit site (L) and walkway width for street furniture (W_s), the necessary width of the walkway of a unit site (W) can be finally estimated from equation 4-3:

$$W = ((P_*S)/F + S_1)/L + W_s \qquad\qquad (4\text{-}3)$$

W – the necessary width of the walkway of an unit site (m)
L – the length of the unit site (m)
W_s – the walkway width for street furniture (m)

Evaluation of Pedestrian Space

Taking Ginza commercial area as a prototype to exemplify the evaluation of pedestrian space, three layouts of the area and facilities compositions of each site (block) are worked out and presented in Figures 6.7, 6.8 and 6.9. Case 1 (Figure 6.7) is close to the existing condition, whereas Cases 2 (Figure 6.8) and 3 (Figure 6.9) are assumed as conditions only for the illustration purpose of the evaluation. The estimated results of spatial distribution of pedestrians in the three different cases by the simulation modelling discussed above are also illustrated in Figures 6.7, 6.8 and 6.9.

It can be seen that from these figures that, owing to the layout of the area and the facilities composition of each unit site are fairly different from case to case, the spatial distribution of pedestrians in each case varies considerably and can be well imitated by the simulation model. Table 6.8 is the result of the required sidewalk width of each site in the three cases respectively, calculated on the service level of a constrained unit flow of 33 p/m/min. It is clear that since the built environment of each site is different from one another, the required pedestrian space of each site is fairly different, which is compatible to the number of pedestrians determined by the surrounding built environment of each site.

If all sidewalks are of the same width as in each case above, the service level of sidewalks can be immediately evaluated. As an example, the situation

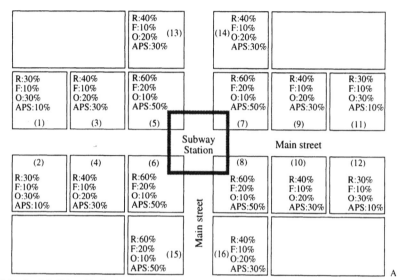

| | | R:40%
F:10%
O:20% (13)
APS:30% | R:40%
F:10%
(14)O:20%
APS:30% | | |

R:30%
F:10%
O:30%
APS:10%   (1)

R:40%
F:10%
O:20%
APS:30%   (3)

R:60%
F:20%
O:10%
APS:50%   (5)

Subway Station

R:60%
F:20%
O:10%
APS:50%   (7)

R:40%
F:10%
O:20%
APS:30%   (9)

R:30%
F:10%
O:30%
APS:10%   (11)

Main street

(2)
R:30%
F:10%
O:30%
APS:10%

(4)
R:40%
F:10%
O:20%
APS:30%

(6)
R:60%
F:20%
O:10%
APS:50%

(8)
R:60%
F:20%
O:10%
APS:50%

(10)
R:40%
F:10%
O:20%
APS:30%

(12)
R:30%
F:10%
O:30%
APS:10%

Main street

R:60%
F:20%
O:10% (15)
APS:50%

R:40%
(16)F:10%
O:20%
APS:30%

A

* R – Retail facilities   F – Food service   O – Offices   * Total facilities in each block: 100,000 sq.m.
  APS – 5 of attractive/popular shops in length along sidewalk   * Each block: 150 × 150 m

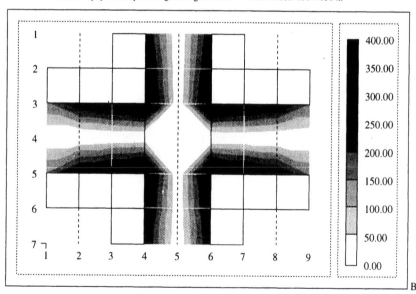

Figure 6.7   The simulation of pedestrian distribution – Case 1 (6:00 p.m.)

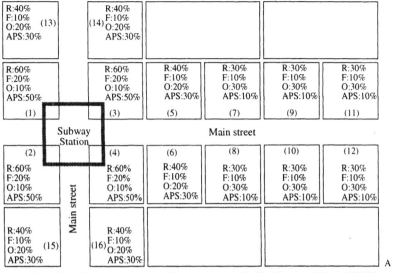

* R – Retail facilities   F – Food service   O – Offices   * Total facilities in each block: 100,000 sq.m.
  APS – % of attractive/popular shops in length along sidewalk   * Each block: 150 × 150 m

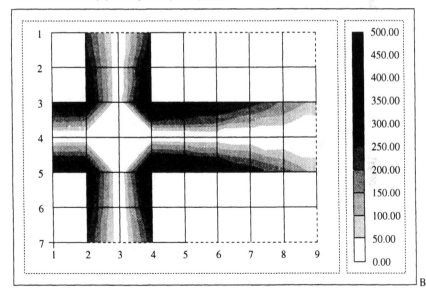

Figure 6.8   The simulation of pedestrian distribution – Case 2

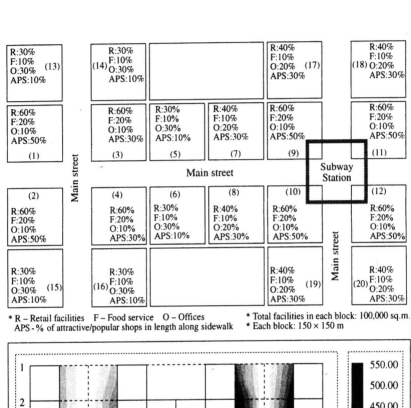

R:30% F:10% O:30% (13) APS:10%	R:30% (14) F:10% O:30% APS:10%		R:40% F:10% O:20% (17) APS:30%	R:40% F:10% (18) O:20% APS:30%

R – Retail facilities   F – Food service   O – Offices   * Total facilities in each block: 100,000 sq.m.
APS - % of attractive/popular shops in length along sidewalk   * Each block: 150 × 150 m

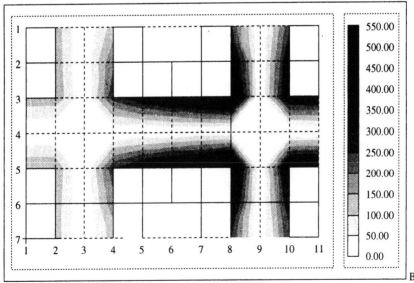

Figure 6.9   The simulation of pedestrian distribution – Case 3

Table 6.8   Each sidewalk width under service level of 33 p/m/min

Case 1:

Block	1 South	2 North	3 South	4 North	5 South	5 East	6 North	6 East	7 South	7 West
Width (m)	2.5	2.5	4.1	4.1	5.6	5.0	5.6	5.0	5.6	5.0

Block	8 North	8 West	9 South	10 North	11 South	12 North	13 East	14 West	15 East	16 West
Width (m)	5.6	5.0	4.1	4.1	2.5	2.5	3.8	3.8	3.8	3.8

Case 2:

Block	1 South	1 East	2 North	2 East	3 South	3 West	4 North	4 West	5 South	6 North
Width (m)	5.6	5.0	5.6	5.0	7.1	5.0	7.1	5.0	5.2	5.2

Block	7 South	8 North	9 South	10 North	11 South	12 North	13 East	14 West	15 East	16 West
Width (m)	3.9	3.9	2.4	2.4	2.0	2.0	3.8	3.8	3.8	3.8

Case 3:

Block	1 South	1 East	2 North	2 East	3 South	3 West	4 North	4 West	5 South	6 North
Width (m)	3.1	2.6	3.1	2.6	3.7	3.1	3.7	3.1	4.9	4.9

Block	7 South	8 North	9 South	9 East	10 North	10 East	11 South	11 West	12 North	12 West
Width (m)	6.4	6.4	8.0	5.2	8.0	5.2	5.6	5.2	5.6	5.2

Block	13 East	14 West	15 East	16 West	17 East	18 West	19 East	20 West
Width (m)	2.2	2.2	2.2	2.2	3.8	3.8	3.8	3.8

of each sidewalk at the width of four metres at the peak period time of 6:00 p.m. has been examined (Table 6.9). If the layouts of the area and facilities compositions of each site (block) are in the condition of Case 1, there will be 12 blocks reaching the situation of *crowded* or *congested* level, 10 blocks in Case 2 and 12 blocks in Case 3.

Implications for Planning and Design Regulation

There are many examples that elaborately designed open spaces do not work very well. The approach and analyses studied above provides a theoretical foundation for design, arrangement and manipulation of public space including pedestrian walkway. If we wish to design a comfortable and popular urban space, then we must design with the knowledge that integration and compatibility are the global variables. The findings discussed above imply that human behaviour and movement is very much influenced by the surrounding built environment in many aspects. Therefore the design of public space must be correlated to the overall plan of functional arrangement, floor space distribution, the layout of an area, the public transit system and so on.

To date, although many cities have tried to control or limit the negative influence of redevelopment on urban environments through regulations like setback and FAR indices, those indices are quite weak in scientific basis or quantitative studies. The present study provides a way to improve the technique. As shown in Figure 6.6, if the required walkway space is correlated with the built surroundings of each site, the necessary setback of a proposed building compatible to the conditions of the site will be estimated simultaneously by the simulation model. In turn, if the walkway space of a site is determined or cannot be changed at a given condition, the maximum FAR of the surrounding buildings can be determined as well by the simulation system. It is believed that the approach studied above, if associated with other necessary considerations relevant to other aspects of urban design, will lead to appropriate guidelines and regulations for redevelopment.

Conclusion

A successful environment should encourage pedestrians to use it by facilitating their movement, behaviour and various activities. It should be a comfortable system as well as a supportive element for the vitality of urban space. To create a pedestrian-friendly environment, the compatible pedestrian space to

Table 6.9   Service level of pedestrian space at four metres of sidewalk width

Case 1:

Block	1 (South)	2 (North)	3 (South)	4 (North)	5 (South)	5 (East)
Space (m²/p)	3.68	3.68	2.25	2.25	1.64	1.84
Service level	Impeded	Impeded	*Crowded*	*Crowded*	*Crowded*	*Crowded*

Block	6(North)	6(East)	7(South)	7(West)	8(North)	8(West)
Space (m²/p)	1.64	1.84	1.64	1.84	1.64	1.84
Service level	*Crowded*	*Crowded*	*Crowded*	*Crowded*	*Crowded*	*Crowded*

Block	9 (South)	10 (North)	11 (South)	12 (North)	13 (East)	14 (West)
Space (m²/p)	2.25	2.25	3.68	3.68	2.42	2.42
Service level	*Crowded*	*Crowded*	Impeded	Impeded	Constrained	Constrained

Block	15(East)	16(West)
Space (m²/p)	2.42	2.42
Service level	Constrained	Constrained

Case 2:

Block	1 (South)	1 (East)	2(North)	2 (East)	3 (South)	3 (West)
Space (m²/p)	1.64	1.84	1.64	1.84	1.3	1.49
Service Level	*Crowded*	*Crowded*	*Crowded*	*Crowded*	*Congested*	*Congested*

Block	4 (North)	4 (West)	5 (South)	6 (North)	7 (South)	8 (North)
Space(sq.m/p)	1.3	1.49	1.77	1.77	2.36	2.36
Service level	*Congested*	*Congested*	*Crowded*	*Crowded*	Constrained	Constrained

Block	9 (South)	10 (North)	11 (South)	12 (North)	13 (East)	14 (West)
Space(sq.m/p)	3.82	3.82	4.62	4.62	2.42	2.42
Service level	Impeded	Impeded	Impeded	Impeded	Constrained	Constrained

Block	15 (East)	16 (West)
Space (m²/p)	2.42	2.42
Service level	Constrained	Constrained

Case 3:

Block	1 (South)	1 (East)	2 (North)	2 (East)	3 (South)	3 (West)
Space (m²/p)	2.97	3.53	2.97	3.53	2.49	2.97
Service level	Impeded	Impeded	Impeded	Impeded	Constrained	Impeded

Block	4 (North)	4 (West)	5 (South)	6 (North)	7 (South)	8 (North)
Space (m²/p)	2.49	2.97	1.88	1.88	1.39	1.39
Service level	Constrained	Impeded	*Crowded*	*Crowded*	*Congested*	*Congested*

Table 6.9 cont'd

Block	9 (South)	9 (East)	10 (North)	10 (East)	11 (South)	11 (West)
Space (m²/p)	1.15	1.31	1.15	1.31	1.64	1.77
Service level	*Congested*	*Congested*	*Congested*	*Congested*	*Crowded*	*Crowded*

Block	12 (North)	12 (West)	13 (East)	14 (West)	15 (East)	16 (West)
Space (m²/p)	1.64	1.77	4.17	4.17	4.17	4.17
Service level	*Crowded*	*Crowded*	Impeded	Impeded	Impeded	Impeded

Block	17 (East)	18 (West)	19 (East)	20 (West)
Space (m²/p)	2.42	2.42	2.42	2.42
Service level	Constrained	Constrained	Constrained	Constrained

each specific site is very important, especially for commercial areas where land price is extremely high. Through the case study in Ginza of Tokyo, it has been clarified that the physical influential factors which significantly affect spatial distribution of pedestrians are retail facilities, offices, food service facilities, distances to stations, facilities distribution in the area, and the spatial configuration of streets.

It has been testified that the situation of pedestrian space in a certain site is not only determined by the number of pedestrians resulting from the distribution feature of pedestrians in a related area, but is also determined by the surrounding built environment of the site. This is reflected by the loss of walkway space due to standing behaviour of pedestrians and by the slow walking speed of pedestrians. Based on these findings, a simulation system has been established, which provides an approach to estimate and manipulate the space issue.

Theoretically, based on the findings in this study, the creation of a pedestrian-friendly environment is no longer just a matter of outside space design, and is neither just a matter of urban beautification. It is a comprehensive issue of the spatial distribution of building-use and bulk, street networks and mass transit planning, as well as compatible space manipulation.

References

Fruin, J. (1971), *Pedestrian Planning and Design*, New York: MAUDEP Inc.

Geographic Description of Japan (1982), *Kanto Part*, Tokyo: Ninomiya Press (*in Japanese*).

Hensen, R.J. (1988), 'Trip Generation Rates for Different Types of Grocery Stores', *ITE Journal*, October, pp. 21–2.

Hillier, B. and Hanson, J. (1984), *The Social Logic of Space*, Cambridge: Cambridge University Press.

Hillier, B., Penn, A., Hanson, J. and Xu, J. (1993), 'Natural Movement: Configuration and Attraction in Pedestrian Movement', *Environment & Planning*, Vol. 20.

Navin, F.P and Wheeler, R.J. (1969), 'Pedestrian Flow Characteristics', *Traffic Engineering*, June, pp. 30–6.

Oeding, D. (1963), 'Verkehrsbelartung und Dimensionerung von Gehwegen und Anderen Anlagen des Fussgaengerverkehrs', *Strassenbau und Strassenverkehrstechnik*, Heft 22 (*in German*).

O'Flaherty, C.A. (1973), 'Movement on a City Center Footway', *Traffic Engineering & Control*, March, pp. 434–8.

Older, S.J. (1968), 'Movement of Pedestrians on Footways in Shopping Street', *Traffic Engineering & Control*, August, pp. 160–3.

Pushkarev, B. and Zupan, J.M. (1975), *Urban Space for Pedestrian*, Boston: The MIT Press.

Takeuchi, T. and Iwamoto, H. (1976), 'A Study of Pedestrian Flow', 31st Academic Conference of Civil Engineering of Japan, Tokyo (in Japanese).

Transport Census of Metropolises (1985), *Capital Region Part*, Japan: Ministry of Transportation (*in Japanese*).

Yoshioka, A. (1978), 'Pedestrian Traffic and Pedestrian Space', *Traffic Engineering*, Vol. 13, No. 4, pp. 25–37; No. 5, pp. 41–53 (*in Japanese*).

7 The Planning and Management of a Better High Density Environment

ANTHONY GAR-ON YEH

Hong Kong – The Highest Density City in the World

Hong Kong is made up of Hong Kong Island, Kowloon peninsular, New Kowloon, and the New Territories (Figure 7.1). In 1991, the total land area was 1,068 km^2 and the population was 5.6 million, giving an overall population density of 5,385 persons/km^2. However, because of the high concentration of people living in the urban areas along the northern coast of the Hong Kong Island, Kowloon, and New Kowloon, the overall population density has greatly understated the population density in Hong Kong (Table 7.1). In the New Territories, where there is less population, the population density is 2,560 persons/km^2, but on Hong Kong Island, Kowloon, and New Kowloon, where the population is mainly concentrated in the limited area, the population density is 26,950 persons/km^2. In the urban area, the density can be as high as 116,531 persons/km^2, such as in the Mongkok district. When the density is calculated at the street block level, some may be as high as 400,000 to 600,000 persons/km^2.

Hong Kong is the city with the highest density in the world. In the study by Newman and Kenworthy (1989), Hong Kong was seen to have the highest population, employment, and activity density among the largest cities in Asia, United States, Canada, Europe, and Australia. According to their study, the population density of the inner area of Hong Kong was 1,036 persons/ha, which is 5.1 times higher than the next highest density city of Singapore and 6.7 times that of Tokyo (Table 7.2). In terms of employment density in the CBD, Hong Kong is also the highest in the world. Its density is 1.3 times than that of the next highest city of Chicago, and 2.6 times that of Tokyo (Table 7.3).

High density development in Hong Kong is a result of its topography,

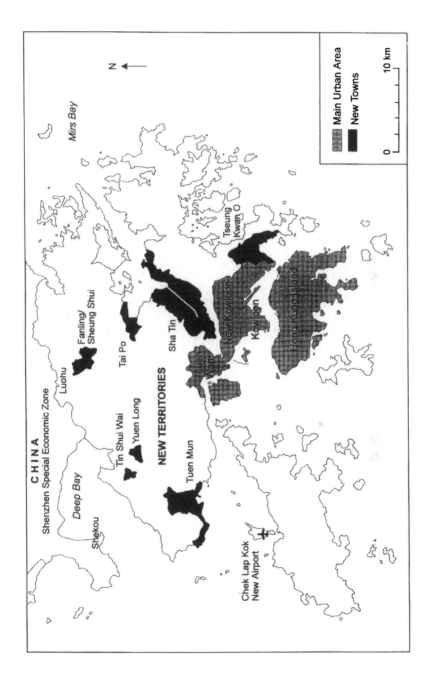

Figure 7.1 New towns in Hong Kong

Table 7.1   Residential population and population density in Hong Kong

District Board District	1981 Population (persons)	1981 Density (persons/ km²)	1986 Population (persons)	1986 Density (persons/ km²)	1991 Population (persons)	1991 Density (persons/ km²)
Hong Kong Island						
Central and Western	283,916	23,448	257,131	20,854	253,383	20,479
Wan Chai	236,149	23,781	200,403	20,182	180,309	18,209
Eastern	474,237	27,150	500,451	27,387	560,200	30,316
Southern	221,354	5,833	243,474	6,380	257,101	6,701
Subtotal	*1,215,656*	*15,695*	*1,201,459*	*15,267*	*1,250,993*	*15,811*
Kowloon and New Kowloon						
Yau Tsim	176,726	54,714	146,496	45,355	111,692	33,232
Mong Kok	247,912	175,612	206,941	142,718	170,368	116,531
Sham Shui Po	467,994	63,190	433,958	56,875	380,615	48,822
Kowloon City	493,325	54,207	432,894	47,156	402,934	41,759
Wong Tai Sin	503,865	53,947	438,417	46,940	386,572	41,331
Kwun Tong	625,552	55,260	690,739	60,826	578,502	52,562
Subtotal	*2,515,374*	*60,164*	*2,349,445*	*55,693*	*2,030,683*	*47,638*
New Territories						
Kwai Tsing	622,387	7,970	420,049	21,464	440,807	21,158
Tsuen Wan	n.a.[1]	n.a.[1]	245,238	4,159	271,576	4,581
Tuen Mun	120,657	1,529	287,539	3,611	380,683	4,711
Yuen Long	189,441	1,397	211,540	1,545	229,724	1,664
North	115,364	844	146,818	1,074	165,666	1,211
Tai Po	74,356	551	140,504	1,033	202,117	1,496
Sha Tin	118,331	1,797	362,033	5,402	506,368	7,378
Sai Kung	42,531	339	46,074	365	130,418	1,026
Islands	45,968	283	47,236	290	47,459	293
Subtotal	*1,329,035*	*1,448*	*1,907,031*	*2,064*	*2,374,818*	*2,557*
All land area	***5,060,065***	***4,878***	***5,457,935***	***5,224***	***5,656,494***	***5,385***

Note

1   Combined with Kwai Tsing in 1981.

Source: Census and Statistics Department (1992), pp. 69 and 70.

Table 7.2   Population densities of selected cities in Asia, United States, Canada, Europe, and Australia in 1980

City	Urban density (persons/ha)	Inner area density (persons/ha)	Outer area density (persons/ha)	CBD density (persons/ha)	Proportion of population in CBD (%) (persons/ha)	Proportion of population in inner area (%)
Hong Kong	293.3	1,036.8	224.4	160.4	0.4	30.0
Asia						
– Tokyo	104.6	152.9	57.8	82.3	1.3	32.3
– Singapore	83.2	201.5	63.1	203.7	6.6	35.2
Average:	*93.9*	*177.2*	*60.5*	*143.0*	*4.0*	*33.8*
United States						
– Los Angeles	20.0	29.6	17.5	29.4	0.1	31.3
– New York	19.8	106.8	12.9	217.1	2.8	39.5
– Chicago	17.5	54.1	11.4	15.6	0.1	42.3
– San Francisco	15.5	58.8	12.9	89.7	1.1	21.3
– Detroit	14.1	48.1	10.6	11.2	0.1	31.6
– Washington	13.2	44.2	10.9	7.5	0.1	21.4
– Boston	12.1	44.8	9.8	125.5	2.7	24.3
– Denver	11.9	19.3	9.8	18.5	0.4	30.9
– Houston	8.9	20.6	7.8	5.5	0.1	16.6
– Phoenix	8.5	19.1	8.3	17.3	0.5	3.7
Average:	*14.2*	*44.5*	*11.2*	*53.7*	*0.8*	*26.3*
Canada						
– Toronto	39.6	56.5	34.0	25.2	0.2	35.7
Europe						
– Vienna	72.1	132.5	59.4	64.9	1.3	31.9
– Brussels	67.4	100.5	49.8	74.1	1.9	51.8
– West Berlin	63.6	83.5	57.2	133.3	0.8	31.8
– Munich	56.9	159.2	48.4	111.2	5.9	21.4
– London	56.3	77.9	48.3	66.4	2.7	37.2
– Frankfurt	54.0	62.5	48.9	65.2	2.5	43.3
– Zurich	53.7	78.9	41.7	44.4	0.9	47.4
– Stockholm	51.3	58.3	46.0	97.0	6.4	49.3
– Amsterdam	50.8	83.3	32.4	108.4	9.7	59.2
– Paris	48.3	106.4	26.0	235.2	5.4	60.9
– Hamburg	41.7	88.4	35.0	26.4	0.7	26.8
– Copenhagen	30.4	59.3	23.6	84.8	2.2	37.3
Average:	*53.9*	*90.9*	*43.1*	*92.6*	*3.4*	*41.5*
Australia						
– Sydney	17.6	39.1	15.8	10.7	0.1	16.7
– Melbourne	16.4	29.3	15.7	24.5	0.2	9.0
– Adelaide	12.9	18.8	12.4	8.2	0.2	11.6
– Perth	10.8	15.5	9.9	8.4	0.7	22.9
– Brisbane	10.2	18.5	9.1	15.3	0.3	21.7
Average:	*13.6*	*24.2*	*12.6*	*13.4*	*0.3*	*16.4*

Source: Newman and Kenworthy, 1989.

Table 7.3   Employment densities of selected cities in Asia, United States, Canada, Europe, and Australia in 1980

City	Employment density (jobs/ha)	Inner area employment density (jobs/ha)	Outer area employment density (jobs/ha)	CBD employment density (jobs/ha)	Proportion of jobs in CBD (%)	Proportion of jobs in inner area (%)	CBD activity intensity
Hong Kong	109.7	478.3	65.6	1,258.6	7.3	45.3	1,419
Asia							
– Tokyo	66.3	114.3	19.8	477.0	26.6	84.8	559
– Singapore	37.1	n.a.	n.a.	339.3	24.3	n.a.	543
Average:	*51.7*	*57.2*	*9.9*	*408.2*	*25.5*	*42.4*	*551*
United States							
– Los Angeles	10.5	13.9	8.9	472.0	4.8	43.3	501
– New York	9.3	53.4	5.8	828.0	22.9	41.9	1,045
– Chicago	8.1	25.5	5.2	937.9	12.3	44.9	954
– San Francisco	7.8	47.8	5.4	713.2	17.0	34.4	803
– Detroit	6.2	19.9	4.8	305.8	6.6	29.6	317
– Washington	8.0	37.5	5.8	584.1	16.1	32.5	592
– Boston	6.2	32.7	4.3	382.8	15.9	34.6	508
– Denver	7.6	16.8	4.9	262.6	11.6	49.7	281
– Houston	5.5	26.4	3.5	442.7	11.6	41.1	448
– Phoenix	4.0	24.0	3.7	66.6	3.9	10.6	84
Average:	*7.3*	*29.8*	*5.2*	*499.6*	*12.3*	*36.3*	*553*
Canada							
– Toronto	19.7	37.7	13.7	757.1	13.4	47.9	782
Europe							
– Vienna	38.4	112.7	22.9	403.2	14.9	50.8	468
– Brussels	42.1	91.8	15.6	591.5	24.6	75.9	666
– West Berlin	26.6	45.9	20.4	333.3	4.8	41.8	467
– Munich	34.2	192.4	21.2	230.9	20.5	42.9	342
– London	30.2	61.9	18.5	396.8	29.7	55.1	463
– Frankfurt	43.2	74.3	24.7	389.1	18.4	64.2	454
– Zurich	32.5	65.8	16.7	422.1	13.6	65.2	467
– Stockholm	34.4	61.5	16.0	279.6	26.3	74.7	377
– Amsterdam	23.2	46.1	10.3	153.1	29.9	71.7	262
– Paris	22.0	60.0	7.6	399.5	20.2	75.1	635
– Hamburg	23.8	105.5	12.0	407.4	20.0	56.0	434
– Copenhagen	16.2	37.9	11.1	325.1	16.0	44.8	410
Average:	*30.6*	*79.7*	*16.4*	*361.0*	*19.9*	*59.9*	*454*
Australia							
– Sydney	7.5	39.1	4.9	433.9	13.2	39.3	445
– Melbourne	6.1	40.3	4.3	646.5	15.2	33.2	671
– Adelaide	5.4	25.1	3.7	250.7	14.4	37.3	259
– Perth	4.6	14.6	2.7	120.7	24.1	51.0	129
– Brisbane	4.1	15.6	2.5	345.8	13.9	45.7	361
Average:	*5.5*	*26.9*	*3.6*	*359.5*	*16.2*	*41.3*	*373*

Source: Newman and Kenworthy, 1989.

historical development, and land policy. Over 75 per cent of the land consisted of hill slopes. A large proportion of the relatively flat areas is under private ownership, making development difficult (Chau, 1981). Most of the development is concentrated in Kowloon, New Kowloon, and Hong Kong Island where most of the flat land is obtained from hill levelling and land reclamation by the government. Population and density in Hong Kong increased drastically immediately after the second world war because of the sudden increase in population and economic activities. There was a large influx of refugees to Hong Kong after the change of regime in China in 1949. Its population jumped from 0.6 million in 1945 to over two million in 1951. This led to a severe housing shortage, squatting, high density, and a poor living environment. Land in Hong Kong is owned by the government. It is formed and subdivided before leasing to the private developers through auction and tender. The government has full control over the timing, location, and amount of land to be leased. The sale of land leases is one of the major sources of revenue of the government. In the property boom period, revenue from land sales can be as high as 30 per cent of the total revenue of the government. The control over the sale of land leases is one of the reasons for the high land price in Hong Kong. The high land price makes office and house prices very expensive, leading to high density development.

The development of Hong Kong has been highly concentrated along the coastal strip of Hong Kong Island and Kowloon along the Victoria Harbour. It was not until 1973, when the government pursued the new town development programme, that urban development was decentralised to the new towns in the New Territories (Yeh, 1987) (Figure 7.1). Through better planning and design, the new towns have better living environments and lower densities than the main urban area. The decentralisation of population into the new towns did not drastically reduce the density of the old urban areas. Their density, although declining, is still very high. In some areas, the density has slightly increased rather than decreased because of urban redevelopment. Despite the continual existence of high density areas in Hong Kong, the urban living environment has been improved through better urban planning and management, making the high density areas more liveable. This chapter examines recent urban planning and management measures that have been taken in creating a better, high density environment.

Density and Crowding

Density is often confused with crowding and has often been used inter-changeably. A distinction should be made. Density is a physical condition and crowding is a psychological experience. Density denotes a physical condition involving the limitation of space. Crowding refers to situations in which the restrictive aspect of limited space are perceived by the individuals exposed to them (Stokols, 1972).

High density is often considered to be undesirable because it is often associated with social pathology. However, the relationship between high density and social pathology is not so straightforward. Studies on the effects of density and social pathology do not show significant relationships. There are many factors affecting social pathology in which density is just one of them. Other factors such as socioeconomic background, education background, and health situations of individuals are more important variables than density in explaining social pathology.

The relationship between density and social pathology is mainly through the intervening variable of crowding. Crowding is a psychological subjective response to density. Social pathology is a result of stress and social conflict caused by crowding. As crowding is a psychological subjective response, high density does not always lead to stress (Mitchell, 1972; Freeman, 1975; Sundstrom, 1978). The effects of high density on crowding may be mitigated by the personal background and interpersonal relationships. The perception of crowding depends on the duration of exposure to high density, expectation of the duration of the exposure, social interaction, the possibility of escaping to an external environment and the adaption ability and tolerance of the persons concerned. The adaptation and tolerance of crowding depends on the socioeconomic background, culture, age, education, and previous living environment of the individual. For a fixed density, people who used to live in a more dense environment will feel less crowded than someone who used to live in a more spacious environment. Different cultures also have different levels of adaptation and tolerance to crowding. Asians and Chinese were often found to have a high adaptation and tolerance to crowding (Anderson, 1972; Schmidt, Goldman and Feimer, 1976), although tolerance to crowding was often involuntary and due to disadvantage not choice (Loo and Ong, 1984).

There are two main types of density and their effects on crowding are different. They are personal space density and external space density, which are related to Stokol's (1972) and Baldassare's (1979) concepts of primary and secondary environment respectively. Personal space density is the measurement of density of the primary immediate working and living environment where the individual has more intimate contact with space for a relatively long time. It can be measured by the number of people per living/ working space. External space density can be measured by the number of people in a certain area. The area can be measured by street blocks, neighbourhoods, districts, and the whole city. It is an expression of the secondary environment where space, facilities, and services are shared among the people within certain areas. Because of the ease of measurement, it is one of the most commonly used indicators for density. In a two dimensional city, there is little difference between residential density (persons/site area) and household density (persons/house or flat). However, in a three dimensional city with high-rise buildings like Hong Kong where there can be more than 30 flats on one site, residential density can be many times higher than household density and therefore residential density is not a good indicator of household density. In fact, in most cases, it has grossly overestimated household density. Crowded primary environments are more undesirable than crowded secondary environments (Stokol, 1972). The effect of density on crowding is mainly through personal space density than external space density. Crowding may not be felt if external space density is increased but not personal space density. For example, the addition of more flats to a building site without reducing the size of each flat.

The effect of density can be affected by the design, layout, open space, degree of sharing, traffic, and community facilities of the external and personal space. For external space, with a fixed density, people will feel less crowded if there is more open space, a freer flow of traffic, and more community facilities. For personal space, with a fixed density, people will feel less crowded if there is less degree of sharing and more privacy. The design and layout of rooms can make a room look more spacious. The subdivision and sound proofing of rooms can increase privacy and reduce crowding. A small room can be made to appear larger by surrounding it with mirrored walls. A cluttered area will appear more crowded than one in which physical objects are neatly arranged. The reduction of the effect of high density can be achieved through better planning and management of the urban space.

The relationship between density and social pathology has been mainly confirmed by experiments on animals. Human beings are more adaptable than animals and the relationship between human density and social pathology has not been well established in non-experimental settings. Many studies have shown that although high density is undesirable, there is little relationship between density and social pathology when other socioeconomic variables are considered (Sundstrom, 1978). The perception of the environment, as well as the experience of crowding, can be mediated by socioeconomic status, cultural tradition, and the nature of activities performed in a given area (Stokols, Rall, Pinner, and Schopler, 1973). Density, though perceived as unpleasant, does not appear to have definite and consistent detrimental effects (Fischer, Baldassare, and Ofshe, 1975).

Hong Kong has the highest density in the world and therefore is a natural setting for carrying out research on high density development. Although Hong Kong has a much higher density than most Western cities where the relationship between high density and social pathology is studied most, similar to the findings of Western cities which have lower densities, there is little evidence of high density being associated with social pathology. Mitchell (1972) found that emotional stress was not related to density. Stress was probably more influenced by inadequate income than density. He also found that forced social interaction between non-relatives as a result of flat-sharing tended to create stress and tensions. It was the degree of sharing that caused stress rather than density. Easy escape from each other by retreating outdoors could significantly assist in reducing such stress. He suggested that there should be more attractive and spacious external environments in high density areas as a means of reducing pressure from overcrowding indoors with a high degree of sharing. Millar (1979) found that relationships between physical density and psychological problems were rather weak once socioeconomic background had been taken into consideration. Many studies also showed that a substantial proportion of urban population did not see high density as a problem (Millar, 1976; Richardson, 1977). On the contrary, many of them enjoyed being surrounded by a large number of people (Millar, 1976; Traver, 1976). Chinese traditions, and the previous poor living conditions they endured when they were refugees, were some of the more probable reasons to explain the tolerance of the Hong Kong people towards high densities and overcrowding (Schmitt, 1963).

Most of the studies on high density living were done in Hong Kong when the density was highest and there was a lack of urban planning and management. Even when the urban density and environment were at their worst in Hong Kong, there is no evidence to support the idea that high density is causing social pathology. The living environment has been greatly improved since the mid-1970s through better urban planning and management. Although improvements still need to be done, the negative effects of high density have been much mitigated.

Advantages and Disadvantages of High Density Development in Hong Kong

There are many benefits from high density development. These include efficiency in the use of land, better public services and provision of public facilities, minimisation of journeys to work and accessibilities to public facilities, energy conservation, avoiding urban sprawl, and conservation of the natural environment (Chau, 1981). High density in Hong Kong has supported and encouraged the use of public transport (Wang and Yeh, 1993). Among the large cities in the world, Hong Kong has been highly regarded as one of the cities which has the lowest energy consumption per capita on transportation (Table 7.4) (Newman and Kenworthy, 1989). High density development also reduces infrastructure costs and promotes closer communities. Jane Jacobs (1961) valued population density as a contributor to the 'exuberant diversity' of an urban area. Le Corbusier's (1933) 'radiant city' considered that high density could facilitate beneficial effects of intense social interaction.

High density development is often considered to be associated with crowding, traffic congestion, and social pathology. However, as discussed in the above section, at present there is little evidence to show the effects of high density on social pathology. It is often found that the effects of high density on human beings vary according to culture and habits of people. It seems that Chinese, especially people in Hong Kong who are mainly refugees or descendants of refugees, have a high degree of tolerance and adaptation to high density environments. People in Hong Kong have adapted to the high density environment by sharing external space outside the personal space environment.

In Hong Kong, the impact of high density is further reduced by its topography. The Victoria Harbour, where the city has developed along its water front, provides a breathing space for the city, making it appear less crowded. The harbour is a centrally-located, natural open space which is within 15–20 minutes walk from most parts of Kowloon, New Kowloon, and Hong Kong Island. The low-rise mountains at the skyline also diffuse the impacts of high density development, providing a glimpse of the natural environment in the midst of the concrete jungle.

The effect of high density is further reduced by better planning, design, and management. Better planning, design, and management have increasingly been used in Hong Kong since the 1980s, particularly in newly developed areas, such as the new towns and reclamation areas. Improvements were also

Table 7.4   Gasoline use, urban density and transportation pattern for selected cities in 1980

City	Gasoline use (MJ per person)	Urban population density (persons/ ha)	Proportion of population in CBD (%)	Proportion of jobs in CBD (%)	Passenger cars ownership/ 1,000 people	Passenger trips per person by public transport per year
Hong Kong	1,987	293.3	0.4	7.3	42.4	466.3
Asia						
– Tokyo	8,488	104.6	1.3	26.6	156.2	471.8
– Singapore	6,003	83.2	6.6	24.3	64.7	353.1
Average:	*7,246*	*93.9*	*4.0*	*25.5*	*110.5*	*412.5*
United States						
– Houston	74,510	8.9	0.1	11.6	602.6	14.7
– Phoenix	69,908	8.5	0.5	3.9	499.0	9.1
– Detroit	65,978	14.1	0.1	6.6	593.8	25.7
– Denver	63,466	11.9	0.4	11.6	666.2	26.9
– Los Angeles	58,474	20.0	0.1	4.8	541.5	59.2
– San Francisco	55,365	15.5	1.1	17.0	543.4	115.0
– Boston	54,185	12.1	2.7	15.9	465.3	79.9
– Washington	51,241	13.2	0.1	16.1	561.4	91.2
– Chicago	48,246	17.5	0.1	12.3	445.0	114.6
– New York	44,033	19.8	2.8	22.9	411.9	121.5
Canada						
– Toronto	34,813	39.6	0.2	13.4	462.5	177.6
Europe						
– Hamburg	16,671	41.7	0.7	20.0	344.4	248.3
– Frankfurt	16,093	54.0	2.5	18.4	386.9	306.3
– Zurich	15,709	53.7	0.9	13.6	374.6	363.3
– Stockholm	15,574	51.3	6.4	26.3	346.5	302.3
– Brussels	14,744	67.4	1.9	24.6	361.1	265.8
– Paris	14,091	48.3	5.4	20.2	338.1	259.1
– London	12,426	56.3	2.7	29.7	287.8	284.4
– Munich	12,372	56.9	5.9	20.5	359.9	306.9
– West Berlin	11,331	63.6	0.8	4.8	268.9	394.5
– Copenhagen	11,106	30.4	2.2	16.0	246.4	200.9
– Vienna	10,074	72.1	1.3	14.9	311.2	312.9
– Amsterdam	9,171	50.8	9.7	29.9	307.8	345.4
Average:	*13,280*	*53.9*	*3.4*	*19.9*	*327.8*	*299.2*
Australia						
– Perth	32,610	10.8	0.7	24.1	474.9	70.8
– Brisbane	30,653	10.2	0.3	13.9	458.1	79.3
– Melbourne	29,104	16.4	0.2	15.2	445.8	94.8
– Adelaide	28,791	12.9	0.2	14.4	475.3	83.2
– Sydney	27,986	17.6	0.1	13.2	411.7	142.3
Average:	*29,829*	*13.6*	*0.3*	*16.2*	*453.2*	*94.1*

Source: Newman and Kenworthy, 1989.

made in old urban districts to reduce the impact of density by providing more open space during urban redevelopment.

Urban Planning and Development Control for High Density Development in Hong Kong

Urban planning in Hong Kong was largely a postwar activity. Although the Town Planning Ordinance was enacted in 1939, it was not until 1947 that a small Town Planning Unit was established in the Public Works Department and in 1953 that a Planning Branch was set up within the Crown Land and Survey Office of the Public Works Department (Town Planning Division, 1984). It was in the 1970s when new towns were in the process of planning and development that the importance of urban planning was beginning to be recognised in Hong Kong. This is reflected by the drastic increase in planners from five in 1960 to 86 in 1980, and 171 in 1990. The recognition of the importance of planning by the community was further manifested by the upgrading of the former Town Planning Office of the Buildings and Lands Department to become a new Planning Department under the new Planning, Environment and Lands Branch on 1 January 1990.

Planning in Hong Kong in the past is conceived to be largely demand-oriented (Bristow, 1981). With the increasing recognition of the importance of planning, this is, at present, less true than in the past. A number of major forward-looking plans were formulated to guide the development of Hong Kong into the twenty-first century. Extensive planning input and coordination have been put into the development of the new towns which on average accommodate 0.5 million people. However, because of the late recognition of the importance of planning, development in the older districts has suffered from the lack of planning. The result is that open space and public facilities are generally underprovided in these districts which also have the highest density. With the benefit of planning, the new towns are better designed and have more open space than the old districts, although the population density is not too much less than that of the old districts. Attempts have been made to apply open space planning standards of the new towns in urban redevelopment in the old districts of the main urban area to improve their living and working environment.

Zoning is the main method of urban planning in Hong Kong. The Town Planning Ordinance, the basis of urban planning in Hong Kong, is not as comprehensive as similar zoning legislation found in other countries (Working

Group on the Review of the Town Planning Ordinance, 1988). It covers only the existing and designated urban areas and has no direct power of development control and plan implementation. The Town Planning Ordinance only empowers the Town Planning Board to prepare statutory Outline Zoning Plans (OZP). Development control has to rely on other legislation, such as the Building Ordinance and lease conditions of land sold or granted (Pun, 1983; Fung, 1988). The intensity of development, such as building height, plot ratio, and sometimes building form, is normally written in the lease condition when the land is leased either by auction, tender, or private treaty grant. Plot ratio and site coverage control are under the Building (Planning) Regulations 1956. As the Kai Tak International Airport is located next to the city centre, building heights for buildings which could be affected by the flight path are under the control of the Hong Kong Airport (Control of Obstruction) Ordinance 1957. Because of this height control, all the buildings in Kowloon and New Kowloon have a similar height, making the skyline appear to look like a cake.

Plot ratio and density control are the major methods of controlling the density of development.

Plot Ratio Control

Before the Building (Planning) Regulations were introduced in 1956, limited by the fire-fighting capability and building technology of that time, plot ratio was no more than three, and limited to five storeys high. The 1956 regulation allowed a normal plot ratio of 6. Because the heights of buildings were related to street width, in areas where site sizes were large, plot ratio as high as 20 could occur (Bristow, 1984). The Ba Man residential complex near the junctions of Jordan Road and Ferry Street is an example of such high density development. The increase in plot ratio gave rise to many opportunities of redevelopment.

The plot ratios were revised in 1962 to reduce the plot ratio which did not come to force till 1966. During the grace period, many urban redevelopment projects occurred. During 1963–66, 93,570 new private domestic units were built, a very high annual production of 23,390 units compared to the 10,000 units per annum over the period of 1958–62 (Pryor, 1978). A sliding scale was used in which the plot ratio was increased as the building height increase, but at the same time the site coverage decreased (Sections 20 and 21, Building (Planning) Regulations).

Because of the sliding scale in which plot ratio and site coverage is related to building height, the taller the building, the more slender it will be. Greater building bulk is allowed for taller buildings to encourage the more intensive and thus more economical use of land. To maintain good quality of light, air, and the environment, the plot ratio is lower for shorter buildings and higher site coverage and site coverage decrease as building height increases. For corner sites (Class B) and island sites (Class C), a greater volume is allowed because they have less obstruction to light and air as compared to inner sites (Class A). For buildings less than 15 m. high, the plot ratio for non-domestic buildings is five and ranges from 3.3 to 4.0 for domestic buildings. For buildings more than 61 m., the plot ratio for non-domestic buildings is 15 and ranges from eight to ten for domestic buildings.

A special formula is used for the calculation of the domestic plot ratio part of a composite building, i.e. combination of domestic and non-domestic functions. The domestic plot ratio of a composite building is the percentage of the unused permissible non-domestic plot ratio multiplied by the permissible domestic plot ratio.

Bonus plot ratio and site coverage can be obtained where part of a site has to be surrendered, or is resumed, for street widening purposes, or where land at ground floor level is retained by the owner but dedicated to the public use. Bonus plot ratio up to three is allowed for 1 m^2 of bonus to every 5 m^2 of public use (Section 22, Building (Planning) Regulations).

Density Zoning

Density zoning was formally introduced in 1965 as an administrative measure to control density (Bristow, 1984). Three density zones were established (Figure 7.2). Zone 1 with the highest density is mainly in the main built-up areas. Zone 2, with medium density, is located in the Mid-Levels on Hong Kong Island and in central Kowloon, and Zone 3 for most of the rest of Hong Kong Island and the foothills of Kowloon. Zone 1 areas will use the plot ratio controlled as stipulated in the Building (Planning) Regulations and development will generally be permitted to the maximum height that is allowed under the Hong Kong Airport (Control of Obstructions) Ordinance. A new plot ratio control is used for Zone 2 (Table 7.5) and Zone 3 areas.

Because of its non-statutory nature, the tighter restrictions of Zone 2 and 3 can only be imposed in lease modifications or new land sales. They are not possible in sites with unrestricted leases.

For large sites, there is a reduction of plot ratio calculation to provide space for internal roads (Table 7.6). For site coverage of less than 3,716 m^2, the full permissible plot ratio is allowed. The plot ratio will decrease

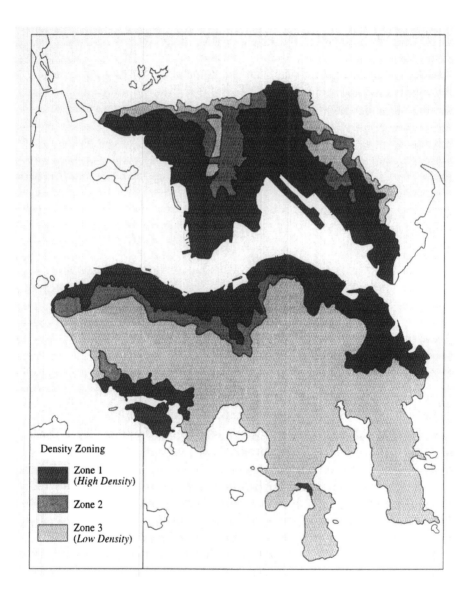

Figure 7.2 Density zoning

Table 7.5 Site coverages and plot ratios

First schedule
Percentage site coverages and plot ratios

Height of building in metres	Domestic buildings						Non-domestic buildings					
	Percentage site coverage			Plot ratio			Percentage site coverage			Plot ratio		
	Class A site	Class B site	Class C site	Class A site	Class B site	Class C site	Class A site	Class B site	Class C site	Class A site	Class B site	Class C site
Not exceeding 15 m	66.6	75	80	3.3	3.75	4.0	100	100	100	5	5	5
Over 15 m but not exceeding 18 m	60	67	72	3.6	4.0	4.3	97.5	97.5	97.5	5.8	5.8	5.8
Over 18 m but not exceeding 21 m	56	62	67	3.9	4.3	4.7	95	95	95	6.7	6.7	6.7
Over 21 m but not exceeding 24 m	52	58	63	4.2	4.6	5.0	92	92	92	7.4	7.4	7.4
Over 24 m but not exceeding 27 m	49	55	59	4.4	4.9	5.3	89	90	90	8.0	8.1	8.1
Over 27 m but not exceeding 30 m	46	52	55	4.6	5.2	5.5	85	87	88	8.5	8.7	8.8
Over 30 m but not exceeding 36 m	42	47.5	50	5.0	5.7	6.0	80	82.5	85	9.5	9.9	10.2
Over 36 m but not exceeding 43 m	39	44	47	5.4	6.1	6.5	75	77.5	80	10.5	10.8	11.2
Over 43 m but not exceeding 49 m	37	41	44	5.9	6.5	7.0	69	72.5	75	11.0	11.6	12.0
Over 49 m but not exceeding 55 m	35	39	42	6.3	7.0	7.5	64	67.5	70	11.5	12.1	12.6
Over 55 m but not exceeding 61 m	34	38	41	6.8	7.6	8.0	60	62.5	65	12.2	12.5	13.0
Over 61 m	33.33	37.5	40	8.0	9.0	10.0	60	62.5	65	15	15	15

Second schedule
Open space about domestic buildings

Item	Class of site	Open space required
1	Class A site	Not less than one-half of the roofed-over area of the building
2	Class B site	Not less than one-third of the roofed-over area of the building
3	Class C site	Not less than one-quarter of the roofed-over area of the building

Source: Hong Kong Government, 1987.

Table 7.6   Density control for density zone 2

Height – no. of storeys used for domestic purposes	Percentage site coverage Class of site			Plot ratio Class of site		
	A	B	C	A	B	C
3	55	66.6	72.5	1.65	2.00	2.18
4	45	54	60	1.80	2.16	2.40
5	40	48	53	2.00	2.40	2.65
6	35	42	46	2.10	2.52	2.76
7	30	36	39.5	2.10	2.52	2.77
8	30	36	39.5	2.40	2.88	3.16
9	30	36	39.5	2.70	3.24	2.56
10	27.5	33	36	2.75	3.30	3.60
11	27.5	33	36	3.03	3.63	3.96
12	27.5	33	36	3.30	3.96	4.32
13	25	30	33	3.25	3.90	4.29
14	25	30	33	3.50	4.20	4.62
15	25	30	33	3.75	4.50	4.95
16	25	30	33	4.00	4.80	5.28
17	25	30	33	4.25	5.10	5.61
18	25	30	33	4.50	5.40	5.94
19	25	30	33	4.75	5.70	6.27
20	25	30	33	5.00	6.00	6.60
More than 20	Any development above 20 storeys shall not have a permitted plot ratio in excess of that permitted for 20 storeys					

Source: Planning Department, 1992.

progressively to 75 per cent for Zone 3 areas and 60 per cent for Zones 1 and 2 for sites larger than 9,290 m². The objective is to ensure that for large development schemes in an area not covered by detailed layout plans, adequate land would be set aside for the necessary community and recreational facilities as well as internal and open space (Planning Department, 1992).

Plot Ratio and Density

Plot ratio is not a good mechanism of density control. Take employment density for example:

Employment Density = Employment/Site Area

Also, employment of a site depends on the gross floor area (GFA) of the building and the workers/floor area ratio:

Employment = GFA x (Workers/Floor Area)
  = Plot Ratio x Site Area x (Workers/Floor Area)

Therefore,

Employment Density = [Plot Ratio x Site Area x (Workers/Floor Area)]/
  Site Area
  = Plot Ratio x (Workers/Floor Area)

With a fixed plot ratio, employment density of a site varies according to different workers/floor area ratios which depend on firm size and work space requirement. For large and less labour intensive offices, their workers/floor area ratio may be low, giving a low employment density. Therefore, increase in plot ratio may not necessarily increase population and employment density. Plot ratio also may not affect personal space density if the work/living space does not increase or decrease.

New Developments in Urban Planning and Management in Reducing the Impacts of High Density

High density may not necessarily be associated with crowding which is the perception of human beings on the environment. Crowding can be affected by the physical design and layout of the internal and external environments of the building in which the person lives and works, and the culture, habit, and socioeconomic background of the person. In Hong Kong, in the last ten years, there have been some developments in the planning and management of office and residential districts which make high density appear less crowded and more liveable and acceptable.

External Environment of Buildings

Better planning and design Through better urban design and layout, buildings are less close to one another and there are more open spaces on the ground level. Newly developed areas, such as the new towns, have more open space and community facilities. There are plans to apply the new open space standards in redeveloping the old districts of the main urban area.

An Advisory Committee on the Appearance of Buildings and Associated Structures (ACABA) composed of representatives from the Architectural

Services Department, Territory Development Department, and Highway Department has been set up to improve the appearance of buildings and structures, particularly the appearance of pedestrian walkways.

Transport management Traffic congestion associated with high density development has been minimised by good transport management to keep the city moving. This involves giving priority to public transport, the development of the Mass Transit Railway, reducing the number of cars on the road through pricing, and using various methods of traffic management (Wang and Yeh, 1993). The development of pedestrian systems also helps to separate pedestrians from vehicles at the street level, making the streets less congested with cars and pedestrians. An extensive pedestrian system has been used in Central to enable pedestrians to move freely on the elevated segregated pedestrian walkways. Most new town centres have an elevated pedestrian system separating pedestrians from vehicles. A bypass has been constructed to divert traffic away from the Central District.

Creation of space from limited space Space is created from limited space by the aesthetic improvement of space and the full utilisation of the limited space. Multistorey car parks, podium block development, and roof top gardens and recreation facilities are examples of the full utilisation of space not only in two dimensions but also in three dimensions (Figures 7.3 and 7.4). Chinese garden concepts in which sceneries change very quickly in short distances, are applied in the design of buildings and parks. There have been aesthetic improvements of urban space by foundation, sculptures, and post-modern architecture, making urban space more interesting and feel less crowded. More recently, there is a trend to converting private space into public space, increasing the amount of space in the urban area. For example, through bonus plot ratio, ground floors of the Hong Kong and Shanghai Bank Building and the New Hang Sang Bank Building are open and provide open space and passage to the public 24 hours a day (Figure 7.5).

Large-scale property development In the past, most of the developments have been small site development. But recently, office and residential developments are often very large-scale, making it possible to have better layout and provision of community facilities. Property developments are increasingly dominated by a few large developers. The ownerships of nearby properties by the same developer can help to make the application of planning and design to overcome high density possible, such as the linking of nearby buildings of the same

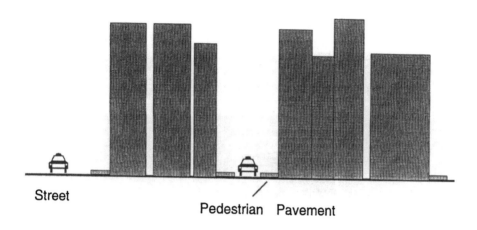

Street

Pedestrian   Pavement

Old form of urban development

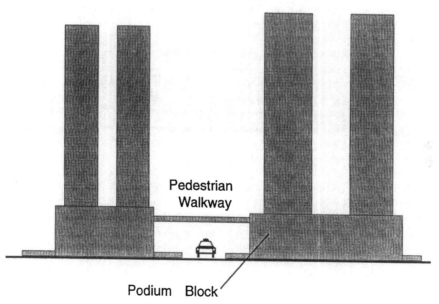

Pedestrian
Walkway

Podium   Block

New form of urban development

Figure 7.3   New form of urban development

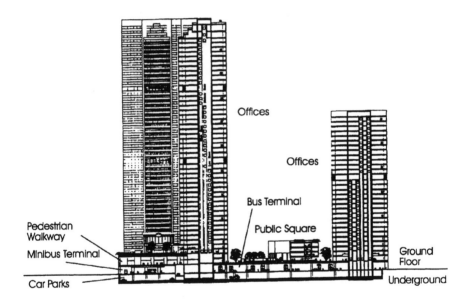

Figure 7.4   Examples of three dimensional land use (Exchange Square, Central)

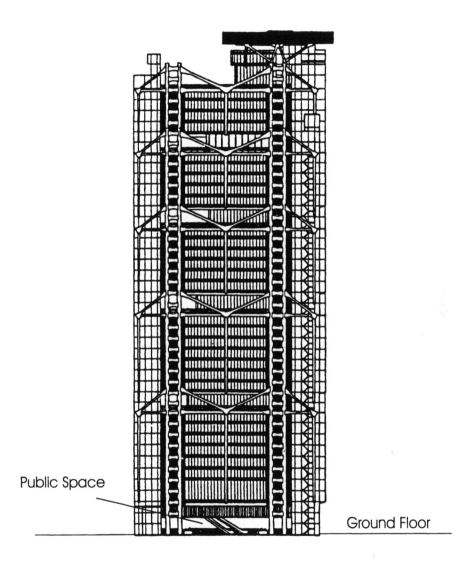

Figure 7.5 **Example of conversion of private space into public space (Hong Kong and Shanghai Bank Building, Central)**

developer by a pedestrian system. Over 40 per cent of the office buildings in Central are owned by Hong Kong Land. This makes it possible to link all their buildings together with an elevated pedestrian system, enabling a freer flow of pedestrian traffic.

Technology New building materials and design help to break the monotony of a district. Glass walls reflecting sunlight and images of nearby buildings help to make the crowded street feel more spacious. The availability of outdoor escalators facilitates and encourages the use of the pedestrian system.

Public education The 'Clean Hong Kong' campaign, coordinated by a joint Urban Council – Regional Council Steering Committee, has frequently been carried out since 1972 to educate the people to make the city less dirty and create a cleaner environment. Posters and TV commercials are also frequently produced by the Government Information Services and City and New Territories Administration to educate people about the danger of letting objects fall from tall buildings. These help to make the high density city a cleaner and safer place to live.

Interior Environment of Buildings

Building management Deteriorating and dirty buildings will make the environment feel crowded. Owners' corporations are encouraged to form to manage and maintain the services and environment of buildings (Fong, 1985). Security guards are hired to make the buildings safer.

Building design Large-scale property developments enable the development of large lobbies and shopping malls. Multistorey large shopping malls inside buildings have increased the amount of shopping space. Interior design with water fountains and Chinese gardens make the internal building environment appear less crowded. Tall ceilings also help to make the inner space of buildings less crowded.

Movement of people Escalators and express lifts have helped to move people quickly inside buildings. The designation of lifts for different floors helps to reduce congestion. All these help to reduce congestion and make the building feel less crowded.

The Experience of Hong Kong – Guidelines for High Density Development

There is a limit to high density development. But, with a given density, good planning and management of the internal and external environments of buildings can lessen the impacts of high density and make it more liveable. In Hong Kong, because of the lack of urban planning in its early days of development, the old urban districts are very dense and crowded. Urban planning did not really take place until the 1970s when Hong Kong actively pursued its new town development problem. The environment in the new towns has improved remarkably because of good planning and management, although its density is only slightly less than that of the main urban area.

The above developments which have helped to reduce the impacts of high density by making it less crowded are achieved mainly through the land leasing system, planning and building regulations, public works, and coordination of different government departments. Much of the efforts in better design and layout of buildings are initiated by private developers who consider improvements in urban and building design and the creation of a less crowded environment to be to their benefit. They can sell their properties at a higher price. The above developments are more successfully implemented in newly-developed areas, such as new towns, than in old urban areas where room for manipulation is limited.

Although the density of some old urban areas has increased through redevelopment, many of the above developments which reduce the impacts of high density have taken place, making them less crowded than before when the density was even lower. Some of the concepts of these developments in creating a less crowded environment have been incorporated in the formulation of the Metroplan (a plan for the future development of the main urban area) (Planning, Environment and Lands Branch, 1990) and the work of the Land Development Corporation which is responsible for redeveloping the old districts in the main urban area through public-private partnership (Yeh, 1990).

Much coordination is needed for the planning and management of the exterior and interior building environment for high density development. In Hong Kong, the main government departments involved in planning and management of the high density environment are:

- Planning Department (*planning*);
- Territory Development Department (*planning and development*);

- Buildings and Lands Department (*land leasing and building ordinance control*);
- Transport Department (*transport policy*);
- Highways Department (*roads*);
- Urban Services Department (*parks, open space, and garbage collection*);
- Environmental Protection Department (*environment*);
- Government Information Services Department (*public education*);
- City and New Territories Administration (*public education, coordination of owners corporation*);
- Independent Commission Against Corruption (*prevent corruption*).

A good urban environment cannot be achieved without uncorrupted and efficient government officials and good coordination among government departments in planning, monitoring, and managing the urban environment.

Guidelines for High Density Development

This chapter is not arguing for high density development. It would be nice to have low density development. However, if high density development is needed, better urban planning and management will make the urban environment more liveable. Hong Kong may be an extreme case of high density development, however, it shows how better urban planning and management can achieve good living environments even under such an extreme high density. The following guidelines for high density development can be developed from the experience of Hong Kong:

- good planning (provision of open space);
- good urban and building design;
- adequate public facilities;
- good public transport;
- good urban management;
- good housing management;
- coordinated efforts of government departments;
- coordinated efforts of citizens;
- public education for high density living.

High density development is very demanding. Because of the large number of people involved and the little amount of space to manoeuvre, there is little room for error. Slight errors in urban planning and management can make the environment unliveable. Urban planning alone cannot achieve a better living environment. It has to be accompanied by good urban management, from the city level to the neighbourhood level. Better *planning, design, and management* can reduce the impact of high density, making the living and working environment *less crowded* at a fixed density. Citizens also have to be educated to know how to behave in public areas in high density zones. Planners, architects, urban managers, communities and citizens all have to work together to make high density living liveable. Experience in Hong Kong shows that high density, if better planned and managed, can provide an interesting and pleasant environment.

References

Anderson, E.N. (1972), 'Some Chinese Methods of Dealing with Crowding', *Urban Anthropology*, Vol. 1, No. 2, pp. 141–50.

Baldassare, M. (1979), *Residential Crowding in Urban United States*, Berkeley, CA: University of California Press.

Bristow, R. (1981), 'Planning by Demand: A Possible Hypothesis about Town Planning in Hong Kong', *Hong Kong Journal of Public Administration*, Vol. 3, No. 2, pp. 199–223.

Bristow, R. (1984), *Land-Use Planning in Hong Kong: History, Policies and Procedures*, Hong Kong: Oxford University Press.

Census and Statistics Department (1992), *Hong Kong 1991 Population Census: Summary Results*, Hong Kong: Government Printers.

Chau, C.-S. (1981), 'High Density Development: Hong Kong as An Example', in R. Yin-Wang Kwok and K.S. Pun (eds), *Planning in Asia: Present and Future*, Hong Kong: Centre of Urban Studies and Urban Planning, pp. 1–14.

Fischer, C.S., Baldassare, M. and Ofshe, R.J. (1975), 'Crowding Studies and Urban Life: A Critical Review', *American Institute of Planners Journal*, Vol. 41, pp. 406–18.

Fong, P.K.W. (1985), 'Management of High-Rise Residential Development in Hong Kong', *Cities*, Vol. 2, pp. 243–51.

Freedman, J. (1975), *Crowding and Behaviour*, San Francisco: W.H. Freeman.

Fung, B.C.K. (1988), 'Enforcement of Planning Controls in Hong Kong', *Planning and Development (Journal of the Hong Kong Institute of Planners)*, Vol. 4, No. 1, pp. 21–6.

Jacobs, J. (1961), *The Death and Life of Great American Cities*, New York: Random House.

Le Corbusier (1933), *The Radiant City*, New York: Orion.

Loo, C. and Ong, P. (1984), 'Crowding Perceptions, Attitudes, and Consequences Among the Chinese', *Environment and Behaviour*, Vol. 16, No. 1, pp. 55–87.

Millar, S.E. (1976), *Health and Well-Being in Relation to High Density Living in Hong Kong*, unpublished PhD Thesis, Australian National University, Canberra.

Millar, S.E. (1979), *The Biosocial Survey in Hong Kong*, Canberra: Australian National University.

Mitchell, R.E. (1972), *Levels of Emotional Strain in Southeast Asian Cities*, Taipei: Orient Cultural Service.

Newman, P. and Kenworthy, J. (1989), *Cities and Automobile Dependence: An International Sourcebook*, Aldershot, Hants: Gower.

Planning Department (1992), *Hong Kong Planning Standards and Guidelines*, Hong Kong: Planning Department, Hong Kong Government.

Planning, Environment and Lands Branch (1990), *Metroplan: The Foundation and Framework*, Hong Kong: Strategic Planning Unit, Planning, Environment and Lands Branch, Hong Kong Government.

Pryor, E.G. (1978), 'Redevelopment and New Towns in Hong Kong', in L.S.K. Wong (ed.), *Housing in Hong Kong: A Multi-Disciplinary Study*, Hong Kong: Heinemann, pp. 266–86.

Pun, K.S. (1983), 'Urban Planning', in T.N. Chiu and C.L. So (eds), *A Geography of Hong Kong*, Hong Kong: Oxford University Press, pp. 188–209.

Richardson, T. (1977), *North Point, Hong Kong: A Case Study of High Density*, Architectural Association and the Royal Institute of British Architects.

Schmidt, D., Goldman, R. and Feimer, N. (1976), 'Physical and Psychological Factors Associated with Perceptions of Crowding: An Analysis of Subcultural Differences', *Journal of Applied Psychology*, Vol. 61, No. 3, pp. 279–89.

Schmitt, R.G. (1963), 'Implications of Density in Hong Kong', *American Institute of Planners Journal*, Vol. 29, No. 3, pp. 210–17.

Stokols, D. (1972), 'A Social-Psychological Model of Human Crowding Phenomena', *American Institute of Planners Journal*, Vol. 38, pp. 72–83.

Sundstrom, E. (1978), 'Crowding as a Sequential Process: Review and Research on the Effects of Population Density on Humans', in A. Baum and Y.M. Epstein (eds), *Human Response to Crowding*, Lawrence Erlbaum Associates, pp. 31–116.

Town Planning Division (1984), *Town Planning in Hong Kong*, Hong Kong: Town Planning Division, Lands Department.

Traver, H. (1976), 'Privacy and Density: A Survey of Public Attitudes Towards Privacy in Hong Kong', *Hong Kong Law Journal*, Vol. 6, No. 3, pp. 327–43.

Wang, L.H. and Yeh, A.G.O. (eds.) (1993), *Keep A City Moving: Urban Transport Management in Hong Kong*, Tokyo: Asian Productivity Organization.

Working Group on the Review of the Town Planning Ordnance, Hong Kong Institute of Planners (1988), 'Issues of Town Planning Legislation in Hong Kong', *Planning and Development (Journal of the Hong Kong Institute of Planners)*, Vol. 4, No. 1, pp. 2–7.

Yeh, A.G.O. (1987), 'Spatial Impacts of New Town Development in Hong Kong', in D.R. Phillips and A.G.O. Yeh (eds), *New Towns in East and Southeast Asia – Planning and Development*, Hong Kong: Oxford University Press, pp. 59–81.

Yeh, A.G.O. (1990), 'Public and Private Partnership in Urban Redevelopment in Hong Kong', *Third World Planning Review*, Vol. 12, No. 4, pp. 361–83.

8 An Ecological Plan for a Better Living Environment in a Modern City: A Case Study of Huizhou, Guangdong Province, People's Republic of China

XIANGRONG WANG AND ZHAOYU QIAN

Introduction

A modern city is a large artificial ecosystem with complicated relationships between various components. The continual increase in population is the principal characteristic of an urban ecosystem. During the process of urbanisation, the settlement of human beings has undergone a process of ecological succession from natural environment to village town, industrial city and modern city. This process has changed and destroyed not only the natural environment, but also the productive condition and living environment of human beings. The results of this succession has brought about a serious challenge to the urban living environment. The historical experiences and lessons have gradually made people aware of the importance of a better urban living environment for public health, urban production as well as the daily lives of inhabitants (Wang, 1992). From the various urban development stages of Western countries in past centuries, we can understand that the development objectives of most Western cities are to strive for a better urban environment. To improve the environment and build a pleasant city have become a theme of planning thought for many years (Wu, 1992). Therefore, it is necessary to approach the ecological planning methods in order to create a better urban living environment and to construct a highly efficient, highly civilised, clean and harmonious eco-city.

Research Progress and the Scientific Significance of Ecological Planning

'Ecological planning' is a kind of planning method which is guided by the national land planning and regional planning. Following the principles of ecology and the regulations of urban capital planning, it puts forward a reasonable strategy of ecological development and construction for an urban ecosystem, which can appropriately handle the relationships between man and nature, and man and environment (Bright, 1988; Lin, 1988).

Although ecological planning is just a new method of solving the problems of urban and village planning, it has a long history as a kind of academic thought. For example, the 'ideal country' of ancient Greek philosopher Plato, the 'ideal country' of ancient Roman architect Vitrruvies, M., the 'utopia' of Thomas More in the sixteenth century, the 'falaji' of Fourier in the eighteenth century, the 'new coordinate village' of R. Owen, as well as the 'garden city' of E. Howard. All of these imaginations have contained the philosophy of ecological planning.

If we consider the theories and practice of Marsh (1864), Powell (1879), Geddes (1915) and (McHarg, 1969) about ecological assessment, ecological surveys and synthetical planning, they established a basis of ecological planning in the twentieth century, Howard's 'garden city' (1902), Sarinen's 'organic scattered theory' and Chicago Human Ecological School's ecological planning for the urban landscape, function, and the open-space system have brought about the first 'high tide' of ecological planning. In the 1940s, the planning works of the Regional Planning Association of America created the second 'high tide' of ecological planning, in which the principal works were focused on the optimum unit and interactions between city and village, and natural conservation, etc. (McHarg, 1969). In the last 30 years, with the rapid development and urbanisation in the world, the presence and aggravation of ecological crises on the areas of environment, resources, population, energy and food have stimulated people's interest in researching the urban ecology. The concept of ecological planning has gone a step further and become a 'hot topic' in urban research today. Many municipalities, such as Washington, Canberra, Stockholm, Frankfurt, Moscow, Hong Kong, Beijing, Tianjing and Changsha, have begun to carry out research on ecological planning.

The 57th regulation of MAB (UNESCO, 1984) plan has pointed out that ecological planning is able to create an optimum environment which merges the techniques and natural human activity according to the natural ecology and social psychology. This optimum environment stimulates and brings out the potentials of human creative power and productive force, and also provides a high level of physical and cultural life. Therefore, ecological planning is not the same as the traditional environment planning and economic planning. Its scientific significance is to stress the activity, coordination, overall interests and administrative levels of planning. It proposes the opening-up of society, a highly efficient economy and the coordination of an eco-environment.

A Case Study of Ecological Planning in Huizhou, Guangdong, China

Analysis of the Urban Living Environment and Ecosystem in Huizhou

Huizhou lies at the junction of 23°05´N and 114°23´E. It is located in the southeast part of Guangdong province and the northeast delta of the Pearl River. It is not far away from Hong Kong and Macao. It has been the centre of politics, economy and culture in the East River region of Guangdong. With its long history, favourable geographic position, abundant scenic resources and cultural relics, Huizhou has many advantageous conditions for urban development. In addition, the reform and open-door policy of China in recent years has brought about great changes in Huizhou. It has become an attractive city for tourism and trade in southern China. Simultaneous with its economic growth, however, was the degradation of the quality of urban living environment and the rapid growth of population which posed some serious problems. In order to face the challenges of the changing urban situation, it is necessary to carry out ecological planning for the city of Huizhou.

According to the theory of pyramidal structures of complex ecosystems, the urban ecosystem in Huizhou may be divided into three sub-ecosystems including natural, social and economic ones. The urban complex ecosystem of Huizhou is situated at the primary stage of ecological succession. Four problems exist in the ecosystem:

a)   degradation of urban eco-environmental quality;
b)   deficiency of infrastructure;
c)   threats of flood in river-lake system;
d)   uncompleted urban open-space system (UOSS).

Ecological Planning of Better Living for the City of Huizhou

Zoning of urban ecofunction district According to the regulations of urban

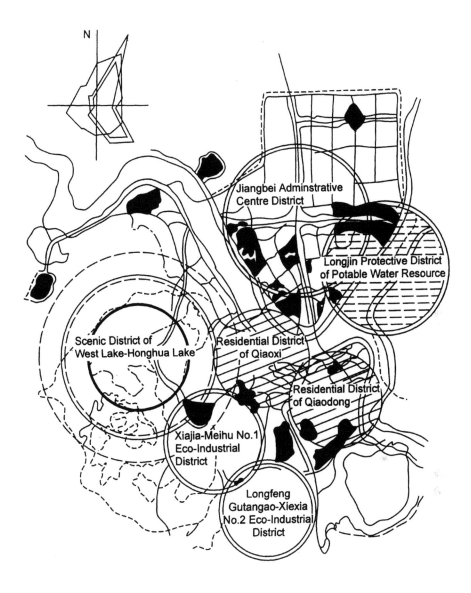

Figure 8.1 Zoning of ecofunction districts in Huizhou

capital planning and ecological conditions of Huizhou, the authors have zoned the city of Huizhou into the following ecofunction districts (Figure 8.1) so as to improve the better coordination between essential ecofactors.

Distribution of urban master structure The dominant wind directions in Huizhou are north and northeast in winter, and southeast and south in summer. The average north wind speed is 2.2 to 2.5 m/s. Since Huizhou is in the monsoon climatic zone of the southern subtropics, the distribution of industrial and residential districts should avoid the blowing direction of winter and summer monsoons. Therefore, natural hills (such as Honghuazhang and Gaoguabang) and the river-lake system are used to form a frame structure of urban community distribution. The urban areas are zoned into two eco-industrial districts, one scenic tourism district, one protection area of potable water resource, two residential districts and one administration district.

Strengthening the systematic engineering of environmental protection The current problems of the living environment in Huizhou mainly come from the unreasonable energy structure and resource waste. Therefore, it is considered the best countermeasure to change the energy structure, renew and reform the technical equipment, and strengthen the systematic engineering of environmental protection.

Control the air pollution Main air pollutants in Huizhou are dust and SO_2 due to the following reasons:

1) direct coal burning;
2) low ratio of removing dust from industrial waste gas;
3) unreasonable urban function zoning.

With regard to the above reasons for air pollution, the ecological planning of Huizhou has attempted to reduce and control the air pollution by renewing the energy structure and technical equipment, using low sulphur fuels, transforming industrial boilers, dispelling smoke, extending urban coal gas, adjusting the zoning of function and increasing greenery.

Treatment for water pollution In order to alleviate the water pollution, some basic engineering works such as protecting the potable water resource, economising water usage, developing the sewage treatment and recycling sewage, are to be carried out. At the same time, industries should set-up objectives and target plans to control and reduce the level of water pollution, or more positively, to improve the quality of the water environment. Moreover, some new eco-engineering techniques, including the oxidation pond, the purifying system of soil-vegetation and the organism purifying system, have also been planned to lessen the water pollution problem. Above all, the authors have suggested that a system of photo-eco-engineering should be set up in shallow water banks of the East River and Xizhijiang River. This photo-purifying system is composed of: 1) submerged anchored hydrophytes (Chara, Najas, Potamogeton, Vallisneria, etc.); 2) floating macrophytes and hydrophytes (Eichhornia, Salvinia, Lemna, Nymphaea, etc.); 3) emergent anchored hydrophytes (Carex, Scirpus, Phragmites); 4) moist herbs; 5) moist trees; 6) landscape forest; and 7) economic fruit forest. The function of this photo-eco-engineering is both to prevent water pollution and to protect the river embankments (Figure 8.2).

Control the noise pollution This plan emphasises the control and management of traffic noise. The residential lands have been categorised according to the noise sources. The multi-level green belts, with widths of 20-30 m, are planned to be placed on both sides of highways and streets. They act as buffer zones to reduce the level of noise pollution.

Planning for the Urban Open-Space System

Planning for the structure and distribution pattern The master structure of urban open-space system is designed to be a radial 'X-form' skeleton which is constituted by three levels of green lands. The first level of green land is constructed by extensive areas of green lands in the scenic districts of West Lake, Honghua Lake and Jinshan Lake. The second level is constructed by line-form green belts of river bank and streets. The third level is composed of spot-form green lands such as public parks, botanical gardens and special-purpose green lands. After a study on the ecological design of the structure of these three levels of green lands, it is discovered that this urban open-space system is not only beneficial in achieving its ecological efficiency which improves the urban living environment, but is also helpful in beautifying the urban landscape (Figure 8.3).

Planning Objectives

The planning objectives were divided into two stages:

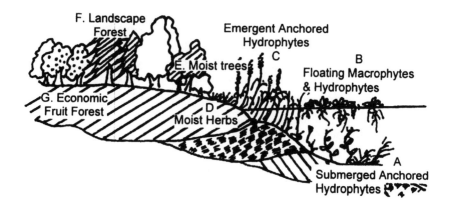

Figure 8.2   Diagram of eco-engineering

a)   1993–1995:

Greenery coverage of city proper area: 40 per cent
Average area of green land/person: 32m^2

b)   1996–2005:

Greenery coverage of city proper area: 50 per cent
Average area of green land/person: 58m^2

Discussion

A satisfactory urban living environment lays a good foundation for urban development. Therefore, municipal officers, planners as well as ordinary people should pay more attention and focus effort on building an optimal urban living environment. In order to achieve this target, ecological principles should be introduced into urban planning such that the traditional planning method can be improved and urban development can be promoted.

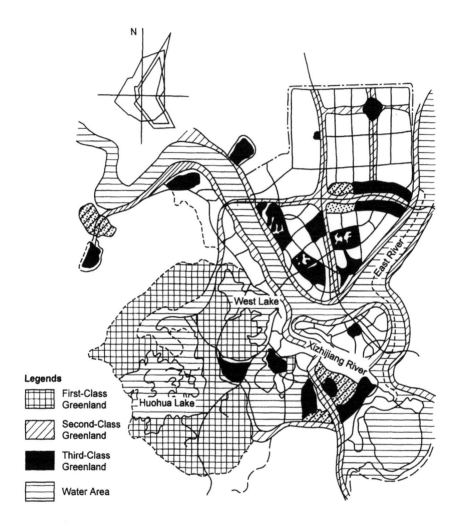

Figure 8.3   Ecological planning of urban open-space system in Huizhou

References

Bright, J.W. (1988), 'Applied Ecology in Action – Design A River Park', *Garden Building*, Vol. 3, No. 2, pp. 29–34.

Lin, D. (1988), 'An Ecological Approach Toward Environmental Planning and Design', *Garden Building*, Vol. 3, No. 2, pp. 8–17.

McHarg, I. (1969), *Design with Nature*, Garden City, NY: Natural History Press.

UNESCO (1984), *Final Report, MAB Report Series, No. 57*, Suzdal.

Wang, X. (1992), 'Garden Building and Ecological Designing', *Urban Planning Forum*, No. 6, pp. 18–21 (*in Chinese*).

Wu, L. (1992), 'On the Capital Planning for Pudong New Area', *Urban Planning*, No. 6, pp. 3–10 (*in Chinese*).

9 The Environment of Densely Populated Areas of Bangladesh: With Special Reference to Dhaka and Secondary Towns

MIR SHAHIDUL ISLAM

Introduction

Bangladesh, with an area of 148,393 km^2 and a population of about 111.46 million (BBS, 1992), is part of the Ganga-Brahmaputra-Meghna delta in South Asia. It is bordered by the Indian territories on the north, the west, the east; Myanmar on the southeast; and the Bay of Bengal on the south.

There are many environmental problems and issues in Bangladesh. According to recent estimates, due to global warming (greenhouse effect), the sea level may rise about 20 cm higher by 2030 and about 65 cm to as much as 1 m by the end of the next century. A one meter sea level change will inundate about 15.8 per cent of the total area of Bangladesh, displace 10 per cent of the present population, lose 13.7 per cent of cropped land, destroy 4,000,000 ha. of mangrove forest, involve a loss in output of 13 per cent of GDP and loss of assets of circa US$ 14 billion. Moreover, the area of high salinity intrusion will increase from an existing area of 13 per cent of Bangladesh to 32 per cent (Report of the Task Force, 1991b). Besides these, several other environmental hazards are responsible for the degradation of the environment in the country.

A visiting multi-disciplinary team carried out an assessment of the chief natural resource management and environmental conservation issues in Bangladesh in 1990. It reviewed the principal areas and causes of environmental degradation, declining soil fertility, lowered water table, river bank erosion and siltation of rivers, degradation of the remaining natural forests, wetlands, coastal environment and fisheries resources. The degradation of the forests is one of the main ecological concerns. Tree coverage in Bangladesh amounts to less than 6 per cent of the total land area, compared to 11 per cent for India (Timm, 1992). The area under government controlled forests continues to dwindle due to encroachments. The area has fallen from 7601 square miles in 1982–83 to 7108 square miles in 1986–87 (Majumder, 1992a). The coastal environment of Bangladesh is very dynamic, though in recent years has begun to show signs of decay. The 710 kilometre-long coastline is now a spectacle of destruction. Unplanned expansion of shrimp culture is leading to the destruction of mangrove forests which are vital to the biological cycle of the coasts (Majumder, 1992b). In the city of Rajshahi an alarming environmental problem is emerging. The ecological balance of Rajshahi is threatened by the unplanned filling of its ponds and low lands. One non-government statistical report shows that in the last five years about 200 natural water reservoirs were filled. As a result, acute water crisis broke out in the city. Due to a disrupted natural environment, the northern region of Bangladesh will start to become a desert with the filling of ponds. The desertification process of the area is thus accelerated, having been also prompted by the Farakka barrage (Das, 1993).

Industrial pollution is also significant, even though the country is still at the primary stage of industrialisation. For a long time, little attention has been given to the discharge of effluent and the disposal of wastes. All industries, big or small, dispose their waste mainly in rivers. In some cases, wastes are dumped into ponds or ditches. Some of the rivers have become excessively polluted due to the discharge of municipal and industrial effluent.

The country is suffering from a number of environmental problems as discussed here. The worst situation occurs in big cities and in a number of medium-sized towns where population density is very high, while services to cater for the population lag behind.

If all the people of the world were crammed into the USA, the population density of that country would be equal with that of Bangladesh (Omar, 1992). Bangladesh depends heavily upon agriculture, but this sector alone can never cater to the needs of the country's growing number of people. A number of industries based on both indigenous and imported raw materials have been set up. Among them are the important ones like jute, cotton textiles, paper, sugar, fertiliser etc. This trend of industrialisation is gradually changing the economic structure of the country towards modernisation. A shrinking land-to-person ratio and the saturated condition of employment in the agriculture sector against the fast growth rate of the national population, has led to the concentration of a large number of people in the big cities of Dhaka, Chittagong, Khulna, Rajshahi and in some medium sized towns like Mymensingh, Sylhet, Rangpur, etc.

Urban and rural developments are supplementary and complementary to each other, but the existing trend of urbanisation through concentration of the larger segment of increased population in a few big cities and towns is certainly mismatched in the context of its linkage with rural population (Asian Development Bank, 1987).

Mass poverty leads to the overuse of natural resources like lands and forests. The below US\$ 200 per capita annual income is one of the lowest in the world. The income is again not distributed evenly. More than two-thirds of the people live below the poverty line.

With over 751 people living in every square kilometre, Bangladesh is the eighth largest populous country in the world and is also the most densely populated country in the world, except for the modern city states of Hong Kong and Singapore. The 1991 census has shown an encouraging trend in population control. The population growth rate has come down to less than 2.17 per cent (*Bangladesh Observer*, 1993a). Still, the growth rate is very high and at this rate the population is likely to be well over 130 million from the present level of 110 million by the turn of the century (Majumder, 1992a).

Rapid population growth has put pressure on housing, sanitation and other civic amenities. According to official records, more than one million people do not have shelter. The shortfall in the number of houses is likely to be 5.29 million by the year 2000 (Majumder, 1993a). According to BBS (1981), only 26 per cent of the urban population had access to piped water supplies and only 11 per cent had access to adequate sanitation. The environmental conditions, especially water pollution problems, arise from inadequate treatment of sewerage, disposal of solid wastes and periodic flooding of poorly drained urban areas. These problems are increasing with the increase in population. Inadequate institutional capabilities and limited resources have constrained the ability of densely populated cities and towns to meet the ever increasing demand.

This chapter attempts to identify some of the major environmental problems of densely populated areas in Bangladesh with a special emphasis on Dhaka City, the capital of Bangladesh, and densely populated secondary towns.

Methodology

Dhaka, the capital of Bangladesh, with a population of 6,959,920 (Bangladesh Observer, 1993a), is one of the most densely populated cities in the world and according to the United Nations Fund For Population Activities (UNFPA) Report 1993, the population of Dhaka is expected to grow to 12.2 million (*Bangladesh Observer*, 1993b). Due to such growth in the population, the infrastructure facilities cannot cope with this load with the result that the living environment in densely populated areas becomes unhygienic.

Environmental degradation due to high density is not only a problem for large metropolitan areas in Bangladesh but also for a number of secondary towns which are facing similar environmental problems.

Information has been collected both from secondary sources and reports of different projects and also published materials. Most of the secondary information is taken from published reports, journals, statistical year books and other similar sources.

Dhaka

Introduction

Dhaka was founded in 1608 as the seat of imperial Mughal Viceroys of Bengal. The city of Dhaka has experienced phenomenal growth in recent years with an annual growth rate of nearly 10 per cent between 1974 and 1981 (Islam and Maniruzzaman, 1991). Considering the population within Dhaka Statistical Metropolitan Area (DSMA) limits, Dhaka has enjoyed a primacy status in Bangladesh: an urban structure with 25–30 per cent of the national urban population during 1961–91.

In 1901, Dhaka had a population of nearly 0.1 million in an area of 10 square miles. In the 1981 census, the population was 3.4 million in 171 square miles (BBS, 1981). The preliminary report of the 1991 census showed an adjusted figure of 3,637,892 people in the municipal area of 344 km^2 (BBS, 1993). That means that the density per km^2 was 10,575. A task force constituted by the Government of Bangladesh (GOB) projected the 2001 population of Dhaka City at 9.3 million (The New Nation, 1991). According to UNFPA's world population report 1993, the population of Dhaka City will stand at 12.2 million by the end of this century (*Bangladesh Observer*, 1993b), making Dhaka the seventeenth most populous city in the world. Currently, Dhaka ranks thirtieth (Islam et al., 1991) and by the end of this century, is expected to rank as the seventeenth (*Bangladesh Observer*, 1993b) most densely populated city of the world.

Dhaka would have to accommodate 20–25 million people in 2015 in an area which would be approximately 20 per cent of the probable urbanised land of the country, at that time (Report of the Task Force, 1991a). As the population of Dhaka City is increasing very rapidly, the service facilities are lagging behind and thus threatening both public health and the natural environment in densely populated areas. The newly constructed flood protection embankment for Dhaka City has added some new environmental problems to the miseries of people of densely populated areas. The environmental problems due to such rapid growth of population and a newly constructed flood control embankment are discussed below.

Physical Destruction of Nature

Some of the early victims of urbanisation are the natural features of the area. Dhaka is characterised by the canals that traverse it. Most of the canals of Dhaka have already disappeared under silt or built structures. Each year, more agricultural land and filled lowland from the flood plains are being devoured by the city. In the late 1970s, the municipal authorities carried out an organised drive that felled hundreds of invaluable old trees in the pretext of road widening. Apart from changing the microclimate and destroying the flora and the habitat of the fauna, these phenomena create more serious problems of waterlogging and hindrances to the recharging of the subterranean water table (Islam et al., 1991).

Urban centres are usually located on strips of elevated land. A dominant feature of the greater Dhaka area is the limited availability of flood-free land. The central part of Dhaka City is developed on hilly land with an elevation of 6.0 m to 8.0 m above mean sea level (Khan, 1993). As the population and the density level within the urban centres rise, the value of the scarce urban land also rises. Islam (1987, p. 2) has observed that the land value in Dhaka over recent years has increased by 60 per cent to 90 per cent per year. As a result, the middle income and the poor, comprising most of the population, are pushed to marginally serviced areas at the fringes of the cities. As urban centres expand, they encroach first on good agricultural land which is more suitable for human settlement. The increased demand for construction materials leads to the conversion of more fringe agricultural lands to brickfields (USAID, 1990).

Problems Due to Newly Constructed Embankment

The flood problem in Bangladesh, because of its unique geography and location, is unusually complex and challenging. Floods damage both the physical infrastructure and disrupt economic activities in the affected areas. Unplanned physical interventions in the natural drainage path cause waterlogging and flood problems.

The problem of extreme waterlogging in the capital city can be blamed on heavy rainfall, the failure to implement a proper sewerage system in the city and the flood control dam. There are allegations that the greater Dhaka flood control dam is also causing the waterlogging to increase. The water which could naturally drain out of the city cannot do so now. It is alleged that this is due to the embankment (Babar, 1993).

Water Supply, Sanitation and Drainage

The Dhaka Water Supply and Sewerage Authority (DWASA) is responsible for water supply and sewerage disposal services in Dhaka City. The level and quality of service are far from satisfactory. DWASA is currently supplying water at the rate of 550 million litres per day (mld) (Sheesh, 1994). Assuming a population of 6.96 million and per capita demand of 180 litres, it appears that DWASA is in a position to supply around 44 per cent of the daily requirement of water.

Viewing the situation from another angle, there are only 1,15,000 (ibid., 1994) water connections. An estimated 3 per cent of the connections are commercial. There are 1,209 community standpipes in old Dhaka (Shankland and Cox, 1979) an apparently very conservative estimate of 115 users per standpipe. The network of 1,250 km of pipes, which are decrepit and worn-out in places, leads to the contamination and leakage of water. These features indicate that DWASA has failed to meet the increasing demand of the city dwellers.

Due to the uncontrolled increase of population in Dhaka City, infrastructure facilities have been pushed to the limits of their capacity. Water supply in parts of Dhaka City is regularly disrupted due to the excessive lowering of the ground water table, especially in the hot and dry season. Studies have also warned about potential subsidence problems due to extensive ground water extraction (Shankland and Cox, 1979, p. 50). Frequent filling of canals and ponds in urban areas also removes an important source of water for the poor.

The sewerage situation is even worse, with only around 40,000 connections and 460 km of sewer line (Sheesh, 1994). Many households, especially those beyond the reach of sewer lines, use septic tanks, although the erection of semi-permanent latrines is prohibited by law. A considerable portion of the city's populations, including most of the slum dwellers, use such latrines. These latrines, located near the settlements and surface water sources, as well as open-air defecation, spread pathogens and pollution and are the major causes for poor hygienic conditions. A study (Islam, 1992) has shown that approximately 15 to 20 per cent of the city's population is currently served by DWASA sewerage system. Of the remainder, about 25 per cent are served by on-site septic tanks, 15 per cent by sanitary pit latrines and 5 per cent by bucket latrines. The remaining 35 per cent to 40 per cent have no sanitary services.

It is evident that 150 km of underground sewage lines of Dhaka are not cleared regularly, so just a little rain causes the streets to go under water. Forty canals within Dhaka City have been marked for re-excavation (Babar, 1993).

Solid Waste Disposal

The city generates around 3,000 tons of solid wastes daily. From the study of different consultants engaged by the Government of Bangladesh, the amount of solid waste generated is increasing proportionately with the increase of the population. There are around 4,300 dustbins around the city for the collection of garbage. The number of dust bins is quite inadequate for a city with 6.96 million people. Dhaka City Corporation is responsible for collection and disposal of solid wastes. The city corporation uses around 100 trucks and 2,380 hand trolleys to collect and dispose of the solid wastes at specific locations of low lying land.

Approximately 50 per cent of solid waste generated within Dhaka City is never collected. Some is deposited in vacant lots, low lying areas, ponds or rivers, and a large percentage is deposited into local drainage ditches, storms sewers and canals (Louis Berger International Inc., 1991). As a result of this uncontrolled disposal of wastes, the degradation of the quality of the environment in densely populated areas has become a concern for the city dwellers.

The method used for solid waste disposal in Dhaka City is 'crude dumping' which is extremely unhygienic and unsanitary. The method of crude dumping being used in Dhaka is nothing but an environmental disaster, causing health risk from flies, rats, air pollution, and water pollution through leaching by rainfall (Paul, 1991). The total, as well as per capita per day, collections of municipal wastes by city corporations are shown in Table 9.1. It should be noted that per capita per day collection of solid wastes is related to the size of the city. The per capita waste collection in Bangladesh for disposal is far lower than the waste generation rates of the United States: 0.51kg in Calcutta, 1.8kg in New York, and 2.42kg US national average (Feroze, 1993). The lower rate of waste generation may be due to the fact that some of the waste escapes municipal collection and is dumped in local low-lying areas. Moreover, extensive recycling of wastes in urban centres has contributed to the reduction in the quantity of wastes for collection and disposal by municipalities. The total waste generated in urban centres is estimated to be 1.5–2.0 times the waste collected by municipalities (Feroze, 1993).

Table 9.1   Collection of wastes in dry and wet seasons by DCC

City	Total collection ton/D		Collection kg/capita/D	
	Dry season	*Wet season*	*Dry season*	*Wet season*
Dhaka	1,000	1,400	0.18	0.25

Source: Feroze, 1993.

The present practice of solid waste collection, recycling and disposal needs improvement in order to reduce adverse impacts on the community, environment and public health, and to maximise resource recovery and utilisation.

Transportation and Vehicular Pollution

The transport problem is expanding rapidly with the very fast growth of population in the city. Dhaka is facing great difficulty with unplanned growth in vehicular traffic, defective planning of roads, road junctions, undisciplined conduct of the road users, an innumerable number of slow moving rickshaws, inadequate public transport and the lamentable management system, despite Dhaka being the main financial and business centre of the country (Islam, 1993). The number of vehicles has increased manifold, but still remains far short of the needs. The cheapest form of public urban mass transport in Bangladesh is the bus. The number of buses in Dhaka City grew from only 1,531 in 1971 to 4,860 in 1988 (Report of the Task Force, 1991a). At either point in time the numbers were small. The number of buses in 1988 (Report of the Task Force, 1991a) was highly inadequate in terms of the demand.

Moreover, in reality not more than half of the registered buses were on city roads at any one time. As a result, one always finds long queues at bus stands and the buses are usually dangerously overloaded with people. In response to the growing needs, the number of other vehicles also increased. Thus the number of auto tempos increased from only 454 in 1981 to 2,148 in 1988 in Dhaka City; the number of auto rickshaws in Dhaka grew from 3,843 in 1971 to 11,066 in 1988 (Report of the Task Force, 1991a). 'Mishuk', or the light bodied auto rickshaw manufactured locally, was introduced to the city roads in 1988 (Report of the Task Force, 1991a). Altogether, in 1988, Dhaka City had over 182,000 motorised vehicles. But, except for the buses, mini buses and tempos, the other and more numerous forms of transports each carried only two to three persons. Altogether more than 1.5 million people in Dhaka are directly or indirectly dependent for their livelihood on the rickshaw. This is nearly 25 per cent of the total population of the city.

The rickshaw, as a mode of transport, is also important from the point of view of environmental balance. It is the most pollution-free mode of transport. On the other hand, the motorised vehicles cause enormous problems of air pollution in the city. It is estimated that fumes from motor vehicles account for 25 per cent of the total emission in the atmosphere each year (Walsh, 1990). Around 50,000 vehicles in the city emit excessive exhausts, but legal action was taken against only 867 vehicle owners in the last 10 years (Sikder, 1991). Moreover, with the current rates of population increase in the city, it is unlikely that the rickshaw can continue with its role for long, considering its inefficient use of precious road space and the consequent congestion that it causes. Therefore, we must be prepared for significant rises in vehicular emissions in the near future.

Health Issues

The living conditions for the larger proportion of urban inhabitants are linked directly to the declining health and nutritional situation in the slums and squatter settlements. In large cities, the poor are often forced to set up their settlements on land filled with garbage. But refuse takes years to be completely purified. Consequently, the poor are vulnerable to many diseases spread through the contaminated ground. Through scanning several studies of (mainly) Dhaka squatters (Hussain, 1982), it was found that the incidence of various types of environment-related diseases is highest among slum and squatter dwellers. The commonly found diseases were gastro-intestinal disorders, cold fever, cough, dysentery, skin diseases, measles, typhoid, cholera, tuberculosis, syphilis, gonorrhoea, malaria, various types of female diseases and chronic malnutrition.

A study (Ashrafuddin, 1987, p. 34) of slum areas near the centre of Dhaka City found that 30 per cent of the respondents were diseased. The study further reported that diarrheal diseases are the leading cause (13 per 1,000) and the respiratory tract infections are the second leading causes of mortality (11 per 1,000). *Tetanus Neonatorum* was the third leading cause (nine per 1,000) and affects new born infants. Most diseases are more fatal to children than adults. The overall infant mortality rate was 152 per 1,000.

Industrial Pollution

Manufacturing industries are a major source of pollution. Though Bangladesh is not highly industrialised, existing industries are heavily concentrated in a few regions, including the Dhaka City region. Data from the census of manufacturing industries of 1982–83 reveal that 50 per cent of manufacturing establishments are located in Dhaka and the adjoining Narayanganj districts. The figure would be much higher if the larger industrial units were considered. The fumes, liquid and solid wastes produced by the industries heavily pollute the air and water in the rivers. It is estimated that the five industrial centres in Dhaka City discharge 49,000kg of Biochemical Oxygen Demand (BOD) into the rivers daily (Islam, 1991). Around 130 tanneries in the densely populated Hazaribagh area produces effluvium and discharges an estimated 680–725kg (1,500–1,600 lb) of BOD and 770–815kg (1,700–1,800 lb) of suspended solids to the Burihiganga river through two drains (Basu, 1979). Chromium, the highly toxic bio-accumulating element, is one of the pollutants that the tanneries discharge. The smokestacks of the industries of Dhaka City, including the numerous brick kilns that are located in the outskirts of the city, readily belch dark fumes into the air. No current data on the level of pollution are available.

Housing Issues

Due to housing shortages, the middle income and the poor are forced to live at excessive densities and without the basic facilities like piped water, sewerage, drainage and garbage removal facilities. The shortage of housing in urban areas has resulted in the growth of slums and squatter settlements which are very unhygienic. In its most recent survey of Dhaka, the Centre for Urban Studies has registered more than 1,000 squatter settlements within the municipal area (CUS, 1988, p. 16). Most housing units in these settlements are made of bamboo, reeds, grass or rags with roofs of the same materials and plastic sheets. Units on average are about 50 ft^2 (for an average family of five to six persons). The average density has been calculated at 162,000 persons per square kilometre or 655 persons per acre. Electricity, water supply, sewerage and drainage facilities are mostly absent. Water is brought from various sources like roadside taps, people with water supply, ditches or small ponds. Makeshift latrines are shared by 10 to 12 families.

Secondary Towns

Introduction

The rapid urban growth of secondary towns throughout Bangladesh (estimated at approximately 7 per cent per annum) is placing severe strain on infrastructure and services. The service facilities are lagging far behind due to increasing demands by the population. Thus, the overall environmental conditions of these towns are declining day by day.

For the secondary towns in particular (i.e. all major urban centres outside Dhaka, Chittagong, Khulna and Rajshahi, which includes towns with populations ranging from less than 30,000 to over 300,000) urban investments have remained generally low due to resource constraint. There have been very few environmental programs because of inadequate institutional and financial capabilities.

According to a report (Louis Berger International Inc., 1990) submitted to the Ministry of Local Government, Rural Development and Co-operatives (MLGRDC) of GOB a number of environmental and infrastructural problems in the secondary towns have been identified. The following towns have been surveyed for study purposes: Sylhet, Mymensingh, Rangpur, Bogra, Pabna, Kushtia, Bhola, Borgura, Feni and Cox's Bazar. Overall features and environmental problems of the surveyed towns are discussed here in brief.

Drainage Condition

The saying, ' there are drains but no drainage system' is true for all 10 of the pourashavas (municipalities). The inventories indicate that out of approximately 526 km of drains, only about 106 km, or 20 per cent are lined, and that most of these are clogged. The household surveys indicate, that on average, only 43 per cent of the population has some form of drainage facility serving their area. The households rated drainage as being poor or very poor. Eight of the 10 pourashavas rated drainage as being the highest priority service for improvement actions, and the remaining two rated drainage as being their second priority.

External flooding is not a major problem for the participating pourashavas. The main problems identified are the inability of the system (or the lack thereof) to drain the areas during the monsoon period, resulting in waterlogging of large urban areas, shallow flooding during peak rainy periods, stagnant and polluted ponds in low lying areas, flooding and accelerated congestion of the roadways, and the generally poor environmental conditions associated with swampy areas during the wet season.

Roads and Footpaths

Roads and drainage systems are complementary to each other and normally go together in an urban environment. In the absence of a proper road network with properly designed drainage facilities, the drainage system cannot be adequately improved.

- Within the 10 pourashavas there are a total of about 763.5km of roads, of which 108.9 km (14 per cent – all main roads) belong to the Roads and Highways Department (RHD), with the remaining 654.6 km belonging to the pourashavas.

- The portions belonging to the RHD are all part of the national highway systems and are generally properly metalled and surfaced. However, even here, inadequacies exist which are a burden on the pourashavas. Out of the remaining roads, which are the responsibility of the pourashavas, about 50 per cent are pucca (metalled), 16 per cent are semi-pucca (brick surfacing), and the remaining 34 per cent are katcha (earthen surfaced). Of the surfaced roads, more than half are in poor condition with potholes or worn-off surfaces; the semi-pucca roads are suitable only for occasional vehicular traffic, and the katcha roads are navigable only during the dry season.

- The average reserve widths varies from 6.0 to 9.5 meters and the carriageway from 2.0 m to 6.5 m. Most are too narrow to meet present traffic demands, and many are suitable for one-way traffic only. Because of the narrow rights of way, widening opportunities are limited without acquiring additional lands.

- Roadside drains are generally absent and mostly nonfunctional due to lack of gradients, clogging, blockages or connection to outfall drains.

- Sidewalks are extremely rare, and road shoulders are frequently not provided.

- Roadside ribbon development right up to the carriageway is common and extremely acute at intersections.

- Bridges and culverts over canals are inadequate.

- Railroad tracks pass through densely settled areas, and railroad crossings are narrow, congested and hazardous.

From the household surveys it was found that an average of 63 per cent of the population felt that the road access was poor or very poor in their areas. In spite of this, however, improved access was considered to be the priority in only one pourashava, the second priority in two towns, the third priority in another two towns, and the last priority in the remaining five towns.

Water Supply

The communal water supply systems within the ten study towns are inadequate to meet the requirements of the pourashavas, nor is it feasible to expect them to do so without major system expansions in all cases. It was found that the present position of water supply is generally as follows:

- The communal piped water supply systems for the ten pourashavas cover only about 22 per cent of the total population, of which 17 per cent is serviced by private connections and the remaining 5 per cent by public standpipes. Of the remainder, 45 per cent derive their water from private hand tube wells, 29 per cent from shared hand tube wells, and 4 per cent from other sources including dug wells, ponds, springs and rivers.

- The average delivery capacity within the service area ranges from between 60 litre per capita per day (lcd) to 150 lcd. Much of this is lost, however, due to system losses and wastage which is estimated to average about 40 per cent overall.

• The average monthly operation and maintenance cost of water supply is approximately 40 to 60 taka ($1 = 50 taka) per serviced household, and is highly dependent on the operating periods for the pumped service.

In spite of the above deficiencies, however, it was found that the overall level of satisfaction was highest for the water supply situation in all the pourashavas, with 54 per cent of the surveyed households indicating that they were generally satisfied with the existing modes and quality of the supply (including all modes). In terms of priority for service improvement, three towns rated water supply improvement as their second priority, two as the third priority, and five as the fourth priority (op. cit., 1990).

Sanitation

Public sewer systems in the pourashavas are almost nonexistent, and in the two towns where they do exist, they serve only 2 per cent to 4 per cent of the total populations. The provision of sanitation facilities is accordingly a householders' responsibility.

• On average, 50 per cent of the surveyed population reported that they used septic tanks or water sealed vaults for disposal, 30 per cent reported that they used pit latrines, 17 per cent reported that they used service latrines, and 3 per cent that they had no facilities and used open spaces.

• On average, 43 per cent felt that their facilities were fair to good, and 57 per cent felt they were poor or very poor.

• Improvement of existing slaughter houses by the construction of sanitary disposal systems, provision of improved water supplies, installation of fly netting etc., is essential. Moreover, the construction of new slaughter houses should be implemented for those pourashavas which currently do not have any.

Solid Waste Disposal

Although the collection and disposal of solid waste is one of the prescribed responsibilities for the pourashavas, they have had few resources with which to do this in the past. From the consultants' investigations it was found that:

- on average, only about 16 per cent of the town populations receive some form of collection service from the pourashavas;

- for the 84 per cent of the population who receive no service, about 61 per cent dispose the wastes onto open lands or into drains and ponds, and the remainder dispose of it by burning or burying. It is accordingly estimated that about 50 per cent of the total waste generated is accumulated in the local environment, equal to about 80 m^3/wk (32 tons/wk) for the smaller towns and up to about 500 m^3/wk (200 tons/wk) for the larger towns.

From the household surveys it was found that 87 per cent of the population felt that the present services were poor or very poor. In order of priority for service improvement, one town rated solid waste collection as the first priority, three as the second, four as the third and two as the fourth (op. cit., 1990).

Conclusion and Recommendations

Attempts have been made to portray the environmental problems and issues of densely populated cities and towns in Bangladesh, with special emphasis on the capital city of Dhaka and secondary towns. A better living environment is not just a demand, it is a fundamental right of all people. According to the nature and magnitude of the problem, a number of recommendations are suggested here. In the case of the Third World countries, funding is often a constraint in development projects. It is suggested here, therefore, that monetary involvement should be minimised, as far as is possible, through proper planning, management and awareness of people.

The formulation of policies and regulations to minimise the destruction of natural features of the country, such as low flood plains, water bodies and greenery, suggests that special care should be taken in the development process of peripheral fringe areas of the city. In order to reduce industrial pollution, proper land use planning should be followed. Moreover, new laws should be formulated for controlling pollution from industries and vehicular exhaust.

Convenient access to safe, potable water must be ensured for each citizen and household. There should be sufficient community water hydrants for those who cannot afford house to house connections. As a short-term and temporary solution, the DWASA may use its tankers to commercially supply water to the slums on a regular basis. Community latrines should also be provided in the densely populated poorer section of the city. A charge may be extracted from the users to meet, partially at least, the cost of maintenance and upkeep.

The problem of garbage disposal deserves more attention than it receives at present. The present system of waste collection, recycling and disposal needs improvement in order to reduce adverse impacts, not only on public health but also for the overall environment. Research and study on the recycling of waste should be carried out for resource recovery and to ensure a healthy environment. Community participation in waste management should be emphasised for the improvement of the waste management system. Topographical, geographical, geological and accessibility factors should be carefully reviewed when selecting sites for waste disposal. Crude dumping of waste on lowlands should be prohibited and sanitary landfill encouraged.

Adequate road width is a primary need for easy access. With the exception of some highways and roads in planned residential areas, most of the road areas are facing traffic congestion due to their narrow widths. Even on wide roads, due to a lack of parking spaces, roadside parking reduces the efficiency of roads. In many urban areas the unplanned development of roadside shops and commercial activities cause transportation problems. Thus, it is necessary to practice bye-laws and regulations properly, if necessary new laws may be incorporated. Rather than replacing slow moving vehicles, it is better to segregate them from the speedy vehicles. In fact, improvement of the solid waste disposal system and drainage system is mostly dependent on better access. While framing transport policy, Environmental Impact Assessment (EIA) should be taken into consideration and policy measures for infrastructural facilities should be adopted taking cognizance not only of benefits, but also probable use-conflicts.

The housing problem is a major issue in densely populated areas. Experience from the failure of a number of rehabilitation program sites and service schemes carried on by government and donor agencies, demonstrates that the government alone cannot solve the housing problem; rather, large scale community participation and private sector involvement should come forward in this respect. Residential space standards should be adopted by relevant agencies and existing space standards for government/private housing should be reduced. It is a recent trend of the country to build high rise apartments; in this regard, effects on the density of that area and pressure on various services should be carefully analysed.

A general dearth of information and data relating to the environmental aspects of Dhaka has been observed. The Department of Environment, City Corporation, Urban Development Directorate, Rajdhani Unnayan Kartiphaka (RAJUK, Capital Development Authority) and municipalities should carry out research work and maintain relevant up-to-date data on various aspect of the cites and towns. Moreover, all groups of city dwellers should be given lessons about personal and community hygiene and every citizen should be made aware about their environment through mass media and educational institutions.

Acknowledgments

The author gratefully acknowledges his indebtedness to Mr Iftekhar Enayetullah, final year student of Master of Urban and Regional Planning, for his help at the time of preparation of the draft and also his daughter Ms Farzana Mir, student of first year Architecture for Computer Composition.

References

Asian Development Bank (1987), *Urban Policy Issues*, Manila, Philippines, pp. 286–7.

Ashrafuddin, M. (1987), *Study on Morbidity and Mortality among the Residents of Geneva Camps, Mohammadpur, Dhaka*, Mohakhali, Dhaka: National Institute of Preventive and Social Medicine.

Babar, S. (1993), 'Dhaka under water', *Dhaka Courier*, May, Vol. 9, Issue 43, 44, p. 10.

Basu, A. (1979), 'Environmental Aspects', working paper for Dhaka Metropolitan Area Integrated Urban Development Project (DMAIUDP), Dhaka.

BBS (1981), *Statistical Yearbook of Bangladesh*, Dhaka: Statistics Division, Ministry of Planning.

BBS (1992), *Statistical Yearbook of Bangladesh*, Dhaka: Statistics Division, Ministry of Planning.

BBS (1993), *Bangladesh Population Census 1991 Zila: Dhaka (Community series)*, Dhaka: Statistics Division, Ministry of Planning.

Bangladesh Observer (1993a), 'Dhaka to be Megacity by Century End', 7 July.

Bangladesh Observer (1993b), 'Census Report Released Growth rate 2.17 pc', 24 December.

Centre for Urban Studies (CUS) (1988), *Survey of Urban Squatters*, Dhaka: Department of Geography, University of Dhaka.

Das, U.K. (1993), 'Northern Suicide', *Dhaka Courier*, June, Vol. 9, Issue 46, p. 17.

Feroze, A. (1993), 'Municipal Waste Management in Bangladesh with Emphasis on Recycling', paper presented in regional workshop on Urban Waste Management in Asian cities, Dhaka, 10–12 April.

Hussain, S. (1982), 'Urban Poor in Bangladesh', unpublished report submitted to Geography Department, London School of Economics and Political Science.

Islam, A.K.M. (1991), 'Bangladesher Pani Dushon: Karon o Pratikar', *To Make Power Janye*, Dhaka: Engineering University Teachers Association (*in Bengali*).

Islam, M.N. (1992), 'Environmental Impact of Greater Dhaka Flood Protection Embankment With Special Reference of Domestic Waste Disposal', unpublished MSc Engg Thesis, Dhaka: Department of Civil Engg, B.U.E.T.

Islam, M.S. and Maniruzzaman, K.M. (1991), 'Urbanisation and the Environment: The experience of Dhaka' in D. Bandhu (ed.), *International Workshop on Human Settlements for Sustainable Development*, New Delhi, India: Indian Environmental Society.

Islam, N. (1981), *Land and Housing Development in the Metropolitan Fringe: A case study from Dhaka, Bangladesh*, Centre for Urban Studies (CUS), University of Dhaka.

Islam, S. (1993), 'Transport in Bangladesh', *Guardian*, January, special issue, pp. 51–3.

Khan, Z.H. (1993), 'Effects of greater Dhaka town protection embankment on the changes in the trend of settlement pattern and landuse in the fringe areas of embankment', unpublished MURP (Master of Urban and Regional Planning) thesis, Dhaka: Department of Urban and Regional Planning, B.U.E.T.

Louis Berger International Inc. (1990), *Final Report on Secondary Town Infrastructure and Services Development Project (Bangladesh)*, Vol. 1, Dhaka.

Louis Berger International Inc. (1991), *Draft Final Report of Dhaka Integrated Flood Protection Project of Flood Action Plan 8B*, Dhaka.

Majumder, M.K. (1992a), 'Environment: A Bangladesh Overview', *Dhaka Courier*, June, Vol. 8, Issue 45, pp. 8–11.

Majumder, M.K. (1992b), 'Coastal Pollution: Question of Life and Death', *Dhaka Courier*, June, Vol. 8, Issue 45, p. 11.

Omar, I. (1992), 'Industrialisation and Pollution in Bangladesh', *Dhaka Courier*, June, Vol. 8, Issue 45, pp. 13–14.

Paul, K. (1991), 'Solid Waste Management in Bangladesh with Special Reference to Situation in Dhaka, the Capital of Bangladesh', country paper presented in seminar on Solid Waste Management held in New Delhi, India, 8–11 October.

Report of the Task Force on Bangladesh Development Strategies for the 1990s (1991a), *Developing the Infrastructure*, Vol. 3, pp. 440, 463, Dhaka: University Press Ltd.

Report of the Task Force on Bangladesh Development Strategies for the 1990s (1991b), *Environment Policy*, Vol. 4, pp. 7, 161, 165, Dhaka: University Press Ltd.

Sheesh, I. (1994), 'Mohanogor Pani Sorboraho abong Poyo Nishkashon Baybostha-Dhaka WASA Prakit', article published in the Souvenir of 38th Convention of the Institution of Engineers, Bangladesh, Dhaka (*in Bengali*).

Shankland and Cox (1979), 'Engineering Infrastructure and flood protection', working paper for Dhaka Metropolitan Area Integrated Urban Development Project (DMAIUDP), Dhaka.

Sikder, S. (1991), 'Paribesh Dushon: Garir kalo Dhuai Dhakabashir Ayo Kome Jachche', *Bichitra*, August (*in Bengali*).

The New Nation (1991), '35 lakh new urban housing units needed by 2000', 23 September.

Timm, R.W. (1992), 'The Environment and Related Problems', *Dhaka Courier*, June, Vol. 8, Issue 45, pp. 20–1.

United States Agency for International Development (USAID) (1990), *Bangladesh: Environment and National Resource Assessment*, Washington DC.

Walsh, H.P. (1990), 'Motor Vehicles and Global Warming' in J. Leggett (ed.), *Global Warming: The Greenpeace Report*, Oxford.

10 The Evaluation of the Five-year Two Million Housing Unit Construction Plan (1988–92)

CHONG-WON CHU

Introduction

Since the 1960s, Korea has implemented an economic policy which aimed at giving prior investment opportunities to the export industry to enable high real-production within a short period of time. Although this policy enabled Korea to achieve high economic growth within a short period of time, investment in the housing sector was neglected and insufficient (Hah, 1991). The reason was because the investment for it took a relatively long time in capital return and did not induce the growth of productivity directly. As seen in Table 10.1, the housing investment ratio to GNP for the five year period was 5 per cent, less than 6 per cent indicated by the UN, and the figure was relatively low in comparison to the economic strength, housing supply ratio and of the household growth rate of Korea.

Table 10.1   The trends of housing investment (1983–87)

	1983	1984	1985	1986	1987
Housing investment (in billion won*)	3,702	2,398	3,431	3,987	4,351
Ratio to GNP (%)	5.5	4.7	4.4	4.5	4.4

*   800 won is equivalent to about 1 US$.

Source: Lee, 1992, p. 2.

The household size has decreased as the result of a natural growth in population and nuclear family formation as shown in Table 10.2. The average

172

growth rate of households was 3.2 per cent a year, which was higher than that of house stock a year, at 2.8 per cent. Under these situations, the shortage problem of housing stock became more serious (Ministry of Construction, 1993).

Table 10.2 The trends of household size (1970–90)

	1970	1975	1980	1985	1990
Size of household (in no. of persons)	5.4	5.1	4.6	4.2	3.8

Source: Economic Planning Board, 1970–90.

To cope with this situation, the government established the Five-year Two Million Housing Unit Construction Plan (TCP). This plan was aimed at alleviating the shortage problem of housing stock by expanding housing production on a massive scale for five years from 1988 to 1992. In addition, it included a series of measures to supply the low-income and working classes with permanent rental houses, which made up a considerable portion of the two million houses supplied during the five years. The government published the development plans for five new towns in the Seoul Metropolitan Area to support this policy.

In this chapter, the backgrounds, goals, and achievements of TCP are introduced, and its effects are evaluated. In addition, the pervasive effects of five new towns, which are under construction in the Seoul Metropolitan Area, are evaluated. Finally, the conclusion of this chapter includes a comprehensive evaluation of TCP and five new town developments, and some prospects for future housing policies of Korea.

The Backgrounds of the Five-year Two Million Housing Unit Construction Plan and Five New Town Developments

A series of policy measures were attempted to alleviate the housing problem. When the real estate market was in recess, it was stimulated by means of reducing tax. On the contrary, when overheated, it was regulated by means of raising tax. But in spite of these policy measures, the chronic housing problem was not alleviated satisfactorily.

On the other hand, the supply proportion of large-sized houses were inclined to increase continuously, which was not a positive sign in relieving the housing supply problem. Indeed, in 1985, the supply of small-sized houses under 15 pyeong (one pyeong is equivalent to 3.3 m^2) shrank by 30 per cent when compared to that of 1975. But, the supply of houses over 20 pyeong was raised by two or three times for the same period of time and particularly, these tendencies were more conspicuous in urban areas (Chu, 1990). The reason for the above tendencies is that houses larger than medium-sized are more effective in returning the amount invested and more lucrative in profit per unit area than small-sized houses. Further, the sale price of small-sized houses was regulated to a lower level by the government in spite of its high construction cost per unit area and therefore, developers have preferred larger-sized houses to smaller-sized houses.

Although the government had made efforts to provide housing and to alleviate the housing problem continuously, the housing market was still un-stable, aggravated by the short supply of housing, natural growth of population and households, and inconsistency in execution of housing policy. As the result, the housing supply ratio had fallen continuously as seen in Table 10.3.

Table 10.3   The trends of housing supply ratio (1970–88)

Region	1970	1975	1980	1985	1988
Whole country (%)	78.2	74.4	71.2	69.8	69.2
Urban areas (%)	58.8	56.9	55.5	57.8	60.2

Source: Korea Housing Bank, 1990.

Under this situation, the housing price had risen as a consequence of the shortage of housing stock resulting from the housing market failure; also, per capita income, nuclear family formation and speculative demand had increased. The housing price spiral was particularly remarkable in urban areas such as Seoul, as seen in Table 10.4.

Table 10.4   The trends of housing stock price index (1981–88)

Region	1981	1983	1985	1987	1988
Seoul	100.0	140.6	142.4	138.8	151.5
Five direct control cities	100.0	107.9	113.4	125.0	146.0
Urban areas	–	100.0	102.8	107.2	121.4

Source: Korea Housing Bank, 1989.

The Korean national economy has grown rapidly from the 1970s to the

1980s, and the overall balance of payment deficit in international trade has turned into surplus since 1986. As a result, floating capital was concentrated in the stock and real estate market, and the price of stock and real estate began to jump. Particularly in the real estate market, the price of apartments in Seoul Metropolitan Area has risen rapidly in a short time. Indeed, the price of apartments in the Kangnam area in Seoul (south of Han River) has risen by 30–50 per cent during six months, from late 1988 to early 1989, and the housing price spiral in the Kangnam area spread all over the country as well as in Seoul Metropolitan Area (Ministry of Construction, 1993).

At that time, the social atmosphere was stimulated by the desire for the political and social democratisation, such as the equity among social classes and social welfare. The housing price spiral led to a rise in rental price, and the working class had more difficulty in being able to afford housing costs, which played a disastrous role in deepening the social conflict. Therefore, the housing problem was highlighted as one of the urgent domestic problems.

The government considered two types of policy to alleviate the housing problem. One was the demand control policy. It was to control the excess demand by imposing the related tax (e.g. real estate transfer income tax) since the housing price spiral resulted from speculation. The other was the supply expansion policy. It was to expand housing production on a massive scale by eliminating negative factors in housing supply, for the housing price spiral resulted from the short supply of housing. The government selected the latter and established the 'Housing Construction Planning Office' in the Blue House Secretary Office. From that time, the various policy measures were implemented to construct two million houses in five years (1988–92).

In order to achieve the goals of TCP, the government had to begin large-scale housing projects. The new town development was inevitable because the mass supply of cheap land was limited due to the increase in land prices and the drainage of developable land in the urban area. These new towns were planned to improve the urban problems of Seoul (e.g. traffic jams, congestion, and so on) and provide a good living and educational environment equal to that of Kangnam area, since the housing price spiral at that time originated in Kangnam area.

Finally, the government provided two new town development plans in April 1989. They were Bundang and Ilsan, whose populations would be about 300,000–400,000. Besides them, three new towns were also to be constructed; Pyeongchon in Anyang City, Sanbon in Gunpo City, and Jungdong in Bucheon City whose populations would be 170,000. The location of these five new towns are illustrated in Figure 10.1.

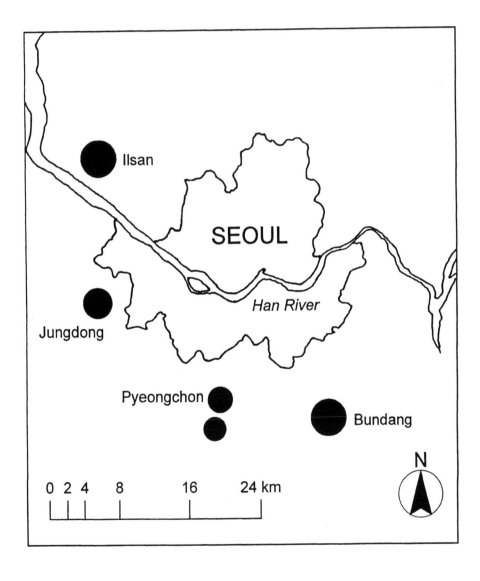

Figure 10.1   The location of five new towns in Seoul metropolitan area

The total housing supply of five new towns is 293,100 dwelling units, which corresponds to 14.7 per cent of the supply goal in TCP. These new towns have not only a practical benefit of providing a mass supply of cheap land in Seoul Metropolitan Area, but it has also a symbolic meaning reflecting the government's strong will for TCP (Ministry of Construction, 1993).

Main Framework for Two Million Housing Unit Construction Plan

Goals and Administrative Support System

Through TCP, The Korean housing policy has reached its turning point. This plan is not a passive treatment, like anti-speculation measures, but a means of positive intervention in the housing market by expanding housing production on a massive scale. The goals set up for the execution of this plan are as follows.

Firstly, housing supply on a massive scale for the stabilisation of the housing market. This goal was set up because it was recognised that the housing problem of Korea had resulted from the shortage of housing stock.

Secondly, encouraging housing price stabilisation for real demand, not speculators. For this purpose, housing production was to be expanded on a massive scale and the speculative housing demand was to be regulated at once.

Finally, creating housing supply for the low income class. For this purpose, rental houses and houses for the labouring class were to be supplied and the role of the public sector was to be reinforced.

According to the supply programme, 400,000 dwelling units a year were to be constructed at the average of 2,000,000 dwelling units during 1988–92. The public sector was to construct 900,000 dwelling units (45 per cent) and private sector was to supply 1,100,000 dwelling units (55 per cent). These supply programmes are illustrated in Table 10.5.

This plan focused on expanding housing production on a massive scale, supported by the fact that an average of 240,000 dwelling units were supplied each year, from 1983 to 1987 (Lee, 1992). The group in charge of supply was composed of public and private sectors based on their sources of financial supply. The public sector provides permanent rental houses for the low income class, welfare houses and rental houses for the labouring class, and long-term rental houses for the middle-income class. On the other hand, the private sector provides houses for the upper and middle income classes.

Table 10.5   Housing supply programme of TCP (revised in 1990) (in 1,000 dwelling units)

		Total	1988 –89	1990	1991	1992
		2,000	779	450	400	371
Public sector	Permanent rental houses	250	43	60	70	77
	Welfare houses for labouring class	150	–	40	–	60
	Rental houses for working class	100	–	20	–	50
	Long-term rental houses	150	91	25	80	14
	Small-sized houses for sale	250	142	55	40	13
	Total	*900*	*276*	*200*	*210*	*214*
Private sector	Subsidised houses for sale	601	241	120	120	120
	Houses for sale	499	262	130	70	37
	Total	*1,100*	*503*	*250*	*190*	*157*

Source: Bae, 1992, p. 8.

An administrative support system, such as arrangements of land, financial funds, and regulations, was necessary for the execution of TCP.

About 18,810 hectares of land was required for this plan. About 80 per cent of the needed land, 15,510 hectares, was to be provided by the public sector, such as Korea Land Development Corporation, Korea Housing Corporation, and local governments. The remainder was to be provided by the private sector by means of purchasing existing raw land (Ministry of Construction, 1993).

Required financial funds were estimated at about 25 trillion Wons for this plan. These were to be provided by a governmental fund, the National Housing Fund, and the Private Housing Fund. In the governmental fund, 42,683 billion Wons have been invested in construction of permanent rental houses and improvement of urban blighted housing. This increase in governmental funds is 567.3 per cent when compared with 7,566 billion Wons invested during 1983–87. The National Housing Fund, 97,956 billion Wons, has been invested in the construction of small-sized houses for sale and long-term rental houses, and Private Housing Fund, 109,582 billion Wons, has been provided by Korea Housing Bank.

Because the private sector took charge of the 55 per cent of planned supply quantity, its role was important. To activate the housing construction of the private sector, the sale price ceiling system, which had been considered as an obstacle in housing supply by the private sector, was modified to reflect cost increase in land, labour, and material in 1989. 'The Housing Redemption Bond' system was adopted to raise housing funds. In addition, the government has eased some regulations restricting developments in residential areas, and improved administrative system to supply houses for needed demanders by excluding speculative factors in the housing market.

The Development of Five New Towns in Seoul Metropolitan Area

Five new towns cover an area of about 4,420 hectares and can accommodate 293,000 dwelling units for 1,180,000 people. The development direction of each new town is outlined below.

Bundang was planned to have commercial and business service functions like Seoul and to develop as a self-contained city, where information-intensive industry would be located.

Ilsan was planned to develop as the city of 'peace and reunification' which would have the functions of art, culture, tourism, and international conferences.

Pyeongchon, Sanbon, and Jungdong were planned to have harmonious urban forms, linked with existing cities. The area, dwelling units, and population of each new town are shown in Table 10.6. Of the total 293,000 dwelling units supplied in these new towns, detached houses account for 11,300 dwelling units, row houses for 11,500 dwelling units, and apartment houses for 270,300 dwelling units respectively (Ministry of Construction, 1993).

Table 10.6 The area, no. of houses, and population of five new towns

	Area (ha)	No. of houses (households)	Population (person)
Bundang	1,965	97,500	390,000
Ilsan	1,568	69,000	280,000
Anyang-Pyeongchon	510	42,000	170,000
Kunpo-Sanbon	418	42,100	170,000
Bucheon-Jungdong	536	42,500	170,000
Total	4,417	293,100	1,180,000

Source: Ministry of Construction, 1993, p. 247.

The government executes Transportation Impact Assessment to alleviate the traffic problem caused by construction of these new towns. Based on the assessment, 38 line roads (213.7 km) and three line subways (62.5 km) are under construction.

Achievements and Effects on the Housing Market of the Two Million Housing Unit Construction Plan

Achievements

The two million unit target was achieved one year ahead of the scheduled five year period. Achievements of public and private sector for four years can be seen in Table 10.7. Public sector supplied 710,000 dwelling units, 79.0 per cent of the 900,000 dwelling unit target during four years (1988–91). Houses for the working class were supplied relatively less because they were introduced recently. The private sector supplied 1,432,000 dwelling units which was much more than the 1,200,000 dwelling unit target. These satisfactory results were achieved by the government's efforts to supply cheap land on a massive scale and the relaxation of various regulations.

Table 10.7 Numerical achievements (1988–91) (in 1,000 dwelling units)

	Goals	Achievements					Ratio
	1988–92 (A)	1988	1989	1990	1991	Total (B)	B/A (%)
Total	2,000	317	462	750	613	2,143	107.2
Public sector	900	115	161	270	164	711	79.0
– Permanent rental houses	190	–	43	60	50	153	80.5
– Houses for working class	250	–	–	61	37	98	39.2
– Long-term rental houses	150	52	39	65	15	171	114.0
– Small-sized houses for sale	310	63	79	84	63	289	93.2
Private sector	1,100	202	301	480	449	1,432	130.2

Source: Kim, 1992, p. 4.

While the ratio of housing investment to GNP was 4.6 per cent on average for the period of 1980–87, it has risen since 1988. Thus, it reached 8.8 per cent in 1991. The ratio of housing investment to GNP for the scheduled period is as shown in Table 10.8.

Table 10.8 Achievements of housing investment (1988–92)

Year	Average (1980–87)	1988	1989	1990	1991
Housing investment (in billion won)	3,304	5,264	6,466	10,440	12,005
Ratio to GNP (%)	4.6	4.7	5.3	8.0	8.5

Source: Kim, 1992, p. 5.

In the case of the five new towns under construction, 202,900 dwelling units were supplied up to 1992 which corresponded to 69.3 per cent of the 270,300 dwelling unit target. The achievements of the five new towns are shown in Table 10.9.

Table 10.9 Numerical achievements of five new towns (in 1,000 dwelling units)

	Total	Bundang	Ilsan	Pyeongchon	Sanbon	Jungdong
Goals	293.1 (100%)	97.5	69.0	42.0	42.1	42.5
Achievements up to 1992	202.9 (69.3%)	66.4	41.8	39.3	29.5	25.9
Supply Plan hereafter	90.2 (30.7%)	31.1	27.2	2.7	12.6	16.6

Source: Ministry of Construction, 1993, p. 249.

Effects on the Housing Market

While TCP caused some negative side-effects in the course of concentrating a good deal of capital in the housing sector during a short period of time, it still brought considerable positive effects, such as the improvement of 'market failure', stabilisation of house prices, and so on. The major effects on the housing market by TCP are as follows.

Firstly, the housing supply ratio, which had fallen since 1983, has risen considerably since 1987, as illustrated in Table 10.10. As a result, the housing supply ratio reached 74.5 per cent in 1991.

Table 10.10 The trends of housing supply ratio (1987–90)

Year	1987	1988	1989	1990	1991
Housing supply ratio (%)	69.2	69.4	70.9	72.4	74.5

Source: Ministry of Construction, 1993, p. 214.

Secondly, because of the supply of housing on such a massive scale, the sale and rental price of housing has fallen since 1990. The resultant stabilisation of housing prices has played an important role in restraining speculative demands. The trends of housing stock price index are shown in Table 10.11.

Table 10.11 The trends of housing stock price index (1990–92)

Year Region	1990.12.	1991.4.(A)	1991.12.	1992.3(B)	B/A(%)
Urban areas	100.0	105.8	99.5	98.5	93.1
Seoul	100.0	106.8	97.8	97.0	90.8

Source: Kim, 1992, p. 8.

Thirdly, TCP has provided the foundation of housing supply for low-income class. It has relieved the difference in housing levels among each income class by supplying permanent rental houses, houses for the working class, and long-term rental houses. In addition, it has provided house buyers with a fair chance of purchasing houses by improving the administrative measures related to housing supply.

Finally, the technology of housing production, that is, the standardisation of housing production, prefabricated housing, and so on, has been developed rapidly through mass housing production.

On the other hand, the five new town development plan was successful in that it achieved its two million unit target by alleviating the shortage of land for housing. It has had other positive effects, such as the stabilisation of housing price, the increase of GDP, and an increase in domestic employment. From the development system's point of view, the joint development system, which allowed the private sector to participate in the development, was adopted. It provided house buyers with houses of various sizes and quality.

The high quality of housing and the creation of various urban spaces in these new towns are the results of urban design methods based on the experiences and technologies accumulated while planning a large-scale housing complex during the 1980s (Korea Land Development Corporation, 1991). In residential districts, a well-arranged site layout made it possible to provide enough outdoor spaces and parking lots, and landscape-colour control made it possible to provide eye-catching scenery. Commercial districts were designed to focus on the functional and aesthetic harmony of its components, such as the urban skyline, building shape, and the interconnections between functions.

Conclusion and Some Prospects for Future Housing Policy

TCP was a 'supply-oriented' policy which aimed at reducing the serious shortage of housing stock and the inflation of house prices (Kim, 1991b). To achieve the plan's goals, the government furnished the resources and measures which were available at that time. As a result, the 'market failure' in the housing market was improved more or less and the housing supply ratio has risen since 1987. Moreover, the government gave its attention to the housing problems of the low-income and working classes, and made efforts to alleviate the problems. However, some negative side-effects were brought about since the plan was not only prepared and executed in a short period of time, but also too much of it was oriented to the achievement of goals.

TCP had brought some effects to the national economy. As the ratio of housing investment to GNP had risen from five per cent to eight per cent during that period, a considerable sum of money was concentrated in the construction sector, which led to the construction boom and the shortage of construction material and manpower (Roh and Lee, 1990). From a macroeconomic aspect, these elements led to inflation, disparities in industrial structure, and the payment deficit in international trade.

Also, the plan is thought to be deficient in detailed measure arrangements, which resulted from concentrating only on housing supply. Though the housing shortage problems in Korea result from the shortage of housing investment and natural growth of population, its major cause is the growth in size of households by nuclear family formation. To cope with these situations, other policy measures are needed to be considered apart from the policy which is oriented to quantitative supply goals. This type of policy may include the measures which give incentives to a family residing with elderly parents, or they promote the construction of houses where three generations can reside together. From this point of view, TCP is thought to be still inadequate in the arrangements of detailed policy measures (The Korea Housing Bank, 1991). In the case of houses for the low-income class, the criteria for selecting its occupants and post-occupancy management problems were still not considered enough, while the quantitative supply goals for them were achieved.

On the other hand, the policy direction seems to have some defects of itself, that the government decided to expand housing production on a massive scale considering the primary cause of the housing problem to simply be the shortage of housing supply. The improvement of obsolescent houses must be considered to be important, as well as new housing construction. In the case of Korea, the total number of newly constructed houses were 3,002,000 dwelling units for the 1976–88 period, while that of obsolescent houses was 1,006,000 dwelling units for the same period, which were equivalent to 35.5 per cent of that of newly constructed houses (Kim, 1991a). This means that the policy measures backing the management of existing houses, must be implemented side by side with new housing construction, which is essential in the efficient use of total national resources.

The two million unit target was achieved one year ahead of the scheduled five year period owing to the strong will of the government, such as the five new town construction. These new town constructions were inevitable in order to stabilise the housing market and for supplying amenities for the housing environment in the Seoul Metropolitan Area. However, the following problems accompanied their construction.

Firstly, there is a concentration problem of the population and economic strength in the metropolitan area. In spite of all the efforts of the government authorities, concentration in the metropolitan area has been very serious until now. Within the radius of 50 km from the CBD of Seoul, there is a concentration of 40 per cent of the national population and 67 per cent of GDP (Hah, 1991). If new towns are constructed within the radius of 20–30 km of the central part of Seoul under this condition, the concentration problem will be more serious and the balance of national land development will have a setback.

Secondly, new town development with large-scale housing construction is realised based on the Housing Land Development Promotion Law and the Housing Construction Promotion Law. This type of development has some merits in that it enables it to achieve the quantitative supply goals and provide house buyers for houses of various sizes and quality. However, it has a tendency to neglect various other aspects that must be considered in the process of new town development. For example, the supply of public facilities and employment opportunities are not considered sufficient because new town development is regarded as a tool for housing supply.

Thirdly, it takes a long time for new towns to have self-sufficient functions and, in this case, the heavy commuting traffic toward Seoul may increase.

Thus, discomfort owing to traffic congestion is anticipated given the time lag between the construction of housing and traffic facilities. The transportation plan of new towns mainly puts emphasis on the traffic network expansion as far as the perimeter of Seoul. Congestion of the main arterial road in Seoul City is also anticipated.

Finally, new towns aimed at providing a good living and educational environment equal to that of the Kangnam area for the purpose of the stabilisation of housing prices. However, in the process of development, the consideration of those public and educational facilities was not made, which resulted from policy attaching weight to quantitative supply goals. In addition, it is anticipated that residents in the new towns will take endure discomforts because of the time lag between their occupation and the construction of public facilities.

Though TCP caused some negative side-effects as it had not established a clear direction path or policy measures, although the problems of housing shortages in Korea have been alleviated to some degree by it. However, the number of housing stock is still short in absolute quantity, and particularly, this situation is more serious in large cities owing to the nuclear family formation and the concentration of population into urban areas. The government has published the housing supply programmes hereafter in the *Seventh Five-year Socioe-conomic Plan* (1992–96) and the *Third National Spatial Development Plan* (1992–2001). In the *Seventh Five-year Socio-economic Plan* (1992–96), the major goals of the housing supply programme are to stabilise housing prices, increase housing supply, and promote the equity between various social classes. According to this plan, 2,500,000 dwelling units are to be supplied for five years (1992–96) and the housing supply ratio is to be raised to 81.4 per cent by 1996. Also, 5,380,000 dwelling units are to be supplied for ten years (1992–2001) and housing supply ratio is to be raised to 92.8 per cent by 2001 according to the *Third National Spatial Development Plan* (Kim, 1991c).

Hereafter, the population growth rate of Korea will be decelerated to below one per cent, but the number of households will rise steadily due to the nuclear family formation (Choi, 1991). Moreover, as per capita income is increased, housing demand will change in both quantity and quality. Hence, it is suggested that a series of policies such as these, which present the supply of average 550,000 dwelling units a year, is thought to be adequate. However, in the light of the latest experience of TCP, the above policies must be implemented with long-term appreciation and consideration of various pervasive effects as well as detailed administrative measures. Many changes are anticipated outside the housing market, such as the opening of the construction market and the establishment of local self-governments. To cope with various housing demand in this atmosphere, the government need to intervene less in the housing market and promote market mechanisms.

Along with the quantitative housing supply, housing for the low-income class must be considered steadily with respect to social equity. In order to induce the participation of the private sector, more financial support is required, such as tax exemption and rent subsidy. For the supply of more rental houses, the land for them should be supplied more cheaply than the land of the house for sale. Until now, dwelling has been produced in real cost, but from now on, to improve the housing problems of the low-income class, housing policy should be served by affordable prices. Also, methods of developing hilly land should be studied by considering topographic characteristics of hilly land. Many scholars have disputed that the high-rise housing is not fit for human scale. However, I think it is necessary to construct high-rise housing in order to reduce the encroachment of fertile land for cultivation because of limited arable land and the dense population in Korea.

References

Bae, S.-S. (1992), 'Housing Supply and Support System of Five-Year Two-Million Housing Unit Construction Plan', paper prepared for Policy Forum on the Evaluation of Five-Year Two-Million Housing Unit Construction Plan and Future Direction of Housing Policy, Korea Research Institute for Human Settlements, Seoul.

Choi, C.-S. (1991), 'Housing Sector in The Seventh Five Year Plan', *National Land and Construction*, Vol. 8, No. 11, pp. 76–81.

Chu, C.-W. (1990), 'A Study on the Korean New Town Development', *Municipal Administration Review*, Vol. 10, pp. 71–100.

Economic Planning Board (1970, 1975, 1980, 1985, 1990), *Population and Housing Census Report*, Seoul.

Hah, S.-K. (1991), *Housing Policy*, Seoul: Parkyoung Publishing Company.

Hah, Y.-I. (1991), 'An Analysis of the Spreading Effect of New Town Construction', *National Land and Construction*, Vol. 8, No. 8, pp. 108–9.

Kim, J.-H. (1991a), 'Population and Housing Problem of Seoul Metropolitan Area and the Methods for Alleviation of Them', *Land Planning Review*, Vol. 2, No. 4, pp. 25–36.

Kim, J.-H. (1991b), 'The Improvement of National Housing Level in the Third National Spatial Development Plan', *National Land and Construction*, Vol. 8, No. 10, pp. 51–3.

Kim, J.-H. (1991c), 'Changing Perspectives for Korean Housing Policy', paper prepared for the International Conference on Dynamic Transformation of Societies, Seoul.

Kim, K.-Y. (1992), 'The Evaluation of Five-Year Two-Million Housing Unit Construction Plan', paper prepared for Policy Forum on the Evaluation of Five-Year Two-Million Housing Unit Construction Plan and Future Direction of Housing Policy, Korea Research Institute for Human Settlements, Seoul.

Korea Housing Bank (1989), *A Report on the Trends of Urban Housing Price*, Seoul.

Korea Housing Bank (1990), *Housing Economy Data Book*, Seoul.

Korea Housing Bank (1991), *A Survey of National Consciousness on the Housing Policy and Housing Problem*, Seoul.

Korea Land Development Corporation (1991), *A Propensity Analysis on Housing Environment of Bundang New Town*, Seoul.

Lee, M.-S. (1992), 'Economic Spreading Effect of Housing Investment', *Bimonthly Economic Review*, Vol. 25, No. 11, pp. 1–24.

Ministry of Construction (1993), *The Administrative White Paper of the Ministry of Construction (1988–1992) – The Change of National Land*, Seoul.

Roh, J.-H. and Lee, D.-K. (1990), 'Input Output Analysis of Five-Year Two-Million Housing Unit Construction Plan', *The Korea Spatial Planning Review*, Vol. 14, pp. 147–56.

11 Housing Attainability: A Concept Towards Solving the Urban Housing Crisis in Bangladesh

MAHBUBUR RAHMAN

Introduction

Urban areas of Bangladesh, accommodating 22 per cent of 111 million people, face an acute housing problem due to the widening gap between the cost of housing and low household affordability. One way of narrowing the gap would be to stem the high rate of escalation in the cost of scarce raw materials, to bring the overall housing cost down. However, there would still be a threshold cost below which housing of acceptable standard could not be provided. The difference between the threshold cost and the affordable cost can be met by subsidies. However, the government cannot extend the provision of subsidy to a large number of households for an indefinite period. The extension of affordability can be seen as another way of reducing the gap between cost and affordability. However, ability and willingness, the two equally important components of affordability, have either been wrongly assessed or ignored, individually or altogether. Therefore, the poorer households still do not have proper housing even after their affordability has been extended. Moreover, what measures the government has taken were inadequate and did not actually reach the needy households. As a result, the housing crisis remains unabated; and this demands an immediate solution, by using all available resources.

This paper put forward the concept of housing attainability, which extends affordability beyond the apparent limit. Housing-deposit schemes, rental-supplement schemes, spare-plot schemes are some of the attainable means of enhancing a household's ability and willingness to pay for housing. This paper examines the levels of attainability established by the study group (SG), through schemes under various credit terms. The findings are based on analyses

188

of information gathered from 751 lower-middle households (monthly income Tk 1,500–3,000; 1 US\$ = Tk 40) and middle-income households (Tk 3,000–5,000) in six areas of Dhaka. Dhaka is the capital of Bangladesh and a fast-growing city, with a population of seven million in 1994. This paper also suggests strategies which the government would have to adopt to create an environment conducive to the proper implementation of the proposed schemes.

Trade-offs and Housing Modes

Since housing costs can be reduced by a number of methods – lowering building standards, applying minimum finishing, adopting more basic services, reducing land area per dwelling, using more local materials, employing self-labour, etc. – different combinations of such methods will result in different housing packages. These can cater for a wide range of households, with various income levels and of various sizes. The different combinations involve several trade-offs: a) between single-family and multi-family structures; b) between vernacular form and modern form; c) between self-help methods and multistorey formal construction; d) among different land areas per household; e) among different liveable spaces; f) among different service levels; g) among different locations; and h) among different qualities of construction and finish. A choice can also be provided between building a complete unit at the outset and incremental building through gradual upgrading of size, finish and service as funding permits.

The point of departure for calculating the costs of different packages is a 42 m^2 house, with moderate finishes and individual services, built on 63 m^2 of peripheral land serviced at an average standard. The cost of such houses, built at a density equal to the average density of the city, is Tk 154,200. The price of the cheapest unit (threshold price), a 21 m^2 structure of just-acceptable quality on a 42 m^2 moderately serviced plot, is found to be Tk 66,375. Several packages in between these two, termed *housing modes*, are developed in order to compare the levels of affordability established by the SG (Appendix 11.1).

Cost, Affordability and Attainability

A rough indication of affordable housing cost is the rule of thumb often used in the developed countries: 2.5 times annual income. On the other hand, in order to maintain the limit of a quarter of the income as the maximum a poor family should spend on housing, direct expenditure should be kept below 15 per cent of income. Proportions of income of various income subgroups within the SG required to meet the cost of housing of various modes under various credit terms are calculated (Table 11.1). It is found that the majority of the SG are unable to afford the threshold price. The result – an unabated housing crisis – is not difficult to predict.

Table 11.1 Effective affordable housing costs by different income subgroups

Monthly Income (Tk)	Percentage of urban population	Credit terms (25 years repayment)	Monthly housing contribution	Income equivalent (month)	Total cost (Tk)
1,500–1,999	14.23	10.5% simple p.a.	1/6 of income	21.6	37,800
2,000–2,499	10.67	10.5% simple p.a.	1/6 of income	21.6	48,600
2,500–2,999	8.59	10.5% simple p.a.	1/4 of income	32.4	89,100
3,000–3,999	7.77	13.5% simple p.a.	1/4 of income	27.8	97,300
4,000–4,999	6.01	13.5% simple p.a.	1/3 of income	37.1	166,950

Source: Rahman, 1991, Table 4.07, p. 136.

The concept of *attainable cost* (AC), developed in this context, is the cost that can eventually be attained after adopting financial or physical mechanisms to extend the apparent affordability limit. The mechanisms are based on methods of repayment, maturity period, length of instalment periods, calculation of interest, adjustment for inflation, channelling of general savings, down payments, grace periods, etc. Here, three schemes and their attainabilities will be discussed by assuming that dwelling units are provided through a system where the cost can be redeemed in instalments.

Housing Deposit Scheme (HDS)

The capability to sustain payments of a housing cost, and therefore, to participate in a housing scheme, depends partially on the household ability to make a down payment (DP) which will reduce the amount that has to be borrowed. It also reduces the regular repayment amounts, and therefore, increases the area of coverage or the standard. The SG can make DPs of significant amounts if they make savings regularly. It is expected that a housing savings scheme will mop up domestic savings for future housing. However, saving part of income for housing could be an additional burden on top of housing expenditure. Nevertheless, the prospect of a secured tenure will increase the motivation and present a genuine reason for increasing expenditure on housing (Tym, 1984; Rahman, 1991). Some SG members accumulate wealth in the form of rural property and jewellery which they are prepared to spend on housing if opportunity arises (Rahman, 1991). Helaluzzaman (1984) claimed that the middle-income government employees could generate DP up to the equivalent of 100 months' income out of assets held by them. However, the volume and availability of such wealth could not be determined.

No statistics or literature are available on the savings by the SG. Moreover, assets are strongly preferred for investment as inflation tends to erode the profits from other forms of investment (Quyum, 1978). However, SG households are found prepared to devote their meagre resources to housing in the form of both additional monetary expenditure and commitment of nonmonetary resources, in particular the spare time of household members. Most of the SG members start earning under the age of 18 and form families between 23 and 30 years of age. Therefore, a 10-year Housing Deposit Scheme (HDS) starting from the first steady employment year will mature when the participant's age is in the 30s, which is a suitable age to enter into a 25-year housing mortgage commitment. Government employees contribute 7.5 per cent of their basic salaries regularly to a provident fund, while national figures suggest only a 5 per cent domestic saving. However, a scheme to channel savings for housing will encourage even greater savings. A saving of 10 per cent of income over 10 years will make available a DP equal to a year's income. This will meet 30 per cent of the cost of a house if the cost is set at 40 months' income equivalent.

A 10 per cent initial investment is acceptable to many commercial banks. Therefore, the SG can possibly claim and secure institutional loans. If the loan is redeemed at a 10.5 per cent simple rate of interest over 25 years (the credit terms of housing loans provided by the only housing finance institution), the monthly repayment will be less than a quarter of income. If an interest equal to the bank's interest on savings accounts (11.25 per cent per annum in 1991) is given in HDS, the savings will amount to an equivalent of 20 months' income at the end of the maturity period; 15 per cent dividend/interest will increase the amount to an equivalent of 26 months' income. Half of the costs of housing of different modes can be met by these savings; the regular repayment on the balance will be between 11 and 22 per cent of income under various credit terms (Table 11.2).

To examine whether the effects of inflation will minimise the above expectations, a recalculation of the variables is made. It is assumed that

Table 11.2   Monthly instalments as a percentage of income of different income subgroups participating in housing deposit/dividend schemes

Income sub-group	Deposit scheme amount[1]	Loan amount (Taka)	Total housing cost	Deposit/ cost in percentage	Monthly repayment as percentage of income[2] 10.5%s	13.5%s	13.5%c	Dividend scheme amount[3] Taka	percentage
1,500–1,999	33,994	33,506	67,500	50.4	14.8	17.2	22.5	45,325	67.1
2,000–2,499	43,706	38,244	81,950	53.3	13.1	15.3	19.9	58,275	71.1
2,500–2,999	53,419	42,981	96,400	55.4	12.0	14.0	18.4	71,225	73.9
3,000–3,999	67,988	57,312	125,300	54.3	12.6	14.7	19.2	90,650	72.3
4,000–4,999	87,413	66,787	154,200	56.7	11.4	13.3	17.5	116,550	75.6
Average	57,304	47,766	105,070	54.5	12.7	14.9	19.4	76,405	72.7

Notes

1   11.25 per cent interest paid on 10 years' regular saving at 10 per cent of income.
2   25 years' repayment period.
3   15 per cent dividend/profit paid on 10 years' regular saving at 10 per cent of income.

Source: Rahman, 1991, Table 4.09, p. 140.

domestic income will rise by 3.5 per cent per annum, inflation will not exceed 10 per cent (inflation rates in recent years have been reduced to only 2.5 per cent), and in accordance with the past trends, housing costs will increase at a rate higher than the inflation rate. As a result, the value of the deposit scheme amount diminishes by only 20 per cent. Nevertheless, it still meets a considerable part (41 to 46 per cent) of the increased housing costs, and the monthly repayment amount is attainable under all credit terms (Appendix 11.2).

Attainability Through Subletting

Subletting or sharing a house is widely practised by the SG (Rahman, 1991). Rent paid by the sublettees can extend the repayment capability of the main tenant. Subletting in government staff houses is a well known phenomenon (Helaluzzaman, 1984; Samiruddin, 1993). Therefore, a large real estate developer has incorporated the possibility of subletting for the purpose of enhancing the buyers' paying capability. However, this may impose pressure on service provisions and deteriorate the environment by increasing density (England and Alnwick, 1982). Therefore, renting a smaller but independent unit instead of subletting is a better option. Despite its undesirability, this practice has some similarities with the spare plot mechanism experimented in Mexico (Appendix 11.3) as it provides double the requirement to the primary participant, half of which will be used to subsidise repayment liability.

There are three modes of housing for subletting (Appendix 11.1). These units can be let out at minimum monthly rents of Tk 900–1,100. This supplementary income, which is to be spent on cost repayment, can significantly increase attainability of housing. Table 11.3 shows that the part of income to be spent on repayment, after the repayment requirement has been reduced by the supplementary income, is well within the recommended affordable limit of under 10.5 per cent annual interest rate. The monthly repayment requirement for the lower-middle income households will be only 14.1 per cent of income while the middle-income group can obtain a bigger house by spending only 13.5 per cent of income.

Table 11.3 Percentage of income required as monthly repayments to meet housing costs under rental supplement schemes

Appropriate mode and cost in Taka	Income supplement	Monthly income	Percentage of income as instalments		
			10.5%s	13.5%s	13.5%c
Mode C' (Tk 149,600)	900	1,500–1,999	14.6	25.3	48.9
Mode B' (Tk 176,000)	1,000	2,000–2,499	16.0	25.8	47.4
Mode B' (Tk 176,000)	1,000	2,500–2,999	13.1	21.1	38.7
Mode A' (Tk 211,000)	1,100	3,000–3,999	15.1	22.7	39.3
Mode A' (Tk 211,000)	1,100	4,000–4,999	11.8	17.6	30.6

Note

*   25 years' repayment period.

Source: Rahman, 1991, Table 4.12, p. 143.

The Rental Supplement Scheme (RSS), described above, may undesirably create a large number of tenants. Households unable to contribute regularly, or those hindered by unemployment or illiteracy from participating in HDS, will have to depend on rental houses. Instead of spending highly on rent, they cant the scope to enter RSS by paying a rent equivalent to 15 per cent of the original participants' monthly income. The rent will be utilised to extend the original participants' affordability to 30 per cent of their income. The secondary participant (renter) may continue to pay rent and share other expenses until the loan is fully redeemed. After the full redemption, he has only to share the maintenance cost and taxes (7 to 10 per cent of income). As the secondary participant will be from the same income group same as that of the primary participant (they might well be related), both of them will be spending an affordable sum on housing. The ownership and other rights may not be granted to the primary participant.

Spare Plot Mechanism

As a general premise, the success of a spare plot mechanism (SP; Appendix 11.2) rests upon the existence of a relatively dynamic and buoyant land market, and a substantial increase in the value of the plot during the grace period, to offset the interest element on the original loan. In fact, the value should be at least doubled to gain an optimum advantage as the secondary participant is charged twice the initial cost of the spare plot. A steadily increasing minimum wage is also a precondition of its success. Although the scheme potentially has widespread application, it has drawbacks too. For example, there is always a danger of gentrification by the upper-income groups. It is liable to inflation and indeed all formal mechanisms will act towards creating a high inflationary situation in land prices to extract the maximum benefit. The government too is in a position to enhance the betterment of the land, through its control over services and utilities, which it is likely to do. Densities will double soon after the grace period expires. This will affect the cost of the land and the quality of the environment.

Table 11.4 is an outline of the SP scheme. Features of the scheme are: a) interest at a simple rate, equal to the inflation rate during the grace period, charged on the land costs, the full redemption of which starts after the expiry of the grace period; b) increase in land value assumed to be at least 30 per cent a year (Rahman, 1991); c) the loan to be repaid in 10 years from the fifth year onward at 13.5 per cent simple interest, which is the most common rate imposed by commercial banks in Bangladesh; d) the costs of the structure and partially of services being repaid within the grace period; and e) the usual land taxes, particularly the Capital Gain Tax from the transaction, being levied. Therefore, the scheme bears no exemptions except some flexibilities in the payment schedule which increases its financial viability by avoiding the status of a subsidised project. The scheme additionally accommodates incremental construction and splits the repayment schedule. Moreover, cost realisation takes place within a short period. Lastly, no down payment is required since it is generated within the scheme. Nevertheless, it encourages inflation, which

Table 11.4 Outline of a spare plot scheme in Dhaka

Average monthly income	Strctr. cost (Taka)	Equivalent months income	Year 1 land cost	Loan in fifth year[1]	Land selling price	Down-payment[2]	Effective loan[3] amount	Monthly repayment amount[4]	Repayment/income% 10 yrs[5]
1,750	21,900	26.4	95,000	133,000	176,370	99,000	34,000	476	24.8
2,250	28,850	27.1	102,850	143,990	190,980	107,250	36,740	515	20.8
2,750	40,500	31.1	111,000	155,400	206,680	115,970	39,430	552	18.3
3,500	55,800	33.7	125,900	176,260	233,760	131,275	44,985	630	16.3
4,500	76,900	36.1	142,300	199,220	264,180	148,360	50,860	712	14.4
Average	44,790	15.2	115,410	161,575	214,390	120,380	41,195	577	19.6

Notes

1 Value of the serviced land cost (adding inflation).
2 Land selling price – 50 per cent gain tax + 10 per cent transfer + other fees.
3 Fifth years' outstanding loan plus down payment.
4 13.5% annual interest, 10 years' repayment period.
5 Percentage of increased income in the fifth year.

Source: Rahman, 1991, Table 4.13, p. 145.

is not desirable even though only at the expense of the wealthier group. The project will fail if the expected rise in land price cannot be capitalised. Therefore, stand-by mechanisms also have to be developed to guard against such failure.

Attainability Under Different Schemes

The attainable techniques will greatly increase affordability if applied together. The HDS and RSS together can reduce the required repayment to 9.2 per cent and 2.7 per cent of income for the lower-middle and middle-income groups respectively, given an interest rate of 13.5 per cent, while the ultimate attainable cost will exceed six years' income equivalent. Under a 10.5 per cent simple interest term, households having an average minimum monthly income of Tk 2,750 and Tk 4,500 will have to pay no monthly instalments for a Mode B' and a Mode A' house respectively, as rent supplement will cover the repayment liability once DP has been made (from HDS). Even the maximum amount that has to be paid under commercial credit terms (13.5 per cent compound interest) by households at the bottom of the SG is below the national average housing expenditure level (Tables 11.5 and 11.6).

Estimation of ultimate attainability is based upon the following parameters: a) DP made from a 10-year HDS with an 11.25 per cent simple interest on

Table 11.5   Repayment in combined deposit and rental-supplement schemes

Housing mode and cost in Taka	Income sub-group	Income supple-ment	Deposit scheme payment	Balance cost amount	Income percentage as monthly repayment (various terms*)		
					10.5%s	13.5%s	13.5%c
Mode C': 149,600	1,500–1,999	900	34,000	142,000	4.4	12.7	30.5
Mode B': 176,000	2,000–2,499	1,000	43,700	132,300	0.1	8.3	24.6
Mode B': 176,000	2,500–2,999	1,000	53,420	122,600	none	3.7	16.0
Mode A': 211,000	3,000–3,999	1,100	68,000	143,000	0.1	5.2	16.5
Mode A': 211,000	4,000–4,999	1,100	87,420	123,600	none	0.2	7.8

Note

*   25 years' repayment period

Source: Rahman, 1991, Table 4.14, p. 146.

Table 11.6   Maximum affordable and attainable costs in various schemes

Average monthly income	Effective[1] affordable cost 16.7I	Thumbrule affordable cost 30I	Attainable cost with[2] deposit 36I	Attainable cost with[3] dividend43I	Reasonable attnable maximum cost=deposit+sup-plement 53I[4]	Attnbl.cost [5]	[6]
1,500	25,065	45,000	54,000	64,500	79,500	94,300	
1,750	29,240	52,500	63,000	75,250	92,750	207,200	232,100
2,250	37,595	67,500	81,000	96,750	119,250	246,050	276,600
2,750	45,950	82,500	99,000	118,250	145,750	271,900	307,050
3,500	58,480	105,000	126,000	150,500	185,500	323,800	366,700
4,500	75,190	135,000	162,000	193,500	238,500	375,650	427,600
5,000	83,545	150,000	180,000	215,000	265,000	401,500	
Average	44,780	80,400	96,480	115,240	142,040	268,325	285,950

Notes

1   Monthly repayment equivalent to 15 per cent of income over 25 years (I for income).
2   11.25 per cent interest on 10 per cent of income deposited monthly for 10 years.
3   15 per cent dividend/profit awarded on 10 per cent of income/month invested for 10 years.
4   Deposit scheme + 15 per cent of income as repayment + 15 per cent of income as rent supplement.
5   Deposit scheme + market rent as supplement, 10.5 per cent interest, 25 year repayment.
6   Dividend + market rent as supplement, 13.5 per cent interest, 30 year repayment period.

Source: Rahman, 1991, Table 4.15, p. 148 and Tables V and VII.

regular savings of 10 per cent of income; b) the balanced cost of housing packages redeemed at a simple rate of interest of 13.5 per cent in 25 years; c) net housing expenditure equal to 15 per cent of income; and d) both participants spending 15 per cent of income in a RSS. Housing costs can be set to three years' income equivalent following the above parameters. With a 15 per cent dividend in HDS, the attainability will increase to 43 months' income equivalent. Tables 11.5 and 11.6 and Table 11.7 present the attainable housing costs for different income subgroups through various schemes.

Table 11.7 Maximum affordable cost through deposit and supplement schemes

Income sub-group	Down-payment (Taka)	Income supplmnt (Taka)	Monthly instalment (Taka)[1]	Maximum attainable cost under various terms			
				deposit schm+25y rpymnt		dividend schm+30y rpy	
				10.5%s	13.5%s	10.5%s	13.5%s
1,500–1,999	34,000	900	1,337.50	207,200	183,000	232,100	201,300
2,000–2,499	43,700	1,000	1,562.50	246,050	217,800	276,600	244,840
2,500–2,999	53,500	1,000	1,687.50	271,900	241,500	307,050	272,720
3,000–3,999	68,000	1,100	1,975.00	323,800	288,000	366,700	326,160
4,000–4,999	87,500	1,100	2,225.00	375,650	335,400	427,600	381,785

Notes

1   25 per cent of income + monthly supplement.
2   25 years repayment period.

Source: Rahman, 1991, Table 4.16, p. 148.

DP for a 10-year HDS, cost redeemed at simple interest rate of 13.5 per cent over 25 years, and Tk 1,000 income supplement added to the 15 per cent of income as the monthly repayment, will attain an ultimate cost of 54 months' income equivalent plus Tk 111,400. Softer terms (30 years repayment period, 10.5 per cent interest rate) will increase the attainability up to 61 months' income equivalent plus Tk 153,500. Tables presenting the attainable costs resulted from the combination of HDS and RSS show that housing of various modes is within the attainability of the SG. Appendix 11.2, presenting the effects of inflation on attainability, shows that payments to be made by the middle-income population after 10 years' saving will be a reasonable part of their income. However, repayments under the same terms will exceed a quarter of income of the lower-income group. This amount can be lowered by reducing the costs of housing packages and softening the credit terms.

Findings from the Survey

The above comparisons are based on assumptions and observations on the SG's housing expenditure, saving capability and affordability. However, since few statistics on them are available, a questionnaire-based survey was conducted in the winter of 1989–90 to gather data on total household expenditure, housing expenditure, regular savings and savings potential, down payment capacity both out of cash savings and assets, possible HDS contribution etc. The survey aimed at making realistic estimations of attainability. Key findings will be discussed in the following sections.

Table 11.8   Percentage of income spent on various types of dwellings

Physical condition of the house	Old city core	Old city intrmedt.	Old city periphr.	New city core	New city intrmedt.	New city periphr.	Row av.
Temporary – not improvable	31.80	26.07	26.45	–	28.51	27.23	27.69
Temporary – improvable	31.43	23.82	28.81	39.25	33.48	31.69	31.20
Permanent – needs improvement	39.51	29.51	32.49	36.28	36.89	33.42	33.73
Permanent – needs no improvement	42.94	39.41	35.56	43.25	40.60	37.24	38.62
Column average	35.56	27.42	30.45	38.77	34.47	32.34	32.38

Source: Rahman, 1991, Table 8.05, p. 249.

It is discovered that the SG spend more than a third of their income on housing (Tables 11.8 and 11.9) which varies with the physical condition of the house, location and tenure type. This is very high by any standards since the national average urban-area housing expenditure by the SG is only 15.3 per cent of income (BBS, 1992). It is also found that they are less habituated to saving. Nearly three-quarters of them had no savings while only a quarter are used to saving irregularly (Table 11.10). Average monthly savings of the saving households (28.5 per cent) is Tk 178.44, less than 2 per cent of income. On the other hand, most households (90 per cent) are willing to increase expenditure by 5 per cent of income on average in return for home-ownership. The average of the extra expenditure, taken as the saving potential (dormant savings), occurred more frequently while the actual savings are mostly irregular and confined to fewer households in the richer range. Moreover, the savings potential and the average savings of each income subgroup are found to be close (Table 11.10).

The respondents intend to utilise the savings (ordered from the highest to lowest preference), for general household needs, domestic appliances, transport, housing-related needs, and income-generating activities. However, no definite pattern can be established. In 95 per cent of the cases, the potential affordability (the figures quoted by the respondents) is found to be equal to

Table 11.9   Percentage of income spent on housing under various tenure types

Rent to income (%)	Percentage of income spent on housing by different tenure pattern									Row percentage
	Owner		Gen. Tenant		Sublettee		Partial Tent.		Staff	
Up to 25%	73.2	100.0	6.0	4.4	2.0	9.5	14.6	57.7	4.2	72.2	14.1
25.01%–30%	–	–	63.2	17.3	11.2	22.2	25.6	37.2	–	–	20.1
30.01%–35%	–	–	89.5	41.0	6.2	20.7	2.4	5.8	1.9	22.2	33.5
Above 35%	–	–	84.6	37.3	14.9	47.7	–	–	0.5	5.6	32.2
Column average	6.58		34.04		34.59		24.36		20.33		32.4
Number of respondents	Chi-square value			Significance		Cells with EF<5			Eta value		
701	305.010			0.0000		None			0.64966		

Note

Column percentage follows the row percentage.

Source: Rahman, 1991, Table 8.06, p . 249.

Table 11.10   Distribution of respondents by frequency of saving and income

Nature of saving	Income subgroups (by Taka/month)										Row percentage
	1,500–1,999		2,000–2,499		2,500–2,999		3,000–3,999		4,000–4,999		
None	14.8	98.6	34.4	92.8	24.1	80.7	16.5	55.2	10.3	36.8	71.5
Irregular	0.6	1.4	6.8	6.7	14.1	17.2	36.2	44.1	42.4	55.1	26.1
Regular	–	–	6.3	0.6	18.8	2.1	6.3	0.7	68.8	8.1	2.4
Av..savings (Tk)	75.00		100.00		122.50		157.50		218.51		178.44
Dormant savngs (Tk)	77.59		99.36		118.87		161.03		339.95		179.06
Number of respondents	Chi-square value			Significance		Cells with EF<5			Eta value		
679	113.5342			0.0000		25%			0.49782		

Note

Column percentages follow the row percentages;

Source: Rahman, 1991, Table 8.23, p. 263.

the summation of present expenditure and savings (or the dormant savings). The findings support the postulation that households will spend most of their savings on better housing. They are willing either to make potential/dormant savings in the face of a home-ownership promise or to make extra efforts to meet the payment requirements in a housing scheme.

It is found that only 6 per cent of the respondents have ready cash savings to make a down payment or initial investment. However, nearly half of the respondents (all having monthly incomes above Tk 2,000) can generate down payments from sources such as rural property (mostly by government employees), household assets like bicycles and appliances (by lower-middle-income group), and ornaments (by traders, craftsmen and service-holders). Nearly 12 per cent of the respondents (mostly businessmen and service holders) own either urban or rural property, part of which will be mortgaged or sold if the respondents are going to participate in an urban housing scheme. The average of the available amount from all probable sources is Tk 15,471. Nearly 30 per cent of these households can generate an amount which is above the average. Since only a 10 per cent down payment or initial investment is the minimum to induce institutional credits or to join a housing scheme, most SG households can afford a dwelling unit of any mode.

An estimation of the probable HDS amount assumes that the respondents will save the dormant amount, at a market interest rate, starting from their regular earning age (established from the survey) up to the age of 30. The average estimated savings are Tk 19,435. If this amount is combined with the amount obtainable from other sources, the ultimate down payment capability, on average, will be Tk 25,582. Two-thirds of those who think they will never be able to buy an urban plot, and more than half of those who do not have any visible source of finance, can indeed generate more than Tk 20,000 in the above process, which is then enough to secure small housing loans (Table 11.11).

The direction and linearity of the relationship between potential affordability and income can be described by the following expression: monthly housing affordability = 0.37 x income – Taka 81.00 (Table 11.12). It follows that the SG can afford to spend around a third of their income on housing. However, it can be observed from Appendix 11.2 that the middle-income service holders are ready to spend more than 43 per cent of their income on housing while workers with the lowest potential affordability are able to spend, on average, 37.4 per cent of their income. These figures, if they do not indicate the ability to spend much on housing, at least suggest the willingness of the SG members to increase housing expenditure for improved services and ownership.

Table 11.11 Distribution of tenants by down payment and income only

Ultimate down payment amount (Taka)	Distribution by income subgroups Tk1,500–2,249		Tk2,250–2,999		Tk3,000–3,999		Tk4,000–4,999		Row percentage
Up to Tk 10,000	45.7	22.	29.6	10.7	24.7	14.5	–	–	13.5
10,001–20,000	43.4	42.9	44.6	33.0	9.6	11.6	2.4	4.4	26.8
20,001–30,000	23.2	23.2	44.0	33.0	23.2	28.3	9.5	17.8	27.1
30,001–40,000	9.3	6.0	31.5	15.2	30.6	23.9	28.7	34.4	17.4
Above Tk 40,000	10.3	6.0	18.6	8.0	30.9	21.7	40.2	43.3	15.6

Number of respondents	Chi-square value	Significance	Cells with EF<5	Eta value
620	175.594	0.0000	None	0.47899

Note

Column percentages follow the row percentages.

Source: Rahman, 1991, Table 8.32, p. 272.

Table 11.12 Affordable housing costs by income and occupation group

Occupation type	Average maximum affordable cost by income subgroups 1,500–2,249	2,250–2,999	3,000–3,999	4,000–4,999	Row av.
Traders + self-employed	106,151	136,517	186,001	237,077	170,534
Service holders	101,057	144,220	182,826	237,284	161,680
Transport + industrial workers	110,523	131,456	178,019	215,440	135,129
Domestic + manual workers	103,647	152,910	170,304	223,752	134,938
Column average	105,032	139,660	181,039	234,338	152,996

Source: Rahman, 1991, Table 8.37, p. 277.

Through a combination of the savings scheme and cash generated from other sources, on the one hand, and the matching of loan redemption amount with the quoted affordable amount on the other, the surveyed households can afford a dwelling unit costlier than Tk 400,000. The ultimate attainable costs vary between Tk 42,360 and 468,000. Only less than 2 per cent of the surveyed households are found unable to attain the threshold cost of Tk 67,500. The relationship between attainable cost and income as obtained from regression analysis is: total attainable housing cost = 61.32 x income – Tk 24,900. The standard error of deviation for both the constant and the multiplying factor is low. However, some attainable amounts are found to be very scattered from the least square line in the upper-income range. Nevertheless, the asymmetric standardised scatterplot supports the validity of the relationship.

Suggestions

The survey has proved that the potential for savings exists among the lower-middle and middle-income urban households in Bangladesh, despite the low national saving-rate. Moreover, they are willing to extend their present housing expenditure in the face of a home-ownership promise. However, provision for mopping up the domestic savings through a network at grass-root level has to be made. The objective of the network is to activate the saving potential of the SG to accumulate a sizeable amount of money over the years which can be utilised to secure housing loans. This can be in the form of a deposit account in a housing bank, a housing fund at the workplace to which the participants will regularly contribute a fixed portion of their salary, or a housing cooperative. The fund may be operated under a contractual system like those in some European countries, where a below-market interest rate is offered on the deposit to facilitate a soft-term loan for housing-related activities later. The housing fund at the workplace may be complemented by employers' contributions out of profits, or bonuses in exchange for certain tax holidays.

Given a 5 per cent inflation rate, certain administrative expenses of the financial institutions in Bangladesh, and a minimum 2 per cent profit for the housing bank or cooperatives, housing loans can be made at an annual interest rate under 10 per cent. Between 1972 and 1985, Bangladesh has obtained US$17.5 billion as loans at an average interest rate of 1.81 per cent (Sobhan and Islam, 1990). Since 1976, various international agencies have given monetary and consultative assistance to Bangladesh for undertaking studies, preparing urban-housing projects and master plans, and formulating policies in the country. At present, Tk 6,532 million is being invested in 63 urban-infrastructure and housing-development studies by foreign consultants (UNCHS-UNDP, 1993). None of the above loans or studies are aimed at increasing affordability. In many developing countries, housing banks or urban-development agencies have been aided by the active participation of international funding organisations. Zero to very low interest is charged on the loans provided by these organisations. Unfortunately, Bangladesh has never benefited from such funding. Therefore it should actively seek funding to facilitate a softer-term housing credit for a larger number of households. The international contribution can form part of the fund for running a housing bank, while the unmatched amount will be provided by the government's allocation.

RSS or SPS can only be operated through well-equipped housing agencies under the supervision of the National Housing Authority. The activities of the present housing agencies in Bangladesh are very limited and conventional as they mostly produce sites-and-services plots. However, due to the disadvantages of social stratification and the inherent weaknesses of the policy, the plots are ultimately held by the rich and privileged groups. Highly subsidised staff-houses cannot even meet a quarter of the demand from the government employees alone. The Second Five Year Plan (1980–85) adopted a policy of not pursuing staff-housing. However, the government agencies totally disregarded that policy and continued to build staff-houses. In the late 1980s, attempts were made to transfer the ownership of staff-houses to interested and able allottees through a hire-purchase system. Unfortunately, it was never implemented after the change of the government in 1991. Most public housing provision is either inaccessible to the target groups or inadequate to meet the real need. In the last two decades, 2,000 housing units, on average, were built annually by the government, whereas the requirement in the capital city alone over the same period was at least 20 times higher. Various governmental committees have suggested, more than once, either abolishing or drastically reorganising the existing housing agencies by orienting their activities towards benefiting the SG. Only powerful housing agencies with radical visions can introduce innovative attainable means, develop and produce cheap, durable and indigenous housing resources, and supply these housing through small cooperative groups.

Land is the most expensive component of urban housing in Bangladesh as well as the most readily accepted collateral for institutional credits. Therefore, land in peri-urban location needs to be cheaply developed, at least partially serviced and sold in small plots by these agencies. However, the requirements of collateral and down payments have put the SG at a disadvantage in competing for the meagre public-housing resources. Hence, the matter needs to be rethought if its importance is not to be overlooked. In fact, the widely acclaimed loan schemes by the Grameen Bank, run among the poor households, do not have those prerequisites which can be a model for small scale soft-term housing credits for the SG.

The need for at least one housing mortgage bank and a separate housing authority has long been felt and has been included in government policy papers in different periods, although no effective steps have been undertaken to establish them. A consortium of three real-estate developers proposed the establishment of a housing mortgage bank in 1989, primarily with assistance from international funding agencies like ADB and CDF. It received sympathetic consideration during the last regime, but no adequate response thereafter. A housing bank and a housing authority have been in the pipeline for years. A housing policy for the country has just been adopted in the parliament. It was anticipated that strategies to extend housing affordability and to reduce the scarcity, and acceleration in the price, of housing resources were going to be included in the policy. Unfortunately, this did not happen as input from relevant experts to the documents prepared by the bureaucrats was too little. As a result, it has become another policy exercise devoid of any effective strategies, means or provisions for strict implementation.

Conclusion

It would be premature to end this article with a *conclusion* since the lessons of providing shelter are still being assimilated. Provision of access to housing using conventional wisdom is an inadequate response by the authorities. However, there is a growing awareness of the need to seek pragmatic solutions to the problems. Nevertheless, no attempt has been made to look beyond the cost and income analyses. Such analyses are imperative since the key issue, as identified in the introduction, revolves around these two determinants. Planners or policy makers are largely impotent in making any significant contribution to housing provision for the low-income groups. The preceding discussion is a partial indicator of where and how the economically weaker portion of the population can be housed within its affordability. This is no more than indicative of what could be accomplished in a variety of contexts. The examples illustrate the levers available to the policy makers to stimulate low-cost housing production and to increase the lower-middle and middle-income urban population's participation in the urban housing market, which currently operates with distortions and anomalies where housing economics are concerned.

References

Bangladesh Bureau of Statistics (BBS) (1990), *Statistical Year Book of Bangladesh, 1989-90*; Dhaka: Statistical Division, Ministry of Planning, Government of Bangladesh.

Bangladesh Bureau of Statistics (BBS) (1992), *Statistical Year Book of Bangladesh, 1991–92*, Dhaka: Statistical Division, Ministry of Planning, Government of Bangladesh.

England, R. and Alnwick, D. (1982), 'What can Low-Income Families Afford for Housing', *Habitat International*, Vol. 6, No. 4, pp. 441–57.

Helaluzzaman, K.M. (1984), 'A Co-operative Housing Development and Ownership Project for Lower-Middle Income Government Employees in Dhaka, Bangladesh', Bangkok: unpublished MS thesis, Human Settlement Division, AIT.

Quyum, A.S.M. (1978), 'Financing Urban Housing in Bangladesh', Dhaka: unpublished MURP thesis, Department of Urban and Regional Planning, Bangladesh University of Engineering and Technology.

Rahman, M.M. (1991), 'Urban Lower-middle and Middle-Income Housing: an investigation into affordability and options, Dhaka, Bangladesh', Nottingham: unpublished PhD thesis, University of Nottingham.

Samiruddin, F. (1993), 'Subletting in the Formal Sector Housing in Dhaka, Bangladesh', Dhaka: MURP thesis, Department of Urban and Regional Planning, Bangladesh University of Engineering and Technology

Sobhan, R. and Islam, T. (1990), 'Debt Burden and the Term of External Borrowing' in R. Sobhan (ed.), *From Aid Dependence to Self Reliance- development options for Bangladesh*, Dhaka: University Press Limited, pp. 79–112.

Tym, R. (1984), 'Finance and Affordability' in G.K. Payne (ed.), *Low-Income Housing in the Developing World*, New York: John Wiley & Sons, pp. 201–20.

Ward, P.M. (1984): 'Mexico – beyond Sites-and-Services' in G.K. Payne (ed.), *Low-Income Housing in the Developing World*, New York: John Wiley & Sons, pp. 209–20.

UNCHS-UNDP (1993), *Preliminary Report of the Urban Sector Shelter Study Team*, Dhaka: Urban Development Directorate.

Appendix 11.1   Housing Modes

Housing modes are developed from a detailed cost estimation on market price based on all possible cost reductions in different housing components. Low land in the undeveloped fringe areas is considered, as it can be developed with cost effective methods. The costs include those of land development and servicing, provision of infrastructure and community amenities, project design, implementation and management, and the interest on the fund invested during the construction period. Detailed cost estimations can be found in Rahman, 1991, pp. 347–52.

Mode A:

42 m^2 standard quality structure on 63 m^2 or 52.7 m^2 plot with either standard or moderate servicing/same structure of moderate quality on 63 m^2 standard plot/31.4 m^2 built-up area of standard quality on 63 m^2 plot with either standard or moderate servicing.

   Cost: Tk 154,200 (US $ 3,855; 1 US $ = Tk 40.00)

Mode B:

42 m^2 built-up area of acceptable quality on 63 m^2 plot with either standard or moderate servicing/31.4 m^2 of standard quality structure on either 52.7 m^2 or 42.2 m^2 plot with either standard or moderate servicing/31.4 m^2 moderate quality structure on 63 m^2 plot with any of the service levels or 52.7 m^2 plot with standard servicing.

   Cost: Tk 125,300

Mode C:

31.4 m^2 built-up area of moderate finish on 52.7 m^2 plot with moderate servicing or on 42.2 m^2 plot with either standard or moderate servicing/31.4 m^2 built-up area of acceptable quality on either 63 m^2 or 52.7 m^2 plot with either standard or moderate servicing.

   Cost: Tk 96,400

Mode D:

21 m^2 structure of either moderate (below US$ 40/m^2) or acceptable standard on 42.2 m^2 plot with either standard or moderate servicing.

Cost: Tk 81,950

Mode E:

21 m^2 built-up area of acceptable quality at a rate of below US$ 24/m^2 on a 42.2 m^2 plot with either standard or moderate servicing.

Cost: Tk 67,500

Housing Modes for Subletting

Mode A':

42 m^2 on the ground floor and 39 m^2 on the first floor of standard finish on 63 m^2 plot with standard servicing.

Cost: Tk 105,500 each unit

Mode B':

31 m^2 on the ground floor and 29 m^2 on the first floor of standard finish on 52.7 m^2 plot with standard servicing.

Cost: Tk 88,000 each unit

Mode C':

31 m^2 on the ground floor and 29 m^2 on the first floor of moderate finish on 42.2 m^2 plot with standard servicing.

Cost: Tk 74,800 each unit

Appendix 11.2

Table 11.13 Effect of inflation on income, deposit and instalment payments

Income sub-groups	Housing cost in Taka	Deposit scheme amount	Balance loan amount	Deposit to total cost in %age	Monthly repayment as percentage of income under various terms*		
					10.5%s	13.5%s	13.5%c
2100–2799	141,750	57,800	83,950	40.8	23.2	26.9	35.2
2800–3499	172,095	74,250	97,845	43.1	21.0	24.4	31.2
3500–4224	202,440	91,000	111,440	45.0	19.5	22.7	29.7
4225–5649	263,130	115,500	147,630	43.9	20.4	23.7	30.9
5650–7049	323,000	148,500	174,500	46.0	18.7	21.8	28.5
Average	220,483	97,410	123,037	44.2	18.3	21.2	27.8

Note

* 25 years' repayment period.

Source: Rahman, 1991, Table 4.11, p. 141.

Table 11.14 Effects of inflation on reasonable attainable costs

Income[1] subgroup (10th yr)	Housing Cost in Taka[2]	Deposit scheme Taka	Loan amount Taka	Deposit to cost (percentage)	Monthly supplement in Taka	Percentage of income as loan repayment[3]	
						10.5%s	3.5%s
2,100–2,799	281,600	57,800	223,800	20.5	490	50.5	62.0
2,800–3,499	305,500	74,250	230,750	24.3	557	38.9	48.1
3,500–4,224	328,400	91,000	237,400	27.7	580	22.4	28.4
4,225–5,649	375,200	115,500	259,700	30.8	741	25.6	32.5
5,650–6,999	422,000	148,500	273,500	35.2	949	18.5	23.9
Av. 4550	342,540	107,305	235,190	31.3	683	24.9	31.4

Notes

1 Income increasing at 3.5% per annum.
2 Housing cost increasing at 15% per annum.
3 Loan repayment as percentage of income after supplementation.

Source: Rahman, 1991, Table 4.18, p. 149.

Table 11.15 Affordability by income and occupation

Types of occupation	Affordability by income subgroups (%age of income)				Row %	
	1,500 –2249	2,250 –2999	3,000 –3999	4,000 –4999	Tenant only	Owner incld
Service holders	38.92	40.79	43.17	39.31	40.84	39.05
Transport and industrial workers	37.69	36.83	38.34	39.10	37.42	36.68
Traders and professionals	39.63	39.04	40.82	40.68	39.96	38.75
Manual and domestic workers	36.73	38.47	38.03	41.44	37.68	37.19
Av. column percentage (tenant only)	37.66	38.73	40.59	40.35	39.08	
Av. column percentage (including owner)	37.66	38.46	39.79	36.17		38.06

Source: Rahman, 1991, Table 8.22, p. 262.

Appendix 11.3

Proposed for the first time in Mexico, the spare plot mechanism is a form of cross-subsidy which reduces the land cost by providing a spare plot to each participant, which is to be sold in a later stage at a higher price than it was worth at the time of the initial project. A reasonably inflationary land market and an exemption from repaying to the government the increased value brought about by self-help initiatives of the participants are two important prerequisites of this method. It provides the target population with twice as much land as they would normally acquire. Although the capital of the participants is suspended in the initial years when they build on half of the plot, the project cost will be repaid as they sell the other half at the later stage when the price has gone up because of an uniform growth over the site area. Residents buying in at this later stage are relatively better off and are reluctant to undergo the rigours of living in a community without service during the initial project years, but are wish to acquire a relatively cheap plot (Ward, 1984). This mechanism attempts to visualise and positively utilise the speculative components existing within the urban land market and lower-income settlements of the newly developing nations. It gives the would-be settlers the wherewithal to speculate on the land market and thereby to gain access to some of its profits. Such activity, in the past, has been reserved only for the real-estate developers.

The proposal also offers considerable scope for formal intervention in the land market that would short-circuit many of the economic contradictions of sites-and-services projects. There is always a fear of speculation in such a scheme: the allottees may start immediate speculation while the organised competitors in the land market may artificially inflate the land value before the official land acquisition or deflate the price near the completion of the grace period. Very large sites may reduce the effective demand, which will then affect the expected price. These projects are affordable and low-priced because of the initial absence of full services. Later expenses incurred can be recouped by the 'sweat equity' as capitalised by the eventual sale. Also, a gradual introduction of upgraded services conforms to the practice of the low- and middle-income population. However, gentrification is a common occurrence in such schemes in which the factors responsible for the encouragement of a highly speculative land market are, to a certain extent, exploited. This imposes no significant pressure on the economy to make land available for the economically weaker section of the population. Such a scheme can be developed as a direct cross-subsidy generator. Moreover, leaving the spare plots to individual care ensures the maintenance of the land and maintains a barrier to squatting.

12 Shophouse: Asian Urban Composite Housing

MAMORU TOHIGUCHI AND HON SHYAN CHONG

Formation of Shophouses

Introduction

Bifunctional buildings, where one's residence and workplace are near each other, have played a very important role in urban living and it is very common in Southeast Asian cities. The kind of this bifunctional building, called '*shophouse*', mixes the function of carrying out business activities that are normally on the ground floor or the front portion with the function of providing living space on the floor above or at the back portion of the building.

Research concerning the formation of shophouses has been done in recent years; however, still a lot of points have not been clarified. Shophouses, defined as '[h]ouses built in a row separated from each other by common party wall, with colonnaded verandah, which have mixed uses of business activities on the ground floor and residence on the floor above' (Tan, 1991, p. 12) are widely distributed in Southeast Asian cities like Singapore, Penang, Kuala Lumpur, Melaka, Jakarta, Bangkok, Quanzhou, Chaozhou, Guangzhou and also in Taiwan. The word '*shophouse*' is firstly used by Cameron, a British geographer in 1865, and was formally used for the classification of housing by British Coloniser in Singapore since 1822 (Izumida, 1990). According to Izumida, shophouses were more similar to European townhouses, rather than the Chinese traditional residences when considering the shape of the site and the use of the area of shophouses. As the colonial power expanded, well-ordered shophouses had spread in the British Straits Settlement and also the Federated Malay States under the British rule in the nineteenth century. Furthermore, another kind of traditional housing type, which had colonnaded verandah had existed for a long period of time in Southern China. According to Mogi (1991), this kind of residence type was brought to Southeast Asia from He-nan of China, and returned to China with its well-ordered form and

212

the know-how about urban development of shophouses. Lacking of data, it is difficult to prove either the European townhouses or the traditional residences of southern China was the original style of the shophouse. However, it suited the life-style of Southeast Asian Chinese and it satisfied the requirements of the European colonisers: supplying various services and dwelling units for the Chinese immigrants who provided labour power. This paper aims to investigate the characteristics of shophouses and their evolution in the context of urban living in Southeast Asian cities, especially in Peninsular Malaysia. Not only will the architectural or urban historical landscape, but also the high-rise shophouses and the relationship between shophouses and urban living will be discussed.

Characteristics of Shophouses

From the results of the investigation on shophouses in Penang and Kuala Lumpur, we could find some common characteristics:

a) most of the traditional shophouses are two or three-storey high, and some of the new types are four or five-storey high;

b) the part which faced the main street at ground floor is used for business activities; the back portion of the shophouse is used as a storage room or a residence unit;

c) connective structure was hindered by a common party wall which was extended up to the roof of the house;

d) decorations mixed with Chinese and European design elements express on the façade;

e) verandah way (five foot way) which is a fixed common porch in front of the building is lining with the street, as this was required by the housing law;

f) the space of the shophouse is long and narrow. It provides a gradual increase in privacy for the residents as they move deeper into the building;

g) courtyard, high ceiling, jack-roof are installed for providing natural ventilation in order to cope with the tropical climate.

Space organisation of the shophouse The basic model of space organisation is shown in Figure 12.1. The ground floor is composed of verandah way, business space, front room, kitchen, courtyard and washroom. The first floor is composed of living room, balcony and air well. A stair case is installed in the shop which is not released to the outside. The front room on the ground floor is supposed to be used as living space, but it is actually used as an office or storage room. The backlane is used as service area for loading and unloading. In the past when urban sewage systems had not been developed, backlane was also used as a place for collecting waste. Besides the ventilation function, the air well is also used as a place for drying washed clothes or worship.

According to a research done in the University of Taiwan, traditional shophouses consist of three parts: the main portion, the subportion and the yard. Relationship between the elements and the characteristics of the space division are shown in Figure 12.2.

The new type of shophouses in Malaysia (Figure 12.3) contain business space in the ground. The first floor, second floor and above are residential space. Composite spaces of the traditional shophouse still have great influence on new shophouses.

Verandah way The verandah way is an important setting of shophouses. It is influenced by the Building Control of the British government that specified the width of the verandah way. Hence, it is also called the 'five foot way'.

Verandah way is a colonnaded sidewalk in front of shophouses and it has the following functions:

1   shelter from rain and strong sunlight;

2   a safe side walk;

3   a neutral zone between street and shophouses which can also be used as an extended space for business activities and living place: customers' reception, window shopping, chatting, playing chess, worship, etc.

The connection to the verandah way provides a closer relationship between the community and the neighbourhood. Discontinued verandah ways can also be found in Melaka.

Tropical environment adaptation Verandah ways as described above are an important setting for tropical environment adaptation. Air wells and jack-

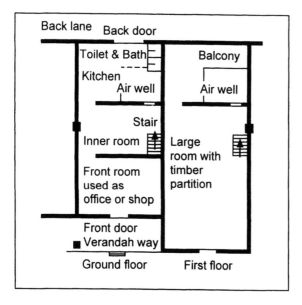

Figure 12.1 The plan of traditional shophouses

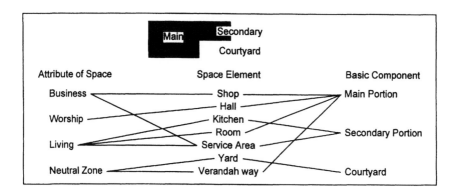

Figure 12.2 Space organisation in shophouse

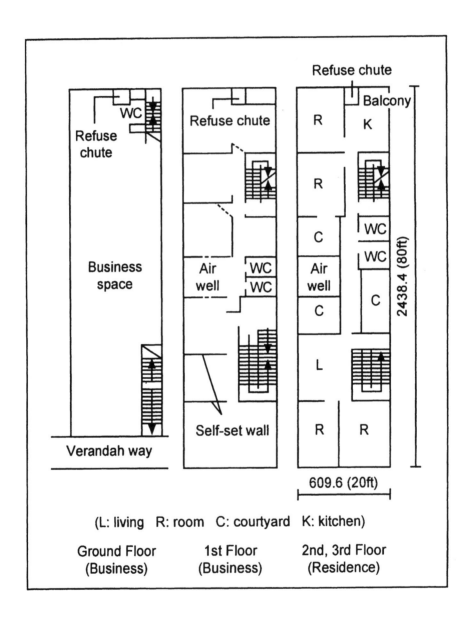

Figure 12.3   The plan of new shophouse type

rooves provide better ventilation. However, the change of the roof materials from the traditional Chinese roof tiles to asbestos or galvanised steel sheets is unfavourable. In Malaysia, the use of traditional construction system and materials is in decline. Traditional environmental adaptation had given way to more mechanical means such as the use of air-conditioning system.

As mentioned before, the shophouse is a kind of building which has a continued semi-public side walk at the front which provides a long and narrow space. Privacy increases as one goes deeper into the building. In short, from the street to the end of the shophouse, public space and private space is linked. The composite space structure, or the combination between the outdoor and indoor expresses one characteristic of Asian architectural form. We could find that some planning principles in the shophouse like place proximately of work and residence, or the combined relationship between public and private space. The comparative study on this kind of Asian residential space or regional structure which is very distinct from Western ones, is very important for the research of Asian flexible planning method.

The Structure and Design of Malaysian Shophouses

Most of the existing Malaysian traditional shophouses are made of brick and mortar, with massive column and thick load bearing wall, timber roof truss and column, timber floor and partition on the first floor. The façades of the shophouses are described as 'a mixed style of English and Chinese', 'a mixed architectural style of Asia and Europe', 'Chinese Baroque', 'a special mix of Roman, Victorian, Chinese and Portuguese' or 'a mix of Confucianism, Buddhism, Islamic and Italian' (Edwards, 1990). The background of a multiracial society and the development of a colonial city are the most important factors that contributed to these mixed representation of the architectural spaces and façades. Also, shophouses in Malaysia express the following elements.

Chinese elements

1   Courtyard, which functioned as a place that provides ventilation and worship facilities.

2   Rounded gable of the pitched roofs and Chinese tiles.

3   Granite in the corbels.

4   Intricately carved doors and panels.

5   Fan-shaped or bat-winged shaped air vents above the windows.

Malay element

Balustrades and eaves with fretwork.

European elements

1   Windows with louvred shutter.

2   Decorative plasterwork.

3   Corinthian pilasters.

4.   Façade in Art Nouveau or Art Deco (1950s, 1960s).

5   Modern style.

Shophouses are normally constructed in blocked form and are connected with other shophouses which were built at different times. Hence, even if they were built in the same row, the representation of the façades expresses a rhythmical landscape. Well-ordered colonnaded verandah way unifies the expression of a uniform scene. In historical low-storied shophouse area, a released human scale is provided. Shophouse townscape presents opposing factors such as continued and divided form, diversified and unified form, the mechanism of which is thought to be useful for new urban and housing design.

Urban Activities and Shophouses

Shophouses and Cities in the Early Days in Peninsular Malaysia

Before the British extended its power to Southeast Asia, bifunctional townhouse had existed in the Dutch ruled Melaka. Immigrants from India, China and Java formed the majority of the population in Melaka in the fourteenth century and their cultural influence had reflected greatly on the Melakan architecture. Especially for the Chinese, shophouses or other housing style could have been brought for their business activities or native living style. However, it is unclear whether any building control or city planning had been carried out. Thus, we cannot conclude whether the unified shophouse townscape had been formed before the European occupation. In 1666, a building judgement was set by the Dutch. The building control included zoning and prevention of conflagration. Bricks had been used by the Dutch in Melaka for the construction together with the 'Melaka tiles', which is a kind of roofing tiles that had a mix of Mediterranean and Chinese style. (Seow, 1983). Mixed European and Asian architectural style could had been created during this period. Chinese architecture in Melaka could be described as follows:

> The Chinese contribution to Melakan architectural styles, as mentioned before, is most strongly exemplified in commercial and religious buildings, namely the shophouse (with its residential development into the terrace house) and the temple. It resulted from the introduction of Chinese building forms, materials, decoration, workmen, system of construction, tools, and craftsmanship (Seow, 1983).

The construction of shophouses in large volumes was undertaken by the British in 1822 for the development of a new colonial city: Singapore. Building controls were carried out which led to the style of shophouses to be fixed. The supply of manpower for the production of urban services in Southeast Asian colonial cities was largely depended upon immigrants. In Peninsular Malaysia, the architectural style provided business activities space and residence. The shophouses were built in the settlements for the purpose of supplying dwelling for new immigrants and the convenience for urban living.

Urban Activities, Housing Estate Development and Shophouses

The development of tin mining was active in the nineteenth century. Kuala Lumpur had been developed as a distributing centre for manpower and materials. This had largely to do with the presence of the Chinese. According to Sidhu (1978), buildings in the inner city of Kuala Lumpur are shophouses. Shophouses in the oldest area were built in the period between 1890 and 1920.

Most of the high-density populated inner cities in Southeast Asia, like Kuala Lumpur, are different from the Western cities which had a 'donut phenomenon'. In the case of Kuala Lumpur, the existence of shophouses in CBD (central business district) was a major factor to evade this problem.

Generally, shophouses were occupied by Chinese traders. They were involved in trading food, clothes, medicine and also josssticks, coffins, etc.

Suburban growth has been one of the most striking aspects of the urbanisation process in Malaysia. The 1970 population census reported that every major town in Malaysia has grown rapidly. To solve the problem of overcrowding in the central area, large scale housing estate development in the suburb has started since the 1970s. The relationship between housing estate development and the shophouse is described as below (Melaka).

Before 1970

Small-scale housing estates were designed merely for residential purposes with no shops, shophouses or other urban facilities, but the location was comparatively close to town centre.

Early 1970s

Large scale (50–150 acres) housing estates were developed which were distinguished from the earliest housing estates by the presence of shophouses.

After mid-1970s

Most of the estates were huge in area (50–400 acres). The developments were sponsored by government bodies or joint ventures between the government and the private sector. These housing estates were multipurpose in function: residential-cum-commercial; residential-cum-industrial; or all three. A number of these estates were furnished with cinemas, markets, and a high ratio of shops or shophouses (Lim, 1984). Self-sufficiency was emphasised in the developments after the mid-1970s.

Furthermore, shophouses are also constructed in centralised settlements villages to encourage Malay entrepreneurs in commercial and social development. Hence, shophouses also play an important role in improving the socioeconomic structure of the Malaysians through the formation of small scale businesses.

However, the way of subdivision, which is specified from the ground floor to top floor as one unit, makes the price of shophouses comparatively expensive. In recent years, 'shop apartments', which were subdivided into each floor, were provided at a more affordable price.

The Activities in Shophouses

Shophouses, including the traditional types are playing an important role in providing space for the retail function in urban living. We completed 97 questionnaires concerning with shophouses in Kuala Lumpur, Malaysia in January 1992. Among these 97 questionnaires, 53 were answered by shop managers, 19 by residents and 25 were answered by people who worked and lived (work-residence proximity) in the shophouses. The results are shown below.

Composition of the inhabitants As a characteristic of urban living, a large number of those who live in shophouses are nuclear families and singles (Figure 12.4). However, there are also various residential patterns like renting out to other people or living together with relatives. About the age structure, generally, they are in their 30s and the 40s, and there are also youngsters under 15 years old. Concerning the occupation, self-employment is the most common, followed by service employees, specialists, office clerks, etc. Figure 12.5 shows the level of satisfaction of the inhabitants of the shophouses.

Ownership Among 44 inhabitants of the shophouses, 13 (30 per cent) are the owners, the others are tenants or subtenants. Among 78 managers, 21 are house owners, and 12 owners live in the shophouses.

Business activities Various small scale businesses are run in the shophouses as could be seen from Figure 12.6.

The majority of the managers are self-employed in small-scale businesses. In fact, 53 out of the 78 managers interviewed (68 per cent) work in family businesses. Chinese traditional business management is greatly reflected, since most of them are restaurant managers or retailers. We can say that nearly all of the urban small scale businesses can be found in the shophouses.

Relationship between the place of work and residence Among the 78 managers, 30 (39 per cent). live in the same shophouse. In addition, among 303 employees, 63 (22 per cent) live in the same shophouse. Recently, this residential style is gradually decreasing.

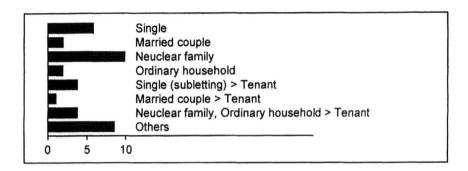

Figure 12.4   Composite of the inhabitants (total: 36; unknown: 8)

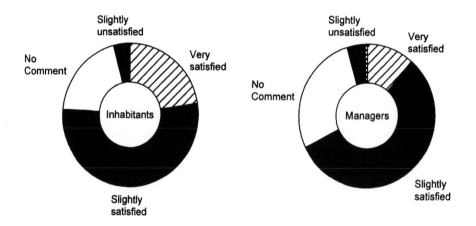

Figure 12.5   Evaluation of shophouse by the users

Inhabitants – total: 43; unknown: 1.
Managers – total: 76; unknown: 2.

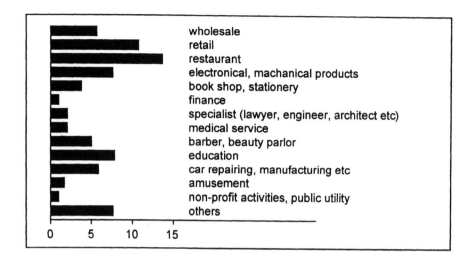

wholesale
retail
restaurant
electronical, machanical products
book shop, stationery
finance
specialist (lawyer, engineer, architect etc)
medical service
barber, beauty parlor
education
car repairing, manufacturing etc
amusement
non-profit activities, public utility
others

0   5   10   15

Figure 12.6   Business activities carried out in shophouses

The Typical Use of the Shophouse

The use of the shophouses expresses various forms that are mixed with various family composition, business activities and the relationship of the place of work and residence as mentioned before. In addition, shophouses are subdivided as a unit from the ground floor to the top floor. Some of the owners use the shophouse for themselves, or they rent out the whole floor or a room. There might be several tenants or subtenants at the same shophouse. Therefore, the relationship between the ownership and the usage could sometimes be very complicated. The following shows five typical cases.

Case 1   The manager owns the shophouse residence (four-storey new type shophouse)

Ownership   One unit (from ground floor to third floor).

Business-type   Electronic goods retailing.

Employees   Six persons, four of them are family members of the manager.

Residence of manager and employees	Ground floor is the shop and first floor is the storage room. Manager and his wife, daughter and another three persons live in the second floor. His son who is also one of the worker lives in another place. The third floor has six tenants; two employees who are not family members live in another place.
Reasons for choosing the place to start business	(1) High-density population of the area. (2) Few business rivals.
Satisfaction	Very satisfied. All the businesses are running smoothly. Preference to continue their living.
Points of satisfaction	(1) Prosperous. (2) Convenient to public transport. (3) Clean. (4) Easy to park vehicles. (5) Good design of façade. (6) Infrastructure is improved. (7) Lot of chance to make friends. (8) Verandah way is good for business. (9) Proximity to residence. (10) Have settled down.
Points of dissatisfaction	None.

Case 2   The manager owns the unit and residence (two-storey traditional shophouse)

Ownership	One unit. Ground floor for business, first floor for residential use.
Business-type	Restaurant.
Employees	20 persons, 15 of them of the same family.

Residence of manager and employees	The proprietress with her son who works in a computer company, elder daughter who works in the shop, twin daughters who are students, younger daughter who is a secretary, In addition, her relatives consist of two boys and two girls who are still students, a man who works in a chemistry company, a lady who works in the restaurant. All of them are living on the first floor, and the ground floor after the closure of the restaurant. Except the proprietress and her elder daughter, all of the inhabitants are part time workers of the restaurant. Another 11 employees live in another place. They consist of six females, three males, some relatives and part time workers. This case showed the complicated living style of the Southeast Asian Chinese in the shophouses.

Reasons for choosing the place to start business	(1) It is convenient to open a restaurant in town centre. (2) Not necessary to worry about the quantity of customers. (3) This kind of shop is necessary for the people around.

Satisfaction	Very satisfied. Preference to continue their living in such an environment.

Points of satisfaction	(1) Prosperous. (2) Convenient to public transport. (3) Lot of chances to make friends. (4) Infrastructure is improved. (5) Have settled down. (1) Noises from cars. (2) Unhygienic. (3) Heavy traffic on the road; difficult to park vehicles.

Case 3   The manager rents one floor for working and dwelling (new type four-storey)

Location and rent	Second floor. M$ 550 per month.
Business-type	Beauty parlour.
Employees	Two persons.
Residence of manager and employees	Two rooms at the front portion are considered as work place and the back portion is for residential use. Manager lives in the back portion and one employee lives in another place.
Composition of the inhabitants	Two men and two women. The manager with her spouse live with their daughter, and a subtenant, who is a student The husband of proprietress works in another place.
Reasons for choosing the place to start business	(1) Familiar with this place. (2) It is necessary for the residents around. (3) It has a convenient location: the town centre.
Satisfaction	No comment. Will move if there are better places.
Points of satisfaction	(1) Convenient to public transport. (2) Lot of chances to take friends. (3) Have settled down.
Points of dissatisfaction	(1) Noises from cars. (2) Unhygienic. (3) Heavy traffic on the road. (4) Lot of business rivals. (5) Verandah way and stair case often occupied by strangers. (6) No playground for children or place for recreation around the area.

Case 4   The manager rents only one floor for business usage, while employees live in the same place (new type four-storey)

Location and rent	First floor only for business usage. M$ 900 per month.
Business-type	Wholesale (incorporated company).
Employees	Six persons. Three of them are family members.
Residence of manager and employees	Manager lives in another apartment house. Two employees who are not family members live in the shop.
Satisfaction	Slightly satisfied.
Points of satisfaction	(1) Residence is nearby. (2) Have settled down.
Points of dissatisfaction	Problems of heavy traffic and parking.

Case 5   The manager rents the whole unit from the ground floor to the third floor for business usage only (new type four-storey)

Location and rent	Ground floor is the shop, the rest are offices. M$ 2,500.
Business-type	Wholesale of industrial materials (cooperation).
Employees	50 persons. Not family business.
Residence of manager and employees	All of them live in another place.
Satisfaction	Slightly satisfied.
Points of satisfaction	Have settled down.
Points of dissatisfaction	Noises of the cars.

Conclusion

Conservation and Development

As described above, the centre of the city such as Kuala Lumpur and Penang is filled with energy. It bears the heart of the city functions and offers places to work and live. What supports living in the city are the shophouses. At present, the accumulation of human and cultural energy attracts market capital and consequently causes various types of development. These have the power to destroy not only the historical and cultural structure of the city, but also the traditional system of society and economy. The rapid growth of the market has added burden to the weak urban infrastructures and worsened the living environment, and the traditionally sustainable economic system of small business.

The survey of the shophouses in Malaysia makes clear that shophouses support urban living and social economic activities, even though their types have changed gradually in history. For example, from timber structure to bricks, from low to medium storey and eventually, the shop-apartments. The traditional shophouses had work places at the ground floor and residences at the upper floor, but now the shophouses have exhibited various types of relationship between residences and work places. They all play an important role in providing convenient places to live and to run small businesses. As for new development of a residential area, shophouses are essential for creating a spontaneous town centre. Urban housing therefore must not be regarded only as containers to live in, but also as part of society and economy. At present, in the Asian cities, the movement of preservation of historical structures and conservation of historical urban areas are becoming stronger. In Malaysia, *Badan Warisan Malaysia* (National Trust in Malaysia) is making a survey of historical structures and areas, and it is also making a conservation plan. In the centre of Kuala Lumpur, buildings which preserved the façades of shophouses are constructed. 'Conservation Guidelines' are made in order to conserve the traditional areas.

In Penang Island, 'Penang Heritage Trust' is set up in order to conserve the traditional areas in the centre of George Town. The study team, including the administrative urban planner with further assistance from the United Nations Centre of Regional Development, is making a survey and a plan. These engagements in preserving the cultural heritage and the traditional sight are expected to lead people to reconsider life and development in Asia.

Universality and Inwardness

Bifunctional urban housing like shophouses is not limited only in Southeast Asian Chinese Societies. In Japan, the scene of row-houses which are called 'machiya' were already drawn in the picture scrolls of Heian era (eighth to twelfth century). These Machiya were used as a kind of urban housing combining a shop and a studio until now. The composition of the space and the usage of the space are very similar to the shophouses, except its wooden structure and its nonexistence of the party wall. By considering the history of the cultural interchange between Japan and China, it could have been influenced by the Southern Chinese house. In Latin-Europe, during the Middle Age period, 'segment house' was used in cities as a tradition. It was a kind of collective housing units with narrow frontage which had shops or studios at the ground floor. It is considered as the origin of *insula* in the Roman Empire, which had a mixed usage of rental housing, five or six-storey apartments with shops on the ground floor and the arcades facing the street. Surprisingly, the composition and usage of space in segment house and *insula* are also very similar to those found in shophouses.

Therefore, shophouse shares the universality of composite housing in urban dwelling and it has also developed and evolved over time its own characteristics in a specific regional situation. Any future planning of urban housing in this part of the world therefore should take the tradition of shophouses into consideration.

References

Edwards, N. (1990), *The Singapore House and Residential Life 1819–1939,* Singapore: Oxford University Press, pp. 17–28.

Izumida, H. (1990), 'Historical Study on the Colonial Cities of Southeast Asia and Their Architecture. Part 1. Singapore's Town Planning and Shophouse', *Journal of Architecture, Planning and Engineering, AIJ*, No. 413 (*in Japanese*).

Laboratory of Urban Planning (1984), *A Study on Traditional Taiwan Shophouse-row*, Research Institute of Civil Engineering, National University of Taiwan (in Chinese).

Lim, H.K. (1983), ' The Suburban Fringe', in K.S. Sandhu, P. Wheatley et al. (eds), *Melaka, The Transformation of a Malay Capital c.1400–1980*, Kuala Lumpur: Oxford University Press, pp. 708–38.

Mogi, K. (1991), 'A Study on The Consistency of Residence with Verandah', *Annual Report of Housing Research Foundation*, No. 18 (*in Japanese*).

Seow, E.J. (1983), 'Melakan Architecture', in K.S. Sandhu, P. Wheatley et al. (eds), *Melaka, The Transformation of a Malay Capital c.1400-1980*, Kuala Lumpur: Oxford University Press, pp. 770–81.

Sidhu, M.S. (1978), *Kuala Lumpur and its Population*, Kuala Lumpur: Surinder Publications.

Tan, T.S. (1990), 'Strategic Areal Development Approaches for Implementing Metropolitan Development and Conservation. Phase Two. Case Study Analysis for Formulating Goals and Objectives: The Case of George Town, Penang, Malaysia', Second International Training Workshop on Strategic Areal Development Approaches for Implementing Metropolitan Development and Conservation, paper for the United Nations Centre for Regional Development, Second International Training Workshop in Jogyakarta.

13 Evaluation of Planned Housing in Java: To Identify the Basic Needs of Interior Space

WAHYU SUBROTO T. YOYOK, KUNIHIRO NARUMI AND NAOKI TAHARA

The Challenge of Urban Housing Development in Indonesia

In developing countries, especially in Asia, the growth of cities is occurring at a much faster rate than ever before. According to the United Nations forecast on population, by the year 2025, approximately one billion people will live in urban centres in Asian countries (Bhattacharya, 1990). This will give rise to some serious problems associated with dense population. Examples include slums, unhygienic environment and general chaos. Each developing country has its own strategy for solving these problems. Basically, the strategy is the promotion of a better settlement that conforms to the local culture.

In Indonesia, the percentage of urban population increases dramatically. From 1961 to 1985, the urban population increased from 14.9 per cent of the total national population to 27 per cent. In 1990, it has reached 30.9 per cent. It is expected that the urban population will increase to 40 per cent in 2005 (Yudohusodo et al., 1991). As a result, the concerns about urban living space as a socially suitable environment which satisfies individual needs have increased.

The Indonesian government has founded *Perumnas* (National Housing Board) since 1974 to promote planned housing development. However, the typical layout in public housing development relies largely on Western techniques and standards which reflect the needs of Western users, their civilisation and diversified social relations. These may not apply to Indonesian communities with unique cultural needs and social customs. So Java, as the city with the highest population density, was taken as a case to be investigated

231

in this paper. According to the census of 1990 (Statistic Centre Bureau, 1990), the population of Java had reached 38.3 million (21.4 per cent of total population of Indonesia). A house for the Javanese is not regarded merely as a real estate or as a commodity but rather a personal property. It grows dynamically in line with a family's socioeconomic condition. A house constitutes a particular world where one can find one's identity. However, in public housing development, very little effort is spent to investigate the living style, the functions of a house and the best way to arrange rooms, particularly in terms of how they are linked to each other. This paper attempts to clarify the linkages among the three factors above. These connections are best illustrated by identifying the organisation of space, the space standard, the living style and the general functions of the space in a house. A case study will be presented to analyse the living style. It includes analyses on activities, the continuity of traditional living style, the house form and systematic interior layout. It is the focus of the case study to find common patterns in functional components and house plans. In the final part of the paper, the implications for policy planning will be highlighted.

Objective

This study attempts to discover the essence of Javanese living patterns and how Javanese integrate themselves successfully with their houses. In order to achieve this objective, one has to thoroughly understand the living patterns of the Javanese, so that the ways in which the house adapts to human activities can be identified.

In the following case study, the interior activity patterns in Javanese urban settlements will be discussed. In addition, the interior physical features, namely the rooms, are classified to gain a better understanding of the basic floor plan. The classification refers to the specific and basic arrangement of the interior space. In fact, the gap between the activity patterns and the physical layout of the house in public housing developments is often proved to be crucial.

Methodology

This study was carried out in Yogyakarta during the period of 1991–92. It used two settlements to represent urban living space in Java. They are a *Kampung*[1] and a public housing project, *Perumnas,* for independent families.

In order to understand how indigenous living style can inform new housing construction, a *Kampung* was studied. In addition, since the majority of the public housing inhabitants originally came from a *Kampung*, the continuity of their living style can be studied in the case of the public housing project.

Spatially, there are 'local rules' governing what kinds of activities should occur, and where they should be taken place. The patterns of specific activities based on 'local rules' are traced in two urban settlements in order to collate and to understand how people arrange their living space spatially.

The primary data from *Kampung* Ratmakan were analysed to uncover the 'local rules'. Samples were obtained through a survey of 14 house units (6 per cent of the *Kampung's* total) consisting of nine single family house units and five 'row-house' units. They were chosen at random throughout nine *RTs*[2] (neighbourhood associations). In addition, a questionnaire and an interview with the respondents and *Kampung* officials were also carried out to obtain data regarding the *Kampung* itself.

In *Perumnas* Condong Catur, the primary data were obtained by surveying 90 respondents. The survey was carried out in an area containing 36m² house type/D.36. A similar questionnaire to the one administered in Ratmakan was also conducted to the inhabitants of Condong Catur. Interviews with local leaders were also carried out in order to secure information about the community itself. Detailed information was collected on different questionnaires. Five respondents were asked to redraw their house plans and furniture. These five samples included three samples in which the house plan has been changed from the original plan. The other two groups had not changed much of their original house plans.

General Characteristics of Java

Java is the main island in Indonesia. The island stretches 1,200 km in length while the width is 500 km. It is geographically situated at the south side of the Indonesian archipelago, at the latitude of approximately 7° and contains some 7 per cent of the total area of Indonesia. The general climatic condition is tropical and humid. The average monthly temperature remains at 27° centigrade in the coastal area and 25° centigrade in the highlands. The average humidity level is high, with an annual average of 75 per cent to 85 per cent. According to the census of 1990 (Statistic Centre Bureau, 1990), Java was inhabited by 110 million people, which is about 60 per cent of the total population of Indonesia.

The term 'the Javanese' refers to the ethnic group who speaks the Javanese language and lives in the central and eastern part of the island, excluding the island of Madura. The Javanese culture could be found in the south central part of Java where Yogyakarta is situated.

Geographical Setting

The area of *Kampung* Ratmakan is by Code River at Ngupasan Sub District in Gondomanan District. It constitutes an area of 5.7 ha. in the centre of Yogyakarta and contains 238 houses. Due to its geography, Ratmakan becomes a strategic *Kampung* since its location is in the central business district (CBD) next to the *Kraton* (palace) and Malioboro Ward.

Perumnas Condong Catur lies about 8 km northeast of the city's centre at Condong Catur *Desa*[3] in Depok District, Sleman Regency. It has an area of 21.7 ha. sitting on barren and unproductive flat land. It contains approximately 1,166 houses (Figure 13.1).

Kampung Ratmakan: General Characteristics

This *Kampung* is a traditional one. It is also an example of an urban *Kampung* where its inhabitants are generally part of the informal sector (52.5 per cent), while some others participate in the formal sector (civil servants and private employees) and semi-informal sector or work as casual labourers (Haryadi, 1991). In 1992, it had a population of 1,422 people and a density of 309 people/ha. This figure represents a much higher density than the average for Yogyakarta. In 1990, population density for Yogyakarta was 157 people/ha. High density has caused inhabitants to change the land use and modify their houses. The purpose of such changes is to provide a living space for relatives, acquaintances or room tenants in their houses. The houses are of various sizes and conditions.

Perumnas Condong Catur: Its General Characteristics

This *Perumnas* consists of 36 m^2 (1,118 units) and 70 m^2 (48 units) single family houses. A typical house consists of standard rooms such as a *ruang tamu* (guest room), a bedroom, a kitchen, and a bathroom. In the area of the 36 m^2 houses, there are passages of 3–5 meters which are used mainly by pedestrians, bicycles and motorbikes. It is inhabited by 1,300 households and

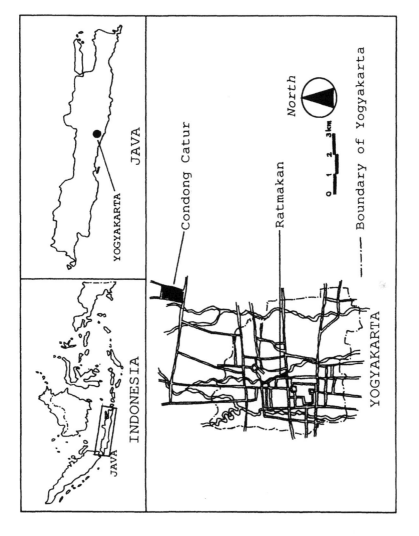

Figure 13.1 The setting of the study area

has a population density of 299 people/ha. (ASIKIN, 1987). According to the survey (1991), the occupancy periods vary, but most of the inhabitants have lived there for more than 10 years. 76.9 per cent of the inhabitants formerly lived in the urban *Kampung* (an indigenous urban settlement) inside the city of Yogyakarta. Therefore, it can be said that the pattern of migration is from the urban area to its fringe. It will be argued that the inhabitants continue the *Kampung* lifestyle as most of them are Javanese.

As could be seen from the social structure of the inhabitants, Condong Catur has an urban characteristic. Most of them are civil servants (61.5 per cent) and belong to the moderate income group (78.8 per cent).[4] This means that most of the inhabitants are middle class people. In addition, 33.6 per cent of them commute and work in the city of Yogyakarta and so Condong Catur serves as a bedroom suburb. This modern lifestyle contrast considerably with the traditional rural lifestyle in the surrounding areas.

Local Life Style: Group Activities in the House

There are many group activities in the neighbourhood that are related to the frequently held rituals and other events typical in Java. They are common in urban communities and offer many chances for people to gather together. This phenomenon stems from the Javanese custom which requires people to establish a reciprocal relationship with their neighbours in order to obtain or exchange any news. Participation in these group activities is to secure one's good reputation and maintain a smooth relationship with others (Mulder, 1989).

At Condong Catur, four of these activities often take place the residents' houses (Yoyok, 1993, pp. 74–9).

Communal Feast

For Condong Catur's inhabitants to reaffirm traditional values, communal feasts (*slametan*) are held at ceremonies and rites to commemorate important events in the life cycle: birth, circumcision, marriage and death. Recognition of these important events remind the Javanese the existence and the relationship between the terrestrial and celestial components. The smallest scale of participation occurs at the neighbourhood level which is designated as *RT* (an average of 46 households in Condong Catur). In cases where the host lives near the boundary of the *RT*, immediate inhabitants across the *RT* are also invited. Adult males attend the functions whereas adult females assist the host's wife with the preparation of the feast. The guests sit on a mat in the 'guest room' (*ruang tamu*) or other rooms such as the dining or family room.

66.3 per cent of the inhabitants observe *slametan*. Of those, 84 per cent invite guests according to the *RT* area. Other inhabitants have other ways in conducting slametan. They invite their relatives and acquaintances, and all together they deliver food to the neighbours. 44.9 per cent of the inhabitants only observe *slametan* on the occasion of the life cycle and communal cyclical events. 40.6 per cent of the inhabitants usually conduct an annual feast on Independence Day.

Savings Club Meeting

In Condong Catur, inhabitants have formed a savings club (*arisan*) which provides credit to its members. The activity is intended primarily to establish harmonious relations among neighbours. Meetings are held in turns at members' houses.

Arisan is dominated by the *RT* and *RW* (community association) domains. Most inhabitants are interested in participating the *arisan*. 97.1 per cent of the families are involved. Various family members form their own group. For example, women can join either a women's *arisan* (30.7 per cent) or an *arisan* for married couples (41.6 per cent). Men can join an *arisan* which is for men only (11.9 per cent). Youth can also join their own group (2 per cent). 88.1 per cent of the *arisan* in Condong Catur conduct their meetings on a monthly basis; others meet more often. 62.4 per cent of the meeting are held in the *RT* area.

Ritual Activities

In Condong Catur, people still stress on the individual's own point of view regarding religious belief and ritual practice. Everyone is free to choose his religion. Yet, the religious adherents (Moslem and Christian) do not merely observe rites in the mosques or churches. Data shows that 50 per cent of the inhabitants also observe communal rites such as the recitation of the *Kor'an* (Muslim) and communal prayer (Christianity) at their houses.

61.5 per cent of these activities are carried out at an *RT* level, 9.6 per cent are held at an *RW* level and 11.5 per cent are held at the area of Condong Catur public housing project. The remaining 17.3 per cent involve in the collaboration of three *RTs* when Christians congregate. Communal religious activities take place monthly (71.7 per cent) whereas others are conducted once every two weeks (26.6 per cent). It seems that these group activities have reinforced the cohesiveness of the religious associations.

House Wife Group for Family Affairs

In Condong Catur, residents (mainly women) are active in the housewife group for family affairs (*PKK*). This is a government program to augment family welfare. Approximately 55.8 per cent of all households are involved in this activity. Meetings are held monthly, usually at members' residences on a rotation basis, although some meetings (10.6 per cent) are held in the *RW* hall. For women, *arisan* can be conducted as a *PKK* activity.

The propensity to participate in those four group activities is also observed in Ratmakan. This emphasis on group activities builds up social solidarity and so conforms to the Javanese character which shy away from individuality (Mulder, 1989). As these activities are carried out in the house units, houses need to have spacious rooms.

The Basic Needs of Interior Space

The Case of Ratmakan

The tremendous demand for space in Ratmakan means that houses have to be rearranged in order to use the space efficiently. So far, this is achieved by living together in a single house. The average density was five people per house unit, and some of the units are as small as 12 m² (Figure 13.2).

A room which is located at the 'front'[5] tends to be multipurpose. In Ratmakan, this room designated as a *ruang tamu* (guest room) is where visitors are entertained.[6] This room can be used for at least nine distinct activities: receiving visitors (either guests or relatives), eating, relaxing, sleeping, praying, studying for children, holding actual group meetings and or ritual group activities (communal feast). This room is always intentionally designated even if the space is restricted to just 4.5 m².

In Ratmakan, the kitchen is almost always constructed as a special room (90 per cent). It is used for five activities: receiving relatives, eating, relaxing, sleeping and cooking. The minimum size of the kitchen is 2.25 m². The bedroom is also usually intentionally created, although there are some houses without a bedroom (10 per cent). Data from random samples reveal that 90 per cent of the houses have one bedroom, 80 per cent have two, 50 per cent

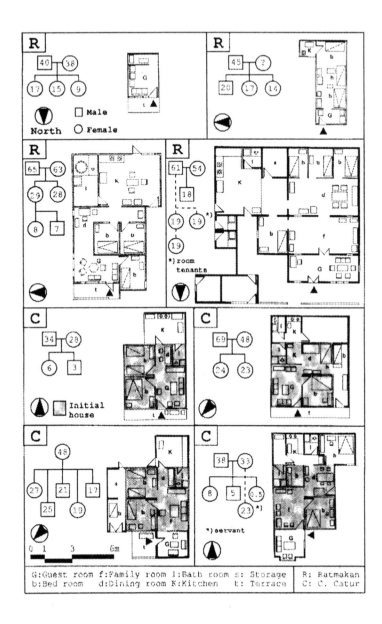

Figure 13.2 Typical plans of housing unit in Kampung and public housing project

have three bedrooms while 20 per cent have four. Various activities take place in a bedroom such as receiving relatives, eating, praying, studying for children, sleeping, holding actual group meetings and ritual group activities.

The limited space in Ratmakan apparently prompted residents to provide a kitchen rather than a private bathroom. People can use public latrines and wells instead of a private one. The congestion in *Kampung* Ratmakan has led people to provide only the most essential room: a *ruang tamu*, a bedroom and a kitchen.

The Case of Condong Catur

Due to the reconstruction of houses in Condong Catur, the average total floor area has increased to 58.4 m^2, which is 62.4 per cent bigger than the original construction. The average number of people per house is 4.8. This means that each inhabitant has an allotment of 12.27 m^2 (Figure 13.2). Similar to Ratmakan, all Condong Catur inhabitants also intentionally designate one room as *ruang tamu*. Seven kinds of activities can take place there: receiving visitors (either guests or relatives), eating, relaxing, studying for children, holding actual group meetings or ritual group activities.

The size of *ruang tamu* ranges from 6.3 m^2 to 10 m^2. 66.7 per cent of the houses also contain a 'second *ruang tamu*' as a family room. It becomes an extra room mainly when there are group activities. Approximately 80 per cent of the houses have a dining room. The inhabitants use this room for private visits (with relatives), praying, sleeping, holding formal group meeting, studying for children and dining. Each house has a bedroom even if it has to be 4 m^2 (Figure 13.2).

55.7 per cent of the households have more than four family members with both sex (76.5 per cent). Hence, there is need to have a bedroom for private activities (studying, sleeping). Yet, congestion necessitates that a bedroom also be multi-functional. Bedrooms, therefore, are also used for various activities such as receiving relatives, praying, sleeping, studying for children. It also serves as an extra space for ritual group activities when necessary. During renovation, there is an apparent tendency to move the kitchen into the backyard in order to increase its distance from the front side of the house. On the other hand, the bathroom is kept in its original place.

Houses in *Kampung* and *Perumnas* actually require similar room types: a *ruang tamu*, a bedroom and a kitchen. The difference is: in *Kampung* rooms are squeezed into smaller spaces while those in *Perumnas* are enlarged.

The Main Body of the House: Its Development Process

An original floor plan measuring 36 m² per house unit in Condong Catur has been changed into various sizes and forms. The reconstruction, where rooms were moved to new locations, occurred mainly in the *ruang tamu* and the kitchen, while the bedroom is more separated from the public space. Residents tend to keep the bathroom in its original place because of difficulties in moving the plumbing system.

It is interesting to note that major renovations involve mainly the *ruang tamu* and the kitchen. A survey conducted in Condong Catur indicated that *ruang tamu* is frequently altered. As mentioned, *ruang tamu* is a versatile space for various activities: a formal space replete with intimate qualities; an informal space for relaxation or recreation; a space for entertaining or receiving visitors; and a haven for family activities, which could be both public and semi-public.

At first, inhabitants kept the house as it was built. The original arrangement is advantageous since the private family activities (entertaining relatives, relaxing, eating and studying) in a *ruang tamu* could be held in the bedroom and the kitchen. However, activities that are related to the public world have apparently interfered with private activities that are also held in the bedroom and kitchen. To overcome this condition, the inhabitants build a terrace at the front of the house so that his visitors can be entertained there. Residents also have a propensity to move the *ruang tamu* into the front yard while the former *ruang tamu* is converted into a family room or a dining room.

It should be noted that these rooms (terrace, family room and/or dining room) are usually not designated since activities carried out there can also take place in a *ruang tamu*. Obviously, extra rooms would be built if the inhabitants can afford it.

In order to reduce the invasion of private space by public activities, the terrace is the first place to be considered for house expansion. The second reason for replacing a terrace with a *ruang tamu* is that it connects with the inner rooms, such as the family room, the dining room or the bedroom. Then, both the family room and the dining room or either one of these rooms can merge with the area of the former *ruang tamu*.

The formation of the three rooms, the terrace, the *ruang tamu* with the family room, and the dining room, is then established in a linear formation. When large group activities are held, this arrangement allows all the three rooms to be used simultaneously. In other words, the three rooms become one spacious room enabling people to meet and talk to each other more easily.

Cooking takes place in the kitchen where women spend most of their time. This activity for the Javanese is considered to be private.[7] The cook feels embarrassed if others see or smell what they are cooking. Clearly then, the back of the house is the best place for cooking because the cook can prepare the meals without being disturbed. In a crowded area, the kitchen activities are protected by a long corridor that runs from the back to the front of the house. As illustrated by the alteration of house plans in Condong Catur, the kitchen has been frequently moved to the 'back' of the house. The juxtaposition of the kitchen and the *ruang tamu* apparently does not conform to Javanese feelings since those two rooms have opposite attributes. There is both topological and visual significance to the position of these rooms. Topologically, the kitchen must be placed at the back. It puts the greatest possible distance between the kitchen and the front of the house. Meanwhile, visually, a filter is created so that visitors cannot directly know, see and smell the cooking in the kitchen.[8]

Through the process of house alteration, *rang tamu* built on the terrace becomes the 'front feature'. In contrast, the new kitchen at the back of the house, can be identified as the 'back feature'.

Bipolar Interior Place: The Basic House Plan

The house and the local life style are ordered by the customs, habits and classification categories of the inhabitants. This practice usually conforms to a consistent set of rules in a specific society. Javanese people perceive the front and the back of the house as the most important aspects of the floor plan. This perception is consistent with the duality of the personal sense of *lair* (outwardly) and *batin* (inwardly). As a result, the house is ordered into sequential space by a set of interior space criteria which conforms to their feelings of social ties. These feelings consist of worth, inhibition, fear, shame and embarrassment (Koentjaraningrat, 1984)[9] and they represent deeply internalised attitudes fostering conformity. Simultaneously, they control behaviour and function as a kind of conscience.

The arrangement of interior space evidently holds a special place in Javanese culture. This phenomenon suggests that it would be useful to uncover systematically the layout of the floor plan. The following will focus on the basic functional components and patterns that house plans have in common.

The traditional Javanese house is arranged in a linear sequence of spaces. This kind of linear formation facilitates organisation and control of privacy, limiting possibilities of invasion by the public world. Contemporary Javanese house plan also bears some similarities to the traditional one (Figure 13.3).

In attempting to uncover the basic pattern of the house plan, the house plans were analysed by adopting the diagrammatic technique developed by Hillier and Hanson (1984) and recently used by Steadman and Brown (1991) in a morphological research. This diagrammatic technique is used to identify the spatial organisation of the Javanese house. By checking the room depth and accessibility, the analysis uses geometric elements to portray the morphology of the interior space which refers to the distance and location of the room's depth.

From Figure 13.4 we can see the floor plan graphs of Condong Catur and Ratmakan which will assist us in the analysis of the access relationship between the rooms. In the graphs, there are some circle symbols indicating rooms (vertices). These circle symbols which are used in the graph are: 1) a vertex (architectural space or room) representing the *ruang tamu* as a solid circle; 2) a vertex representing the kitchen as a circle with dark shade; 3) a vertex representing the bath room as a circle with dim shade; and 4) a vertex representing other rooms (mainly bed rooms) as a hollow circle; while 5) the external spaces at the front and the back are represented as a hollow circle with a cross. A solid line (linking space, door and corridor) indicates that two rooms are accessible or are joined by a door in a shared wall.

The graph is drawn according to the conventions developed by Steadman and Brown (1991): where a vertex denoting the front external space is set at level-0; all spaces accessible from the 'front' exterior space are set on level-1; all spaces accessible from the 'front' exterior space by passing through level-1 space are set on level-2 and so on. The level on which any vertex appears records the depth (in the topological sense, measured by the number of solid lines on the graph) of the corresponding room in the plan. The depth of an individual room refers to the topological distance of each room. The value of the depth measure is represented by high lighting the distance of a space from the main entrance. It indicates how public or private that space is intended to be.

The analysis then focuses on the access and depth of each room from the front where the main entrance of the house unit is located. The data, which were obtained by random sampling, constitute a significant contribution to the understanding of the relationship between the spatial arrangement, the variety of room types of the house unit and culture.

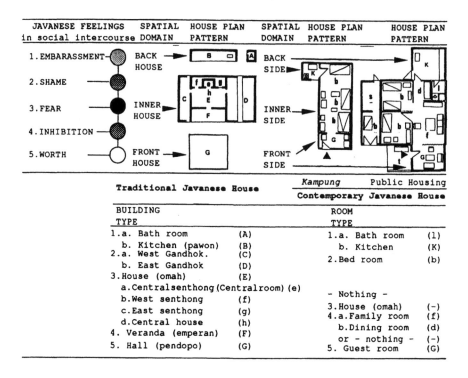

Figure 13.3 **The alteration of floor plan for traditional to contemporary Javanese houses in confirming with the Javanese feelings in social intercourse and the spatial concept of floor plan**

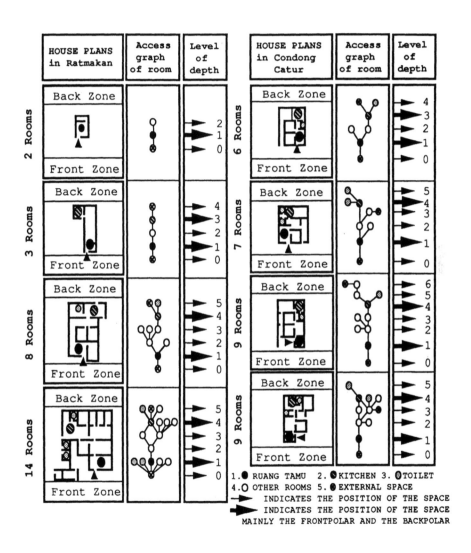

Figure 13.4   Spatial organisation of contemporary Javanese houses

Front Polar: The Main Privacy Filter

The differences between the inside and the outside conform to Javanese feelings and move people to generate or control encounters. It is expressed by physical design elements. An understanding of the relationship between defensible spaces is important in urban areas such as Condong Catur and Ratmakan, where all spaces are essentially man-made, and many of their physical characteristics are affected by social conventions. The data indicate how inhabitants (Javanese) protect their privacy against the possibility of uncomfortable psychological and physical social encounters.

Various floor plans indeed represent a number of different house forms. Yet they all have identical patterns of functional relationship. To identify the basic floor plan then, it is important to make an access graph of rooms. By making a graph correlation between the floor plan and the access flow (Figure 13.4), a certain access relationship of the rooms is indicated. The graph also illustrates that a *ruang tamu* is often reached directly from the exterior of the house, where the terrace[10] is located. It also tends to minimise the distance between the exterior space and the front of the house. *Ruang tamu* as the front interior room then, is represented as the vertex level-1, both in Condong Catur or Ratmakan. Considering the idea of privacy gradients, this room belongs to the public domain.

A *ruang tamu* for Javanese guides the direction of the central interior flow. Its spatial attribute, similar to the spatial position and purpose of an entrance hall in western countries or a genkan in Japan, is explicit and specific. It is intended to regulate the access of people and objects between the private and the public domains. It is required to control visibility between the exterior and the interior. It also constitutes the place where personal appearance can be controlled. The similarity of *ruang tamu* to an entrance hall and a *genkan* is signified by its position as a part of the main building. It is different from a terrace which, although a component of the house, is not effective in controlling encounters. A terrace is apparently similar to a verandah in the Western or an *engawa* in Japan. It is the boundary between the inside and the outside.

A *ruang tamu* with the vertex level-1 is a significant room where the host and the outsiders (guest, acquaintances, relatives) make the first contact. It means that a *ruang tamu* signifies the main filter to control intrusion of the private residence by the public world. It is a room demarcated from other rooms and will be kept tidy. It represents a feeling of *aji* (worth).

Ruang tamu also signifies the most powerful space. It is in the central position to distribute people's paths from the main entrance to another space inside the house. It is also the focal feature of social gatherings for various group activities or private visits. In addition, most family activities occur in this room. These elements comprise the key characteristics of the 'front polar'.

Middle Interior Room: The Sub-privacy Filter

The edges (solid lines) radiating from the focal point of the *ruang tamu* indicate the enlarged private spaces behind a *ruang tamu*. The private domain obviously tends to extend beyond the main entrance of the house. The spatial arrangement of the private domain corresponds to the Javanese inward feelings. This domain, as the inner space, is filled mainly by the bedroom.

In Condong Catur, where house alteration was made possible, a family room which was not available in the original house, tends to be provided in the renovation. Generally, the range of the vertex level is between 2 and 3, which is the middle interior room. Similar conditions also prevailed in Ratmakan where there is a tendency to squeeze in more houses. This area seems to create a sub filter to control the intrusion of public into private life.

Back Polar: Space for Private Activities

The last domain of interior space in a Javanese house is found at the back of the house. For the Javanese, cooking takes place in this space. The kitchen must satisfy particular conditions since the Javanese have their own ways of cooking. They tend to avoid encounters when they cook. It is induced by the feeling of *isin* (embarrassment) and fear of eyes, ears and opinions from others concerning what and how food is being prepared.

Supporting this feeling, a kitchen (and a bathroom) is suitable if it is located at the back, covered by both front and inner spaces. In order to avoid encounters with outsiders in the front space, a kitchen must have its own access from the exterior space, meaning that the kitchen has its own entrance, such as a side entrance. As this space occupies its own particular position at the back, and it also takes on opposite attributes to the front space, this room is the 'back polar' of the main body of the house.

The Terminology of the Bipolar Interior Space

The graphs of Condong Catur and Ratmakan samples show that the position of the two basic rooms, the *ruang tamu* and the kitchen, must have a maximum number of connections and the greatest possible walking distance between them. As a result, the *ruang tamu* must be located at the front, while in contrast, the kitchen is always found at the back.

The graphs also show that the two polar spaces, the *ruang tamu* and the kitchen, play a powerfully distinct role in separating the public and private spaces. They also have the function of limiting multiple entrances and avoiding unnecessary access between the rooms, the front and the back of the house.

The term 'bipolar interior space' refers to the ordering of two interior spaces based on opposing principles. One room must avoid encounter while the other encourages it. One room must maintain strict control of boundaries and interior organisation; the other requires transparent boundaries so that outsiders will feel familiar enough to enter the house. It is expressed in the Javanese house by the interior layout. The *ruang tamu* and the kitchen enclose the inner space, that consists of a bedroom, a dining room and a family room. This bipolar interior space creates the main body of the house.

Bipolar Interior Space in Javanese House Plans

The house for the Javanese is shaped by values, as reflected in the bipolar interior space in maintaining a strong specific categorical order in the domestic interior. A house is set in three distinctive areas: the front, the middle and the back space. They are aligned in a linear formation creating the main body of the house where two filters are arranged. The Javanese are greatly concerned with the public world. On the other hand, they still need autonomous private control of their houses. It is reflected in its room accessibility and the depth of each house. It is also indicated by the two entrances, the main and the side entrances, which support the powerful and functional distinction between the front and the back.

Based on the spatial analysis of the floor plans above, there is no doubt that the bipolar interior space principle exists in a single family house, whether in a *Kampung* or a public housing project. This principle provides a particular link between the interior and the exterior where people can have full contact with other community members within the house. Contact between the 'two-worlds', the private and the public realm, takes place in what local people called a *ruang tamu*, which is demarcated from the interior. Local people can express their feelings of respect toward visitors in this room which also plays a significant role in the control of encounters.

This principle also leads to the existence of the kitchen at the 'back' which assures that people can protect their private life from the public world. It is a unique room because even though this room belongs to the private space, it has a particular link to the exterior through a side entrance.

Implication for Policy Planning

People who live in the *Kampung* and *Perumnas* share the need to modify their houses. As the *Kampung* is an urban settlement with very high population density, people are forced to modify their houses into smaller dimensions to fit in more rooms. In contrast, the original house units in *Perumnas* have only a few rooms and so more rooms can be created.

The effort to derive a standard internal layout from the interior space analysis of the house is becoming a major issue in Indonesia, particularly in face of multi-family housing developments, such as walk-up flats, which require particular life style changes. Therefore, the evaluation of planned housing in Java and the identification of the basic needs of interior space are worthy of intensive investigation.

The analysis of house plans in the two settlements in Yogyakarta may help to identify the basic rooms and where they should be placed when taking the users' point of view into serious consideration. This analysis and interpretation is of particular relevance as even the recent studies only refer to the evaluation of physical features in isolation from other sociocultural factors.

House types in Indonesia is still distinguished by its size type: T 15, T 21, T 36 etc., where the number indicates the house size in square meters. This system does not help people to recognise other attributes such as room types. It is also difficult to identify other features such as the rooms' organisation, topological space and the spatial order of the house.

House types discussed above show that the arrangement of the interior space exhibits two opposing attributes: front-back, public-private, open-closed and social-personal dimensions. These features are important especially when we are designing multi-family housing units which require the observation of some strict rules in space utilisation.

The bipolar model may help to clarify our image of the house type. The bipolar model offers certain advantages, including the identification of individual rooms, the consideration of room organisation, as well as the topological form.

This model offers a brief guidance to the planner, designer, and even inhabitants in constructing a floor plan configuration which consists of the basic components such as the guest room and the kitchen. House standardisation should not only consider the house size, but also the room types and its configuration which should be suitable to the local life style.

The bipolar model attributes offer beneficial input to the planning of a multi-family house, where public facilities are provided. As the bathroom is excluded from the bipolar model, it can be a public facility rather than another room in the house.

In sum, it is clear that when space is limited and there is a high population density, a basic floor plan, at least, should contain a guest room (G), a kitchen (K) and a number of bedrooms (n), which could be stated as 'n-GK'. If this configuration fits the local lifestyle, then it could be considered as a viable alternative to the present classification, which is based exclusively on physical dimensions.

Acknowledgments

We would like to express our appreciation and gratitude to Ms Heather Black who assisted us in editing the English of this paper.

Notes

1 *Kampung*: village-like settlement contained in an urban area.
2 One RT (RT 31) which had formerly been occupied mainly by industrial buildings, in Ratmakan has been abolished and replaced by commercial buildings such as a hotel. When the survey was carried out, therefore, the data were only obtained from the eight remaining RTs remained in Ratmakan.
3 *Desa* is a village-like unit of rural subdistrict.
4 According to a study of the World Bank, the salary of the moderate income group is between Rp. 150,000 and Rp. 450,000 per month (1US$ = Rp. 2,080).
5 The 'front' in a traditional Javanese house is known as *pendopo* (hall).
6 The term 'guest room' was designated by the respondents. It is necessary to confirm it with the respondents since there were many other activities occurring in this room which have no relation to the 'guests'.
7 Cooking for Javanese seems to be an activity which needs privacy in special gradient to control unwanted interaction and information (confidential).
8 The juxtaposition between the *ruang tamu* and the kitchen is established and is a result of congestion. Yet those two rooms tend to become demarcated because each has its own entrance.
9 This is the set of Javanese feelings in a hierarchy which is in effect when social intercourse occurs.

10 Terrace for the Javanese (and also for Indonesians) is an appendage room which usually constitutes a part of *ruang tamu*. It is a place for informal gathering. Also it is intended to provide an additional spatial privacy filter supporting the main privacy filter, the *ruang tamu*.

References

Asikin, M.M. (1987), 'Inhabitants Aspiration in the Growth of Living Demand Order in *Perumnas*', paper presented at a seminar on the occasion of the International Year of Shelter for the Homeless, Yogyakarta, Indonesia (*in Indonesian*).

Bhattacharya, K.P. (1990), 'Metropolitan Cities in Asia – Size, Scale and Its Management: Need for Self Sustaining Cities', *Proceeding of the 4th International Meeting on Urban Problems in Developing Countries*, Toyohashi, Japan, p.161.

Brown, F.E. and Steadman, J.P. (1991), 'The Morphology of British Housing: An Empirical Basis for Policy and Research', *Environment and Planning B: Planning and Design*, Volume 18, pp. 385–415, Great Britain: Pion Ltd.

Haryadi (1991), 'Residents Participation in Planning for the Improvement of Living Environment: Case study of Kampong Ratmakan, Yogyakarta', paper presented at The Third International Training Workshop on Strategic Areal Development Approaches for Implementing Metropolitan Development and Conservation, Penang, Malaysia.

Hillier, B. and Hanson, J. (1984), *The Social Logic of Space*, Cambridge: Cambridge University Press.

Koentjaraningrat (1984), *Javanese Culture*, Jakarta: PN Balai Pustaka (*in Indonesian*).

Mulder, N. (1989), *Individual and Society in Java: A Cultural Analysis*, Yogyakarta: Gadjah Mada University Press.

Statistic Centre Bureau (1990), 'The Outcome of Population Census 1990', in S. Yudohusodo et al. (1991) (eds), *House for All People*, Jakarta: Unit Percetakan Bharaderta (*in Indonesian*).

Yoyok, W.S. (1993), 'Evaluation of Planned Housing in Javanese Living Space: An Attempt to Identify the Activities Pattern in Houses and Neighbourhood. Case study: Urban Settlement in Yogyakarta, Indonesia', thesis submitted for the Master's degree, Osaka University, Osaka, Japan.

Yudohusodo, S. et al. (1991), *House for All People*, Jakarta: Unit Percetakan Bharakerta (*in Indonesian*).

14 Urban Redevelopment with Landowner Participation using the Land Pooling/ Readjustment Technique

RAY W. ARCHER

Introduction[1]

This paper discusses the use of the land pooling/readjustment (LP/R) technique for the redevelopment of established urban areas, with special reference to the Taiwan and Japanese experience. It is presented in four parts. The first part outlines the LP/R technique and its possible application in redevelopment land projects and in land and building redevelopment projects. The second part presents a case study of a redevelopment land project in Taiwan. The third part outlines the wide use of LP/R for redevelopment land projects and for land and building redevelopment projects in Japan. The fourth part considers the lessons of the Taiwan and Japanese experience.

The Land Pooling/Readjustment Technique

Urban LP/R is a land management technique in which a selected group of adjoining land parcels in an urban fringe area are consolidated for their unified design, servicing and subdivision into a layout of streets, open spaces and serviced building plots, with the sale of some of the plots for cost recovery and the redistribution of the other plots to the landowners. The landowners then sell, develop and/or hold these plots. A definition from another perspective is that LP/R is a technique for converting rural land into urban building land whereby a group of neighbouring landowners in an urban-fringe area are combined in a compulsory partnership for the unified planning, servicing and subdivision of their land with the project costs and benefits being shared among the landowners. It is a simple concept and this is illustrated by the diagram in

Figure 14.1 which shows a typical LP/R project to produce serviced building plots.

The LP/R technique can be used to improve land development for urban expansion and provide net benefits for the landowners, the government and the public. The potential benefits of LP/R projects include land assembly for project sites, government land acquisition for public roads and facilities, construction of network infrastructure, timely land development, official plan implementation, full project cost recovery, and the equitable sharing of project costs and benefits among the landowners.

LP/R is widely used in Japan, South Korea and Taiwan and is used in some cities in Australia and Canada. It has also been adopted in Indonesia and Nepal where a number of LP/R projects have been carried out. The technique has different names in different countries, being known as land readjustment in Japan and South Korea, land consolidation in Taiwan and Indonesia, land pooling in Australia and Nepal, and land replotting in Canada. These different names mainly reflect different parts of the LP/R process. There are differences between countries in the form and procedures of their projects but they are all landowner partnerships for the unified design, servicing and subdivision of their land with project cost recovery by the sale of some of the new plots.

There is an important legal difference between land pooling (LP) and land readjustment (LR). In LP projects, the land parcels in the project site are legally pooled/consolidated by the transfer of ownership to the implementing agency which later subdivides it and transfers building plots back to the landowners. In LR projects, the land parcels are only notionally pooled/consolidated for their unified design, servicing and subdivision because the landowners remain the owners of their land parcels throughout the project until they exchange them for their new building plots as shown in the replotting plan.

Most LP/R projects are prepared and implemented by local governments but they can also be undertaken by other authorised government bodies such as highways departments, public housing authorities and urban development corporations, and by landowner cooperative associations. The use of the technique is normally authorised and regulated by a central government law for LP/R which identifies the possible implementing agencies and sets out the principles and procedures to be followed in proposing, preparing, approving and implementing each LP/R project. A scheme is normally prepared for each project by consulting the landowners. This scheme defines and explains the project and, when approved, then authorises and regulates its implementation,

A) The Basic Concept

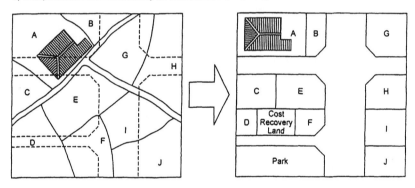

B) Readjustment of Land Boundaries by the Road Layout and Replotting Plan

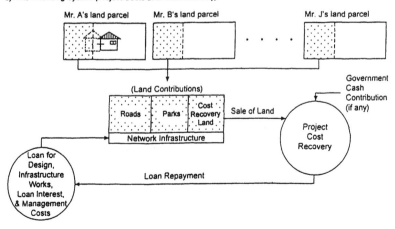

C) The Financing System (Project Costs and Cost Recovery)

Figure 14.1   Urban land pooling/readjustment

so that the scheme can be seen as a form of partnership agreement for the project.

One of the basic principles of LP/R is that there should be majority landowner support for each project. In addition to that, the LP/R law has to authorise compulsory participation or compulsory purchase acquisition of land by the implementing agency, since some landowners may not agree to participate.

LP/R is mainly used as a technique for new urban development by the conversion of rural land into urban building land but it can also be used for urban redevelopment. As redevelopment is a complicated and costly process, the application of LP/R to redevelopment becomes complicated and costly so that most countries make little use of it for urban redevelopment. However, it has been widely used for many redevelopment projects in Japan, and Japan is the main source of Asian experience and lessons of urban redevelopment with landowner participation using LP/R. There is also European experience in countries such as Germany where the technique has been used for more than 100 years. It has also been used for the redesign and resubdivision of old undeveloped subdivision estates in Western Australia and Canada (Archer, 1988; British Columbia Department of Housing, 1976).

Complications of Urban Redevelopment for LP/R

Urban redevelopment is more complicated and costly than new urban development, so that it is more difficult to use the LP/R technique for redevelopment. The complications for LP/R can be identified and grouped under the headings of: the many private ownership interests in the project site, the numerous occupants (many of which oppose redevelopment), the large expenditure required, and a complex management task.

Urban redevelopment is defined as the conversion of an existing urban area of streets, open spaces, plots and buildings to a new layout of streets, utility service lines, open spaces, plots and buildings for a new pattern of urban uses. These new uses usually involve higher density building development for the more intensive and more valuable use of the land.[2] Redevelopment can be carried out by private developers who assemble the land for their project sites through negotiated purchase of adjoining plots and buildings. However, these private projects are usually of limited size (and within a single city block) because many landowners do not wish to sell at the time the developer is buying. These private projects are normally undertaken for the profit generated by converting the land to a more valuable and intensive use. Multi-block redevelopment usually has to be carried out by a government agency which assembles the project site by a combination of negotiated purchase and compulsory purchase of plots and buildings plus the transfer of public streets and spaces and government plots and buildings from other government agencies. LP/R provides the government implementing agency with an alternative to land purchase as the method of land assembly for the redevelopment project.[3] These multi-block projects normally require government financial contributions for the infrastructure and other public-purpose components of the project.

Compared to the new urban development situation the use of LP/R for redevelopment is usually much more difficult because most redevelopment projects involve:

1) many landowners, as the project area is usually divided into many plots, and multistorey buildings may also be divided into separately-owned condominium units;

2) many other persons and organisations with 'ownership interests', such as leases, easements and mortgages in the land and buildings;

3) various government agencies that control, operate and maintain the public roads, utilities and facilities in the area;

4) household occupants of the buildings, whose main objective is meeting their housing needs rather than possible financial gain from the redevelopment project; and

5) some or many of the household occupants of the land and buildings may be illegal (squatter) occupants;

6) business-firm occupants of the buildings, whose main objective is in continuing their business activities rather than possible financial gain from the redevelopment project;

7) the construction of multistorey buildings within the project so as to provide condominium units and floorspace for the landowners in exchange for their land and buildings, plus some units and floorspace to be sold for project cost recovery. Some/many of the tenant- and lessee-occupants of the buildings in the project site may also have to be accommodated in the new buildings;

8) large capital outlays that are necessary to cover the combined costs of: paying compensation for the buildings to be demolished, paying relocation assistance to the departing occupants, demolishing the buildings, reconstructing the network infrastructure, and then constructing multistorey buildings;

9) the need for government funding on the cost of reconstructing the network infrastructure and other public-purpose components of the project;

10) the need for large project loans to finance the large capital outlays;

11) high project costs, including the interest payable on the large project loans required for relatively long periods;

12) the possible need to arrange long-term loans for the small landowners to enable them to pay the extra cost of a new unit or floorspace; and

13) the heavy demand on management and professional manpower to meet the need for complex designs, valuations, negotiations, design revision and community relations.

When some or many of the owners and occupants have to be rehoused within the project site after redevelopment, then they will also need relocation to temporary housing and business premises while the demolition and construction work is carried out.

These complications for LP/R arise when an area of streets, plots and buildings is redeveloped into a new pattern of streets, plots and buildings. These land and building redevelopment projects are three dimensional projects that require three dimensional LP/R. It is also possible to have simpler, land-only redevelopment projects that are two dimensional. These redevelopment land projects cover the clearance, redesign, reservicing and resubdivision of the project area into a new layout of streets, utility lines and plots, most of which are transferred to the landowners for them to build on or to sell after the LP/R project. This two dimensional redevelopment land project requires only the conventional two dimensional LP/R. It has been widely used in Japan for the reconstruction of urban areas which were damaged by earthquake and fire during World War II. LP/R projects for redeveloping land could be used in most countries to redevelop devastated areas to improved layouts instead of reinstating the previous layouts of streets and plots. They can also be used to redesign and resubdivide old undeveloped subdivision estates that have become obsolete (Archer, 1988).

The complications and difficulties of land and building redevelopment projects makes the use of LP/R for these projects difficult. The main reason for using LP/R in these projects is to overcome or reduce the landowner opposition that delays and restricts the land assembly for private and/or government redevelopment projects. The landowners opposing these projects usually do so because they wish to obtain a share of the anticipated redevelopment profits, or they wish to remain in, or return to, that location. The landowner participation in land and building redevelopment LP/R projects also reduces the capital funding required to finance the project and reduces the project risk of the implementing agency.

Redevelopment land projects are the simpler form of redevelopment project as the new buildings are constructed after the project. LP/R can be used successfully for these projects, particularly if the area has been badly damaged by fire or other disaster. It is useful to consider a case of redevelopment land project in Taiwan.

A Redevelopment Land Project Using LP/R in Taiwan

Most LP/R projects in Taiwan are carried out by local governments. The Kaohsiung City Government has been the leading user of the technique. After implementing the first LP/R project in 1958 and another three projects during the 1960s, it adopted LP/R in 1971 as the main system of land development for the city's growth and expansion. As in January 1992, the Kaohsiung City Government had completed 36 projects with another 19 in preparation and implementation. It had also formulated a set of project regulations that have been adopted for general use by local governments across Taiwan.

There is very little use of LP/R for urban redevelopment in Taiwan. Local governments normally use general land acquisition by purchase ('zone expropriation') and public housing projects to redevelop urban areas. The Kaohsiung City Government has used LP/R for urban redevelopment in only three projects and these were early projects, two of which were combined new development and redevelopment projects. These early redevelopment projects were undertaken mainly to reconstruct areas of uncontrolled urban development where there was no public street layout and a jumbled mixture of business land uses and housing, with many illegal businesses and squatter occupants. The only redevelopment project without a new urban development area was the Kaohsiung 'Tenth Stage Project' that was carried out between April 1973 and August 1976 (Department of Land Administration, 1985).

Outline of the 'Tenth Stage Project', 1973/1976

The Tenth Stage LP/R Project was undertaken to redevelop a 3.5 ha. site in central Kaohsiung from an area of mainly low-rise unregulated buildings without public roads into a planned layout of roads and plots suitable for medium and high-rise buildings. A case study of the project was made by Ying-Shy Lai and the following outline is based on his report (Lai, 1988). A map and plan of the site are shown in Figure 14.2.

Project site An inner urban area of 34,875 m^2 with 403 land parcels, three of which were government land (for future roads), occupying 11.27 per cent of the area. Most of the land parcels on the site had no public road frontage. The site was mainly occupied by old low-rise buildings. As many of these buildings were built without permits, they were used for both legal and illegal business activities and housing. The site had about 2,300 residents, including tenants and 690 squatters. The site was zoned for commercial land use in the city plan.

Project plan The project was designed to establish a planned layout of streets for the area and to convert the land parcels into 138 regular building plots with street frontages suitable for walk-up and high-rise commercial and residential buildings. Only 20 of the existing buildings were designated for retention, and the rest were to be demolished.

Project progress

- April 1973 – Meeting of officials with owners and residents to announce the proposed project.
- October 1973 – Cadastral survey and building survey of the site.
  - Preparation of LP/R scheme and negotiations with landowners.
  - Majority landowner acceptance of scheme, by 67.57 per cent (271) of the landowners holding 66.14 per cent of the site area.
- June 1974 – Provincial government approval of scheme.
- August 1974 – Public exhibition of scheme and bar on new building construction within the site.
- December 1974 – Completion of building demolitions.

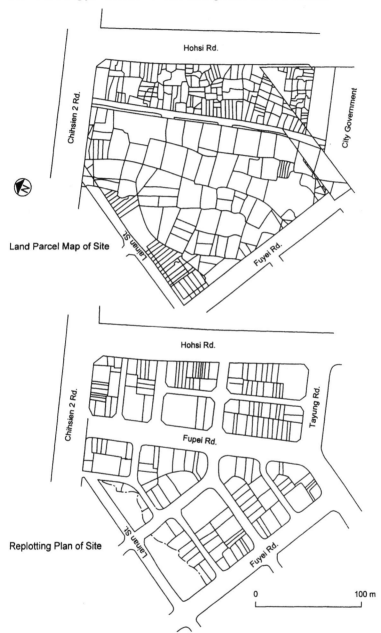

Figure 14.2   Tenth stage land pooling/readjustment project in Kaohsiung, 1973/76

Source: Lai, 1988, p. 51.

- August 1975/April 1976 – Contractor construction of roads, drains, etc.
- October 1975 – Negotiations with small landowners for 'collective consolidation' (i.e., to receive shared plots), for the preparation of the replotting plan.
- March 1976 – Public exhibition of replotting plan.
- August 1976 – Transfer of building plots to landowners.

Project costs and revenues

The project costs totalled NT$ 67.3 million and were made up of:
- compensation for demolished buildings and relocations – NT$ 47.1 mill. (70.0 per cent);
- road and drain construction – NT$ 7.4 mill.(11.0 per cent);
- interest on project loan – NT$ 11.5 mill.(17.1 per cent);
- project management – NT$ 1.5 mill.(1.9 per cent).

The project revenues totalled NT$ 81.3 mill., with NT$ 71.1 mill. (87.3 per cent) from the sale by tender of the 2,691 m² of cost-recovery land in July 1977 and January 1978, and NT$ 10.2 mill. (12.7 per cent) as payments from landowners who had received land above their entitlement. The project financial surplus was therefore NT$ 14.0 mill.[4] (During 1973/1976, US $1.00 = NT$ 38.)

Project land values The assessed market value of the land on the project site before LP/R averaged NT$ 7,986 per m² and ranged from NT$ 3,788 to 12,727 per m² After LP/R, it increased by 170 per cent to an average value of NT$ 21,571 per m², with the values ranging from NT$ 14,545 to 27,455 per m².

Sharing of project costs and benefits The project site of 34,875 m² was redeveloped into 12,712 m² of streets (36.45 per cent of the site) and 22,163 m² of building plots, of which twelve plots of 2,691 m² (7.72 per cent) were designated to be sold for project cost recovery. Hence, 55.83 per cent of the site area was returned to the private landowners. As the city government and the private landowners respectively owned 11.27 per cent and 88.73 per cent of the site before LP/R, the private landowners were returned only 62.92 per cent of their land area at the completion of the project. The percentage of land area returned to each landowner, varied as their land contribution to the project, would reflect the benefit their land obtained from the project. For example, the owners of land with an existing (boundary) road frontage would be returned a 65 per cent area while the owners of interior land that obtained a road frontage from the project would be returned a 40 per cent area.

As many landowners owned small land parcels, they would be entitled to a plot smaller than the minimum permissible site area under the building construction regulations. Most of these landowners were subjected to 'collective consolidation' and were allocated a part-ownership of a building plot that they shared with one or more of the other small landowners. They then had to reach agreement among themselves and with a developer on the building development of the plot, and on the sharing of the ownership and use of the plot and building. The small landowners that lived in the site were also entitled to buy a public housing unit outside the site.

The owners and residents of the demolished houses were paid compensation and/or assisted to relocate. The owners of the legal houses could get compensation at the reconstruction cost of the building, whereas the owner-occupants of the illegal (squatter) houses were paid small amounts calculated by a formula covering the plot, the building, the household size and voluntary demolition. Low-income tenant households were paid small amounts of NT$ 4,000 to 5,000 per household. The small landowners and the tenant- and squatter-households were offered the purchase of public housing by the city government. However, these compensation and resettlement arrangements were not sufficient and there was resistance to the final building demolitions.

Assessment of the project The Tenth Stage LP/R Project was a successful redevelopment project for the Kaohsiung City Government. The local government was able to convert a problem area of uncontrolled urban development and land use into a planned layout of streets and reshaped and serviced building plots ready for medium- and high-rise building construction. The project was prepared and implemented over three years and with full cost recovery, but it was opposed by some landowners and many of the occupants. It was a successful project for most of the landowners as they obtained serviced building plots with a public street frontage and a much higher market value than their original land and buildings. But some of the landowners were not satisfied with the project. They included owners of the land parcels with the 20 undemolished buildings on them, as they had to pay cash contributions to the project but without realising the new redevelopment value of their land. Some of the small landowners that were 'collectively consolidated' were not pleased to become a part-owner of a new building plot.

Most of the other participants did not consider it so successful. Most of the business occupants, squatters and residents of the project site who were relocated could claim that they were not fully compensated for their relocation costs and did not share in the project benefits.

Building development after the project The 126 building plots were transferred to the landowners in August 1976. They included the 20 plots with existing buildings. (The 12 cost recovery plots were sold in July 1977 and January 1978.) During the next 12 years buildings were constructed on 80 of the 118 redevelopment plots. Therefore, the other 38 plots ,which occupy 19 per cent of the total plot area, were still undeveloped at mid-1988. Most of the new buildings were nine- to 12-storey buildings with elevators or five-storey walk-up buildings, with shops and offices on the ground and lower floors and condominium apartments on the upper floors. Shophouses were built on some of the smaller plots.

Limitations of the Redevelopment Land Projects Using LP/R

As redevelopment land projects are simpler than land and building redevelopment projects, they facilitate the use of LP/R for landowner participation. However, they have limitations. The Tenth Stage LP/R Project demonstrated these limitations.

1) *Small plots*: when the project site contains small plots, there is need for a mechanism such as 'collective consolidation' to ensure that the new plots will be large enough for the construction of space-efficient multistorey buildings. However, some/many of the landowners may insist on receiving individual building plots and may not have the funds to pay for additional land above their entitlement. The project reduced the original 403 land parcels to 138 plots plus roads by consolidating multiple land holdings and by 'collective consolidations', and many small plots were produced, including eleven plots smaller than the regulatory minimum plot area of 38.5 m^2.

2) *Slow building development*: as most of the landowners are investors and land users rather than multistorey building developers, the building development of the new plots is likely to be slow. These landowners can either sell their plot to a developer, or become a building developer or join with a developer for a joint project. This third option, a joint building project, is a real possibility in Taiwan as developers without land often make a 'co-construction' (cooperative construction) agreement with a

landowner to construct and finance a multistorey building with ownership of the plot and building being divided between them.[5] However, this shared ownership is not acceptable to many landowners who prefer to own their own shop-house type of buildings.

3) *Building blockages*: building development of some of the new plots can be delayed for many years due to speculative withholding of the plots, or due to development difficulties such as the inability of a group of 'collectively consolidated' landowners to agree on the building development of their plot. At mid-1988, 12 years after completion of the project, some 38 of the 138 plots on the project site (19 per cent of the area) were still without buildings.

4) *Landowner dissatisfaction*: some of the landowners will see the project as disadvantageous to them and are therefore likely to oppose it. They are mainly the landowners whose buildings are retained and who have to contribute land or cash to the project, the landowners who do not want disruption to their business activities and/or their housing, as well as the owners of small plots who cannot afford to contribute land or cash to the project. These are the legitimate grounds for objection. If there are numerous cases of objection, they may outweigh the benefits of the project and its supporters.

5) *Need for rehousing*: the projects cannot reinstate the tenant-occupants in housing and business premises on the project site as new building development takes place after the project. Many of these occupants will therefore probably oppose the project. The project should provide some financial assistance for their relocation costs (or compensation if the local landlord and tenant law makes this necessary), with this expenditure to be included as a project cost to be recovered from the land value increase.

These limitations of redevelopment land projects indicate the desirability of measures to support and improve these projects. The implementing agency should be authorised and funded to buy and consolidate the small plots of those landowners within the project site who are willing to sell their plots. The agency should also be able to purchase the plots and buildings of the landowners who prefer to sell rather than join the project and who cannot find a private buyer. This would mean that the implementing agency could implement projects by a combination of LP/R and negotiated land purchase.

The implementing agency should also undertake programme of measures to promote and assist early building development of the new plots after the project. This should include assisting the collectively consolidated landowners to reach agreement on the building development of their plots and to arrange co-construction projects with building developers for their plots. This assistance could begin with the preparation of an information booklet for landowners which explains the nature and principles of co-construction projects and provides formulas to guide the sharing of the building space.

The likelihood of both landowner and occupant opposition to redevelopment land projects indicates the need for careful project preparation. This begins with the selection and delineation of the project site so as to ensure that the benefits of each project will exceed its costs, both financial and social ones. When the projects do not require majority landowner support, as in the case of government LP/R projects in Japan, there is need for a full assessment of each proposed project with clear criteria for approving them for implementation.

Urban Redevelopment Using LP/R in Japan

Japan is the only Asia-Pacific country regularly using LP/R for urban redevelopment. A large amount of redevelopment has been carried out, using two forms of LP/R. This redevelopment has been carried out by government authorities and agencies, and landowner cooperative associations.

Redevelopment Using Two Forms of LP/R

The first form of LP/R is called land readjustment (LR). It is used for the two-dimensional redevelopment land projects in which an existing urban area is cleared and converted to a new layout of streets, infrastructure and building plots, some of which are sold for project cost recovery and the others are transferred to the landowners. The second form of LP/R is termed 'rights conversion' (RC). It is used for land and building redevelopment projects in which an existing urban area is cleared, converted and rebuilt to a new layout of streets, infrastructure, plots and buildings. In these three-dimensional projects the landowners and occupants that have chosen to remain in the area are allocated the ownership or tenancy of units in the new buildings in accordance with the project RC plan, and the rest of the floorspace in the buildings is sold for project cost recovery. The other landowners and occupants who choose to leave the area are paid compensation for their property and/or given rehousing assistance.

The RC technique is complicated and time-consuming. In order to expedite the redevelopment of larger areas for priority purposes, a government agency is authorised to purchase all the land and buildings in the project area, but it is obliged to sell or lease units in the new buildings to those owners and occupants that choose to return to the area.

There are a number of reasons why Japan has made extensive use of LP/R (including RC) for landowner participation in urban redevelopment projects whereas other Asia-Pacific countries have made very little use of it in redevelopment. The main reasons are as follows:

1) the successful use of LP/R in urban-fringe development projects over a long period, mainly under the *City Planning Act, 1919* and then under *the Land Readjustment Act, 1954*. This provided an institutional framework and skilled manpower resources for its extension into redevelopment projects, first into the redevelopment land projects and later into the land and building redevelopment projects;

2) appropriate and successful use of LP/R in redevelopment land projects to replan, reconstruct and replot of the earthquake-damaged urban areas, and the fire-damaged urban areas from World War II;

3) existence of other urban areas that were suitable for redevelopment land projects by LP/R (combined with some government land purchase), as they were areas with mainly single-storey timber buildings and obsolete street layouts;

4) Japanese cultural preference for negotiation and consensus rather than compulsion, and the landowners' awareness of the LP/R technique and willingness to accept the preparation and implementation of LP/R projects over long time periods;

5) strong landowner and occupant opposition to government land assembly by negotiated and compulsory purchase for redevelopment projects, as was authorised by the 1961 urban redevelopment law which was later repealed;

6) provision of government contributions and subsidies to redevelopment

projects using LP/R, including some subsidies to the private projects by landowner cooperative associations.

Features of Japanese LP/R

Urban LP/R has been used in Japan for most of this century and the technique has been modified over the years, on the basis of the country's urbanisation experience and needs. The LP/R concept was introduced to Japan from Germany. The early urban projects were based on the experience gained with rural LP/R projects. The urban projects were carried out under the *Agricultural Land Readjustment Act, 1909*. Although the later *City Planning Act, 1919* authorised the urban LP/R projects after 1919, some continued to follow the procedures of the 1909 law. This continued until 1954 when *the Land Readjustment Act* was passed.

Most LP/R projects are carried out under this 1954 law with later amendments, except the LP/R projects for land and building redevelopment. They are carried out under the *Urban Redevelopment Act, 1969*, with its amendments in 1975 and 1980. Although Article 93 of the 1954 law authorised vertical replotting (also called stereo replotting and three-dimensional replotting), it did not set out the principles and procedures to be followed in this replotting, so that land and building redevelopment was not carried out under this law. The 1969 law for redevelopment authorised the 'rights conversion' form of LP/R for land and building redevelopment.

The 1954 land readjustment law provides five categories of the implementing agency. They are one or a few landowners (including land lessees), landowner cooperative associations, local governments, central government agencies, and government development corporations. As shown in the Table 14.1 statistics, most LP/R (by area) is carried out by local governments, of which the Nagoya City Government is one of the most active and progressive users of LP/R. The landowner cooperative associations are the second main implementing agency (by area). These associations can be formed by a group of seven or more landowners and lessees. Their project has to be supported by a majority who own and/or lease two thirds or more of the project area. These landowner association projects are mainly for new urban development but some are redevelopment projects, usually for retail and business centres. The Housing and Urban Development Corporation (HUDC) is the largest implementing agency, and it uses LP/R for a wide range of projects.

LP/R projects are undertaken for all types of urban development and they are usually grouped into four broad categories. The first is the redevelopment projects for earthquake and fire-damaged areas, for obsolete and deteriorated areas, and for the creation of new commercial areas. The second category is redevelopment undertaken in conjunction with road and railway line construction projects that have been undertaken in established urban areas. The third one is projects for new urban development by the conversion of rural land to urban uses. The fourth category covers the combined LP/R and land purchase projects to develop new towns, mainly by the HUDC.

Some of the main features of the LP/R under the land readjustment law can be noted, as follows:

1)   the landowners who can participate in the projects are the owners and lessees of land in the project site, but not the lessees and tenants of buildings;

2)   all the projects undertaken by government agencies are designated as city planning projects and have to be approved as such. LP/R is seen as the main technique for the positive implementation of city plans (as an alternative to regulatory controls) and is described as 'the mother of city planning';

3)   as the government LP/R projects are undertaken for public purposes, there is no formal requirement for majority landowner support for proposed projects. Landowner association projects require 66 per cent landowner and lessee support, by both number and land area;

4)   in the government LP/R projects, the landowners are represented by a landowners' council of ten to fifty members for each project who are mainly elected by the landowners and lessees;

5)   a panel of three or more evaluators ('valuers') is appointed to evaluate the lands and leaseholds in each project for the replotting planning, at an advisory capacity to the implementing agency;

6)   each project contributes the land required for public facilities such as roads, drains/canals and open spaces, but the land for social facilities such as schools has to be purchased from the project or covered by government land in the project;

Table 14.1   Urban land pooling/readjustment projects in Japan, 1919–91

Law applied and implementing body	Projects undertaken		Projects finished		Projects in progress at end of 1991 fiscal year	
	No. of projects	Total land area (ha.)	No. of projects	Total land area (ha.)	No. of projects	Total and (average) land area (ha.)
Former city planning law	1,183	49,101.0	1,183	49,101.0	–	–
Implementation by individual(s)	1,108	19,429.0	1,035	17,862.3	73	1,566.7 (21.5)
Implementation by l/r cooperative	4,256	97,291.3	3,356	69,925.8	900	27,365.5 (30.4)
Implementation by local government	2,219	14,596.9	1,517	79,712.3	702	34,884.6 (49.7)
Implementation by administrative agency	320	33,881.0	307	31,434.8	13	2,446.2 (188.2)
Implementation by Housing and Urban Development Corporation	136	18,564.6	81	10,541.6	55	8,023.0 (145.9)
Implementation by Regional Development Corporation	2	573.9	–	–	2	573.9 (286.9)
Local Housing Corporation	4	222.3	2	175.2	2	47.1 (23.5)
Grand total	8,045	284,559.0	6,298	209,652.0	1,747	74,907.0
Total	9,228	333,660.0	7,481	258,753.0	1,747	74,907.0

(Rows indicated by side label: Land Readjustment Law)

Source: Hasegawa, 1991.

7)   central government subsidies and other contributions are provided for government projects, and interest-free loans are available to finance the landowner association projects. A case study of a government redevelopment project showed that only 21.6 per cent of the project cost was recovered by the revenue from the sale of plots (City Planning Bureau, 1982);

8)   the calculation of each landowner's and lessee's plot entitlement is usually based on the estimated values of the original land and the new plots. However, this valuation for replotting is not estimated as a money value but as an index number of land utility. K. Hayashi (1978, p. 36). noted that '... The evaluation of the land is regarded as "top secret" and is never shown to the public. Otherwise, small differences of figures will bring up meaningless disputes'.

The planning of a government LP/R project is done in three stages. The first stage is the preparation of the layout plan together with a feasibility study, which is used to authorise the project as a city planning project. The second stage involves the preparation of the project plan (mainly the engineering designs, the implementation programme, cost estimates, and financial plan) which, when approved, authorises the implementation of the project. This implementation begins with the election of the landowners' council and the appointment of the advisory evaluators. The third or final stage is the preparation of the replotting plan. It is usually first prepared as a provisional replotting plan which is revised and finalised after the project infrastructure construction works are completed and the project costs are known. The provisional replotting plan enables the landowners to relocate to their new plots and continue their land-use in advance of project completion, if they wish.

The LP/R projects take a long time to prepare and implement. K. Hayashi (1978, p. 364) has commented that '... If a typical size of land readjustment (project) could be finalised within ten years from planning, it can be said to be a "successful project" in Japan'. This is for a new development project, whereas a redevelopment project could take a 'tremendous time'.

Land and Building Redevelopment Using LP/R

Although Article 93 of the *Land Readjustment Law, 1954* authorised vertical replotting for land and building redevelopment projects, this authorisation was not used as the principle of land evaluation. Moreover, replotting principles devised for land surface areas could not be applied to multistorey buildings. A new method had to be devised for converting each landowner's share of the project site into a fair share of the plots and buildings produced by the project. The 1961 urban redevelopment law (the *Act Concerning the Redevelopment of Urban Area in Association with Development of Public Facilities*) avoided the issue by authorising full government acquisition of the project site, by negotiated and compulsory purchase. However, there was strong opposition to these redevelopment projects from the landowners and from the business and household occupants on the project sites. The following redevelopment law responded to this experience by providing for both landowner and occupant participation in the projects.

The *Urban Redevelopment Act, 1969* introduced and authorised the 'rights conversion' (RC) method for calculating the landowners' and other participants' shares of the new plots and buildings in the three-dimensional redevelopment projects. The RC concept is illustrated in Figure 14.3. The law provided a three-stage project planning process similar to that under the land readjustment law, except that the third plan to be prepared would be an RC plan for the project instead of a replotting plan. The RC plan covers all the landowners and occupants in the project site as property right holders. It shows how the landowners and occupants who choose to participate in the project would be allocated rights in the new plots and buildings, covering both their space allocation and their legal tenure. The other floors of the buildings would be sold as apartment units and as floorspace in order to recover project costs. After the RC plan is approved, the new plots and buildings are transferred to all the right-holders according to the plan at the same time. This transfer is by the exchange of all the rights in the new plots and buildings for all the rights in the old plots and building. Therefore the RC plan has to be comprehensive and shows the exact entitlements.

The preparation of these RC plans which covers all the right-holders and their exact entitlements for a single comprehensive exchange of rights was very complicated and time-consuming, particularly for larger projects. Hence, the preparation procedure of RC plans is too slow for urgent projects. The redevelopment law was amended first in 1975 and again in 1980, in order to authorise a new system for the larger and urgent redevelopment projects so that government authorities could undertake these projects more expeditiously. These projects are termed 'category 2 redevelopment projects'. In these projects, the government authority purchases all the land and buildings in the project site but is obliged to accommodate those owners and occupants who wish to return.

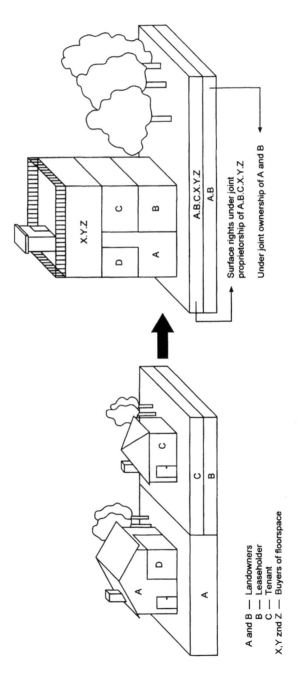

A and B — Landowners
B — Leaseholder
C — Tenant
X,Y znd Z — Buyers of floorspace

Figure 14.3   Rights conversion in a land and building redevelopment project

The purpose of the *Urban Redevelopment Act* is to authorise and promote the land and building redevelopment of selected urban areas towards the objectives of providing a favourable urban environment, preventing fire and earthquake disaster, providing housing in central areas, and modernising and restructuring the urban infrastructure. Redevelopment for these purposes is not seen as a profit-making activity. Subsidies are provided by the national government and contributions are made by the government implementing agencies. For example, the national government subsidies cover one-third of the costs of project planning and design, land development, and the construction of social and safety facilities, and up to two-thirds for some other cost items. The law authorises redevelopment projects by one or a few landowners and by landowner cooperative associations (known as urban redevelopment associations). There are many of these landowner association projects, which are mainly undertaken for profit and are usually smaller projects for shopping and commercial centres. Government loans at nil interest and some other subsidies are available to these landowner projects.

By April 1986, 289 projects for 528.6 ha. had been undertaken under the 1969 redevelopment law, of which 167 projects for 212.6 ha. had been completed. These 289 projects were undertaken as government and landowner projects, as follows (City Bureau, 1990):

- local government   91 projects, 366.1 ha.   (4.02 ha. av.);
- HUDC   8 projects, 10.6 ha.   (1.33 ha. av.);
- local housing corporation   4 projects, 5.2 ha.   (1.30 ha. av.);
- landowner association   136 projects, 128.6 ha.   (0.95 ha. av.);
- other landowners   50 projects, 16.1 ha.   (0.32 ha. av.).s

The 1969 law for urban redevelopment introduced the RC technique as an adaptation of land readjustment for land and building redevelopment projects. It was introduced in response to the opposition to the land and building redevelopment projects by government purchase of the project land that had been authorised by the 1961 law for urban redevelopment. This opposition had come from the landowners and occupants who wished to remain in the redevelopment area, and from the landowners who wished to share in the profits of some of the redevelopment projects. However, the RC form of LP/R was soon found to be too complicated and too slow for the larger and urgent government redevelopment projects. As a consequence, the 1969 law was amended in 1975 and 1980 to authorise the partial return to the government land purchase approach. The amendments authorised some government agencies to undertake some land and building redevelopment projects by purchasing all the project site, but with the obligation to accommodate the owners and occupants that elected to remain in the project area. These land purchase projects were designated 'category 2 redevelopment projects', with the projects using the RC technique designated as 'category 1 redevelopment projects'.

Most of the land and building redevelopment projects are carried out as 'category 1 projects' because the redevelopment law restricts the use of the land purchase approach. The 'category 2 projects' can only be undertaken by local governments and government development corporations, and the project site has to be larger than 1.0 ha. These projects should redevelop areas for priority purposes such as to reduce fire disaster hazard and to install needed public facilities; or the project site should be located at areas designated for redevelopment promotion. In these projects, the implementing agency purchases all the project site but the owners and occupants have a re-entry option by the right to purchase or lease units or floorspace in the new buildings. The compensation, which they are entitled to, is retained as a credit for the purchase or lease of their new unit or floorspace. This land purchase and re-entry approach expedites project implementation because it is not necessary to prepare a single RC plan that covers every right and entitlement in the project as a whole. These 'category 2 projects' cannot be classified as LP/R projects but they do allow a partial form of landowner participation.

Potential for Urban Redevelopment With LP/R

The first part of this paper outlined the potential benefits that LP/R can provide in the conversion of urban-fringe land for new urban development. It also noted the features and complexities of urban redevelopment that make it more difficult to use LP/R for redevelopment, particularly for the larger multi-block redevelopment projects. In spite of these difficulties, there are potential benefits in using LP/R for landowner participation in redevelopment projects, and they are:

1) giving landowners the opportunity to return to the project location and/ or to share in the profits/benefits (including the costs and risks) of the redevelopment of their land;

2) facilitating land assembly and reducing project delay by avoiding or

reducing possible landowner opposition to the project; and

3) reducing the amount of funding required for the project, and spreading the project risk.

Japan is the only Asia-Pacific country using LP/R for urban redevelopment and has made wide use of it for redevelopment land projects since the Kanto Earthquake in 1923, and for land and building redevelopment projects since 1969. This redevelopment has been successful but this Japanese experience does not necessarily indicate that LP/R can be adopted by other countries to improve the efficiency, economy and equity of their urban redevelopment projects. This cautious assessment is supported by the fact that the first redevelopment law in 1961 focused on government land purchase as the government's preferred method of land assembly. Even though landowner opposition to government land purchase forced its abandonment through the 1969 redevelopment law that introduced the rights conversion form of LP/R, this law was soon amended to re-authorise the land purchase approach. This cautious assessment is also supported by the Taiwan experience, such as the Kaohsiung City Government's decision to abandon the use of LP/R for urban redevelopment in favour of the land purchase approach after the completion of its Tenth Stage Project in 1976.

The use of LP/R is most appropriate for redevelopment land projects in which the landowners are returned reshaped plots for their building development after the project. This type of project is most suitable for the redevelopment of urban areas devastated by fire and natural disaster, as the buildings have been destroyed and will be replaced by the landowners. It would usually be preferable to use LP/R to replan and resubdivide these devastated areas instead of reinstating the old obsolete layouts of streets and plots. However, if the area is to be redeveloped with multistorey buildings and many of the original plots are small, then it will be desirable to promote their consolidation so as to provide larger plots for space-efficient buildings. The implementing agency can do this by encouraging private consolidation activity, by its own purchase of small plots, and by adopting a form of plot co-ownership that is acceptable to the landowners. The agency should also encourage and assist the landowners to undertake or facilitate the early building development of their new plots after the project.

In the case of the land and building redevelopment projects, the 'category 2' redevelopment technique based on government land purchase, with a re-entry option for the landowners and occupants, appears to be more efficient and economical than the rights conversion form of LP/R. However, there is need to obtain more information on the experience with these 'category 2 projects' so as to make a proper assessment of their merits and to identify their lessons for the improved management of land and building redevelopment projects. The re-entry option caters for the landowners and occupants who wish to be reinstated within the project. In the case of landowners who are mainly interested in sharing in the project profits, they might be better catered for by forming a landowners' financial syndicate for the project which they could join.

There is also need for further study of the Japanese experience with the landowner association redevelopment projects as they could provide a transferable model for private redevelopment projects, or lessons to guide the organisation and management of landowner cooperative projects in other countries.

Notes

1 This paper forms part of the author's continuing study of the land pooling/readjustment technique for managing and financing urban land development in Asia-Pacific countries. Also see Archer, 1992, 1989, 1985 and 1980.

2 It should be noted that single block redevelopment projects (i.e., redevelopment projects within a single city block) are simpler and easier to implement than multi-block redevelopment projects. The multi-block redevelopment projects usually include the replanning and reconstruction of the street and public utility networks so as to prepare the project site for higher density building development and more intensive land use.

3 The leasehold land system provides another alternative. When the urban land is under a public land and leasehold system, such as in Hong Kong, Delhi and Canberra, then land assembly for multi-block redevelopment can be facilitated by coordinating the ground lease periods of adjoining plots to a common expiry date. This can minimise both the social and the financial costs of land assembly for redevelopment projects (Archer, 1974).

4 This financial 'surplus' obtained from the project by the City Government was less than the market value of its share (55.83 per cent) of the 3,929 m² of land that it contributed to the project.

5 The sharing of the building space between the landowner and the building developer is usually determined in Taiwan using the following formula. The landowner's share of the total floorspace value equals the landowner's share of the project value equals the ratio of the net land value to the project value. The net land value is the estimated market value of the land minus the land value increment tax payable. The project value is the total of the net land value plus the building cost, including loan interest, construction tax, brokerage commission, sales tax, etc. (Lai, 1988).

References

Archer, R.W. (1974), 'The Leasehold System of Urban Development: Land Tenure, Decision-making and the Land Market in Urban Development and Land Use', *Regional Studies*, Vol. 8, No. 4, pp. 225–38.

Archer, R.W. (1980), 'Land Pooling by Local Government for Planned Urban Development in Perth', in W.A. Doebele (ed.) (1982), *Land Readjustment: A Different Approach to Financing Urbanisation*, Lexington, Massachusetts: D.C. Heath and Co., pp. 29–53.

Archer, R.W. (1985), *Bibliography P1794; A Bibliography on Land Pooling/ Readjustment/ Redistribution for Planned Urban Development in Asian-Pacific Countries*, Monticello, Illinois: Vance Bibliographies.

Archer, R.W. (1988), 'Land Pooling for Resubdivision and New Subdivision in Western Australia', *The American Journal of Economics and Sociology*, Vol. 47, No. 2, pp. 217–20.

Archer, R.W. (1989), 'Transferring the Urban Land Pooling/Readjustment Technique to the Developing Countries of Asia', *Third World Planning Review*, Vol. 11, No. 3, pp. 307–31.

Archer, R.W. (1992), 'Introducing the Urban Land Pooling/Readjustment Technique into Thailand', *Public Administration and Development*, Vol. 12, No. 2, pp. 155–74.

British Columbia Department of Housing (1976), *Replotting Assistance Programme*, Vancouver.

City Bureau (1990), *Textbook for the Training Course on Urban Development*, Tokyo: Ministry of Construction.

City Planning Bureau (1982), *Introduction to Land Readjustment (Kukaku Seiri) Practice*, Nagoya: Nagoya City Government.

Department of Land Administration (1985), *A Briefing on the Land Consolidation of the Kaohsiung Municipality*, Taiwan, Republic of China, Kaohsiung: Kaohsiung City Government.

Hasegawa, S. (1991), 'Introduction to Land Readjustment in Japan' in Department of Town and Country Planning (ed.) (1992), *Sixth International Seminar on Land Readjustment and Urban Development, 1991*, Bangkok: Department of Town and Country Planning.

Hayashi, K. (1978), *Japanese Urban Planning, Development and Land Readjustment*, Nagoya: Department of City Planning, Nagoya City Government.

Lai, Ying-Shy (1988), 'The Use of the Land Consolidation/ Readjustment Technique for Urban Redevelopment; A Case Study of the Tenth Stage Project in Kaohsiung, Taiwan', Masters degree dissertation, Asian Institute of Technology, Bangkok.

15 The Urban Renewal Projects in Japan: Nonresidential Projects

TAKATSUNE FUKAMI

Characteristics of Japanese Cities

From the viewpoint of the urban renewal projects, Japanese cities have some common characteristics related to their geography and urbanisation history.

It is noteworthy that no Japanese city had its own city walls. In the past, city area was not demarcated, within which the area might have been planned, nor did the Japanese citizens desire to make a city plan for such an area. Japanese cities grew by building one house after another around city nuclei, e.g. street junctions, temples, shrines, the castle, etc. As a result, a sprawling, built-up area has emerged.

Japan is generally calm in climate, so that we did not have to cover up completely our space by artificial structure. Therefore, Japanese cities did not need to prepare positive plans. Rather, the Japanese city could be understood as being a large rural settlement. Since horse and carriages were not used in Japanese cities before the Meiji Revolution (1868), roads were narrow and the crank shaped street pattern was formed for military reasons. In Japan, water resources had been rich and the water transportation and irrigation systems had been well developed. Rivers and moats were good defensive facilities for street fighting and the construction of bridges was limited. Most buildings were made of wood, as Japan retained ample forest resources. These were the characteristics of a typical Japanese city until the beginning of this century.

After the Meiji Revolution (1868), the Japanese government started to construct trunk roads, bridges, second bypasses and harbours for modern water transportation. They then built a railway network all over the country for the modern surface intercity transportation. The railway tracks were constructed in and through non-built-up areas and the railway stations were built at the edge of the existing cities. In large cities, tram facilities were constructed by the widening of roads. For most cities, the main purpose of constructing major road systems was to encourage industrial activity. Nevertheless, the built-up areas remained almost the same as they were in 1850. New land uses, such as factories and warehouses, were located near harbours or at places that were convenient to the railway service. In 1940, the Japanese population was estimated at 73 million; the number of cities was 125, which composed of 37.9 per cent of the national population. In the cities, the average population density was 31 persons/ha. It had a high density, in spite of the fact that the cities included agricultural land and forest. Hereafter, in this paper, the built-up area in the 1940s is referred to as the 'old built-up area'.

During World War II, the central parts of most cities and towns (215 cities and towns) were heavily damaged: 64,350 ha of the built-up area were burned down, with 2.7 million dwellings destroyed. Of these, 115 cities implemented reconstruction projects through the 'Land Re-adjustment System'[1] (LRS), and these projects produced urban areas characterised by small street blocks, parks and road system. Since 1955, urbanisation (the concentration of the population, industrial activities into the cities), the increase in the nuclear families and motorisation have accelerated the expansion of built-up areas. The new built-up area has been constructed through either unplanned sprawl, or the LRS. In the sprawling built-up area, the road system was very poor. It had narrow roads and they often led to dead ends. At the same time, new large industrial areas were planned outside these new built-up areas and at sea fronts where reclaimed lands with harbour facilities were available. Bypasses were constructed in the outskirts of the built-up area and the widening of major roads was initiated in most cities, with the removal of the tram facilities in some cities.

The model that was adaptable to most Japanese cities is shown in Figure 15.1. Of course, the cities which escaped damage by the war did not need to reconstruct their central areas through the LRS. The old built-up area was usually small. So indeed was the area damaged by war. However, the reconstructed area was even smaller than the war damaged area. At the first stage, the CBD was full of retail facilities, which replaced dwellings. In the next stage, motorisation resulted in the construction of bypasses, along which roadside facilities, such as gas stations, fast food restaurants, Pachinko game houses and car repair factories. As we entered into the highway age, the Japanese industrial society discovered that the coal mining industrial zone and the textile industrial zone required restructuring. By the 1960s, these industrial facilities were closed. Also, the factories located at the fringe of the old built-up area were closed or relocated to the new industrial area. These sites were to be reused for the new urban facilities, such as a shopping centre, a new city hall, a school, a music hall, a library, etc.

The notations in Figure 15.1 are as follows:

A–C: the old built-up area;
D–G: the new built-up area;
A: the LRS implemented area in the old built-up area. (A = B in the case where the war damaged area was not reconstructed);
B: the unimproved area within the old built-up area;
C: the industrial area located at the fringe of the old built-up area;
D: the postwar planned area for industry;
E: the postwar planned built-up area;
F: the postwar sprawled area;
G: the postwar harbour and the sea front industrial area with planned development;
H: the first bypass;
I: the second bypass;
J: the motorway or the highway;
K: the railway;
L: the agricultural land or forest;
N: the new shopping facilities.

The Need for Renewal in Japanese Cities

The basic project for urban renewal in Japan generally involved the construction of roads, often accompanied by the construction of the water supply system, the sewage system, the energy (electric and gas) supply system, and the transformation would normally be carried out by an owner or dweller. The construction of roads, however, involved the removal of some houses, suggesting that road construction is more difficult than the rebuilding. The Japanese urban renewal projects were implemented so as to use surplus floor space, with the construction of roads, the transformation of old dwellings into fire proof structures, and the introduction of new urban facilities. The characteristics of the Japanese urban renewal projects in each of the areas shown in Figure 15.1 are as follows.

1)   This is the main shopping street in the old built-up area where the

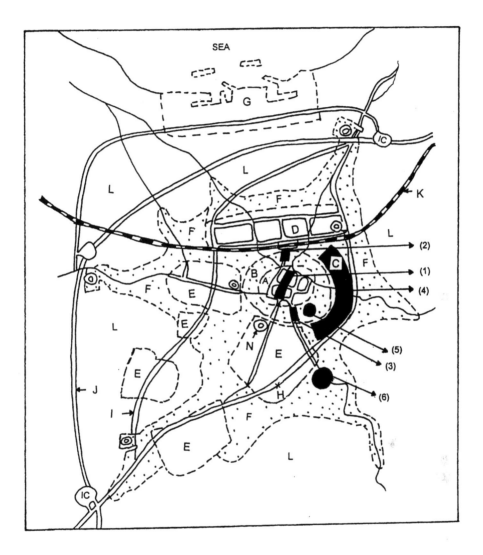

Figure 15.1   The Japanese city structure

transformation into fire proof structures was implemented in the first instance. The first government-aided project was renamed the 'Transformation into Fire Proof Structures Project'. This construction changed 1- or 2-storey wooden shops into 3- to 5- storey fire proof structures with residential use on the upper floors. Moreover, the facade of shopping street was modernised. The government subsidised the difference in the building costs between wooden and fire proof structures. Next, the Shopping Street Modernisation Project was applied to the main shopping street. This is basically a financing project; a special public organisation called the Small Business Support Organisation, supplied the necessary funds for the shopping street associations, authorised by the Shopping Street Association Act. The funds were allocated for building the arcade, organising a cooperative store such as a department store, paving the pedestrian zone, and furnishing the shopping mall. A part of the interest was paid by the government. In some districts, the Urban Redevelopment Projects were implemented, introducing large commercial facilities into the shopping street with the use of government grants and loans. Recently, the government offered a number of project systems and plan-making support systems (e.g. the General Improvement Scheme, the Commercial Complex Project, the Community Mart Project). These systems gave the incentive to revitalise the downtown area, the main shopping street or the central district of the community through financing, investment or deregulation, by legal or political arrangement. The difficulty of the project varied between the reconstructed area and the non-reconstructed area.

2) This area is close to the railway station. The urban renewal projects in this area consisted of the construction of the station square (for buses and taxis) and the wide trunk roads that connect the station with central city zones. There were three types of urban renewal project systems applicable to this area. The urban reshuffling type of the LRS, the Urban Remoulding Project and the Urban Redevelopment Project. All three systems were concerned with the construction of the large station square, wide trunk roads and the large buildings facing the square.

The urban reshuffling type of the LRS should be carried out using the following steps. First, the local authority purchases the sites from those owners who want to move out, by using the fund prepared to compensate the decrease in the total land value expected by the LRS. Then, the local authority carries out the project under the LRS. Finally, the site owners build large buildings that are appropriate in front of the main railway station (for example, large commercial buildings, office towers, or tall hotel buildings).

In the Urban Remoulding Project and the Urban Redevelopment Project, the conductor of the project rearranges the rights of the land and buildings in such a way that each right holder's asset value before the project implementation is equal to the values of the part of the building site and the part of the building floor which is to become his or hers after the implementation. The remainder, derived from subtracting the converted rights for the existing right holders from the total value of the new building, is sold in order to cover a part of the project costs. This means that these projects can only be carried out in the area where floor area ratio (FAR) designated by the city planning regulation is high enough. For example, the main railway station terminal area, the central part of the city or the area along wide trunk roads meets this condition.

3) This is the area that requires urban renewal projects in order to widen the trunk roads. Here, there are shops along the narrow road. When the project is carried out, the narrow road will be widened; however, the project is carried out through direct purchase. Consequently, the shopping function might be wiped out. Therefore, the project most appropriate for these areas is the one that can carry out both the road widening and the shop relocation at the same time. There are two project systems suitable for this purpose: the Urban Redevelopment Project System and the LRS roadside type (one version of the LRS). The former deals with both land and buildings, whereas the latter deals only with land. However, both are carried out only for the corridor area along with the road, and the area behind it remains unimproved.

4) This area often includes vacant sites where factories, warehouses and company houses were once located. There are three types of redevelopment projects for this area.

The first type is very simple: the vacant site is reused as a site for residential buildings, a shopping centre, office buildings or a theme park by the developer who purchased this site. Of course, the new land use is decided by the site conditions, such as its location, size, land prices, etc. The former site owner moves to a new location where he builds a new factory or warehouse with innovative machines or systems on the new site and with a size which is generally larger than the former one. One interesting case can be found in Kawasaki City near Tokyo. There, an industrial incubator was developed on a single 0.5 ha. site that was once occupied by a factory. The developer was in a typical public-private partnership organisation that consisted of a construction company, some private companies and the municipality.

The second type is the renewal project involving an improvement of the area surrounding the site. This is desirable, but persuading the surrounding residents and property owners is always very difficult. Therefore, this case is not popular and the public sector initiatives are important and necessary. Most of the actual cases are carried out by the public agency itself or the public sector involving private developers. In this case, and also in the one mentioned above, the owner who sold his site renewed his factory with innovative facilities at the new location. The renewal area changes from industrial to residential, commercial or business use, according to market and site conditions, such as the city size and the site location.

The final type is the project in which the site owner himself changes the usage of his own site from an industrial use to a more suitable use for its location, such as a tall office tower or an amusement park. Sometimes the owner cooperates with private developers. This type of development has been increasing recently. There are cases, though not many, where conservation or rehabilitation has been attempted. For example, a factory building made by brick was changed to a museum and restaurant in Kurashiki, near Hiroshima.

5) Most of this area is used for residential purposes, and therefore, the physical environment needs to be improved. Old houses should be renewed, roads should be widened, small parks for children should be built, and car parking lots should be constructed. But this area is located at a central part of the city so that high demands cause high land prices, high density and complicated property rights. For this reason, it is very difficult to make the land available for renewal projects. Thus, the public body is sought to enforce the project strongly, but so far only a few projects have been undertaken.

6) Although this area needs to be improved as a high amenity residential area, few projects have been carried out. This is simply because this area is geographically too large.

7) There are some projects in this area because the decline of the shipping

and seafront industries brings about an opportunity for the development of the water front.

As can be seen, it is true to say that the built-up area in Japan generally needs to improve its houses and living environment, but the fact is, there are few projects carried out except in areas (1) and (2).

Case Study: The Revitalisation of a Main Shopping Street

The Short History of the Shopping Street in the Core of City

The features of the revitalisation projects of the traditional main shopping street located at the central part of the city vary between the 1960s, 1970s and 1980s, corresponding to the situation of the central part of the city in each period.

In the 1960s, Japanese life became stable and Japanese household income and expenditure increased. Accordingly, shops changed from the confused wagon market-style to independent stores where shops could deal in sound goods. In the main shopping street particularly, many cooperative commercial buildings (quasi-department store style) were constructed. These fire proof buildings were constructed on the sites where wooden houses had been removed. The changes were supported by the low interest loan system that, initiated in 1963, was to help building the cooperative commercial buildings by the shopping street associations. This system was able to support only the small shop organisations that satisfied the criteria written in the Retail Promotion Act. These organisations consisted of smaller shops in the commercial district or shopping street (e.g. shopping cooperative associations and shopping street associations). They could apply to the prefectural government and the Small Business Support Foundation, provided that the members of the organisation decided to implement shopping street promotion projects, such as, for example, a cooperative shop, lighting, arcade, advertisement control, coloured paving of a pedestrian path, street furniture or mall. The Small Business Support Foundation consulted with the prefectural government and decided on the amount of loan, the level of interest rate (including zero) and repayment conditions, including unredeemed terms according to the contents of the projects. The prefecture paid to the Small Business Support Foundation the difference between the general interest rate and the actual interest rate. This system is called the Shopping Street Modernisation Project. Also, the Ministry of Construction assisted the projects which rebuilt fire proof structures. In the 1960s, the shopping street had a row of the RC shops and the arcade, and was very busy as household income increased.

In the 1970s, supermarkets and chain stores were spread as new types of commercial industry. The city area enlarged and shopping centres were built along bypasses as the number of car-owners grew. Also, the Urban Renewal Act was enacted in 1969 and the main railway station squares were constructed with department stores and large commercial buildings. The city hall, civic halls, the police head office and other public buildings, which had been located in the central part of the city, moved to the outskirts of the city so that these facilities could enlarge their size. As a result of this relocation, the main shopping street in the core of the city faced new competitors, such as a department store in the station district and shopping centres in the fringe zone. Consequently, customers were lost because the public buildings had moved out. Since the competition between large stores and shopping streets was very severe, the government introduced a new act: the Large Store Act. The Large Store Act restrained the shopping centre development in the outskirts of the city, but the commercial facilities in the station districts and the New Town centres developed by public agencies were not controlled.

In this period, the purpose of revitalisation projects was mainly to construct cooperative buildings and arcades, although there were cases where the revitalisation depended upon department stores or large supermarkets. Shopping streets in the core of the city still prospered because income grew and the strong demand continued.

In the 1980s, the period of rapid growth came to an end. The economy became stable and so did population and income. The commercial capital started to reform the distribution system in order to reduce the cost and to target on segmented consumers for restructuring the sales. Speciality shops in the shopping street started to move out to the suburbs, because in the suburbs, they could acquire a large and cheap site and arrange large car parks. As a result, they could set low prices in order to face price competition. The vitality of the shopping street rapidly ceased. In 1985, the government started to deregulate various controls, attempting to depend more on the power of private activities. One of such policy changes appeared as the Private Power Promotion Act. However, the energetic private investment power overheated the Japanese economy and money games, such as the real estate investment and the stock investment spread all over the country. In the shopping street, many shop owners quit the management of the sales and let their space to national chain stores. Therefore, the retail industry itself did not change but the industry that gained the most momentum was the real estate industry.

In this period, the revitalisation projects of the shopping street included constructing car parks and shopping malls. Nevertheless, the shopping streets at the core of the city were undoubtedly affected. Office tower development became very popular in large cities, whereas in local cities suburban shopping centre developments boomed. In 1991, the economic recession began, and the efforts to restructure private enterprises were also affected seriously. The decline of the main shopping street located at the central part of the city accelerated and its revitalisation became more difficult. The large sites, which were left after the medium-sized stores moved to the suburbs, are still vacant. It is evident that the central part of the city now faces serious problems.

The Case of Toyama City

Let us examine the revitalisation of the core of a typical local city. Toyama is one of the war-damaged cities where the reconstruction projects were carried out. It contains 300,000 population. In the station zone there are hotels and an office tower constructed by urban renewal projects; in the suburbs, numbers of commercial facilities have been located along the two bypasses.

In the central part of Toyama, the following can be found:

1) a site developed by an urban renewal project in 1977;

2) a main shopping street covered by the arcade;

3) vacated site since a large shop moved out;

4) a new civic hall on a former public facility site;

5) a pedestrian precinct with street furniture and high quality pavements;

6) a multistorey car park that was developed as a cooperative building by the street association;

7) the prefectural government head office, a tower building on the site where the castle was once located; and

8) part of the old castle, being used as the central park.

The main shopping street, called the central street, is a traditional shopping district and it was prosperous until the 1960s. In the 1970s, its prosperity had been retained due to the construction of large shops and arcades. In February 1972 a fire broke out and the landowners decided to carry out an urban renewal project, but some owners disputed the rights of properties. In 1975, the owners excluding those who opposed the project established an Urban Renewal Association, and in 1977, a department store was constructed. This association consisted of 19 members and completed a building that has 20 storeys on the 3,500m site, with a standard floor space of 2,400 m. The total cost of this project was 2.745 million yen and 13.3 per cent of the total cost (0.366 million) was subsidised.

After this development the new civic hall was constructed as a cultural facility, and streets were modernised to become good pedestrian zones. These improvements inevitably enhanced the image of the central place. Nevertheless, the impacts of the commercial facilities in suburbs and the station zone were so strong that after all, a big shop was forced to close down. This site is still vacant. At the time of writing, another revitalisation plan is being considered. The introduction of a hotel and a general merchandise store poses a problem, because the access to the central part of the city is limited and therefore there will not be enough traffic capacity. Another problem is that owners and managers in the city core neither care to nor are they capable of providing attractive services and facilities.

Case Study: The Redevelopment of an Industrial Site in the Inner Area

There are three patterns of renewal projects on industrial sites within the inner area, as we have mentioned before. Industrial sites usually include railway sites and harbour lands. The location of these sites became very convenient as the city area was enlarged. Also, these sites are arge by themselves.

Two typical and famous cases are the Minato-Mirai 21 (MM21) project in Yokohama and the Ookawabata project in Tokyo. The MM21 was carried out on 186 ha. land, where there had been a large factory, a large railway yard and harbour facilities, including warehouses. The plan supplies office towers, residential towers, a museum, convention facilities, and a tall hotel. At this moment, the Inter-Continental Hotel, the Yokohama Museum, the convention hall and several office towers, including the tallest building in Japan (about 300 m high) called the Landmark Tower, have been completed. The Ookawabata project is a development along the River of Sumida, on the sites where factories and warehouses were once located. The project includes redevelopment of surrounding small sites. Here, however, we shall focus on a smaller renewal project which challenges the improvement of a built up area.

The Case of Kamiya District in Kita-Ward in Tokyo

Kita-Ward, located along the River of Sumida which has been the major water transportation for industry, was one of the industrial concentrations in Tokyo. Due to the immense number of immigrants to Tokyo, factories in this area were replaced by public housing facilities. Consequently, this environment became less suitable for industrial activities.

In 1981, a metal factory on 2.8 ha. site in Kamiya district moved out, and subsequently, this site has been bought by the Japan Housing and Urban Development Corporation. The Corporation started to prepare a plan for housing, but the physical shape of the site was found to be unsuitable for high-rise buildings because an access road from the trunk road could not be arranged. The Corporation and Kita-ward officials organised a committee for the Kamiya district redevelopment, and the committee studied a renewal plan including the surrounding small sites. This block is surrounded by the river of Sumida, the canal that is now a subway depot in the underground and a park on the surface level, and two main trunk roads. The block has 5.5 ha.

Figure 15.2 shows the building structure before the project. There were only two roads whose width were more than 4 m. One road with 6 m width is used as the access road from the trunk road to the factory; the other is the main street in the district. As can be seen in Figure 15.2, almost all of the buildings were made from wood. The committee did not favour a housing project only on the factory site, but emphasised the renewal of a whole area, or at least, with the rest of the non-wooden factories. At that time, however, there was no institutional framework that could implement the committee's plan, with which the Corporation wanted to start the house construction. As a result, the project had been delayed.

The Corporation proposed to divide the metal factory site into three zones: the Corporation housing zone, the small factories apartment building zone, and the residential area improvement zone. The Corporation started to build its own housing complex in the Corporation housing zone, but a subsequent legal change allowed the Corporation to carry out a residential area improvement project. In 1984, meetings with the residents began, and finally in 1986 the project started. In the new plan, a new block had been created by a new road. A new building which was to accommodate small factories was

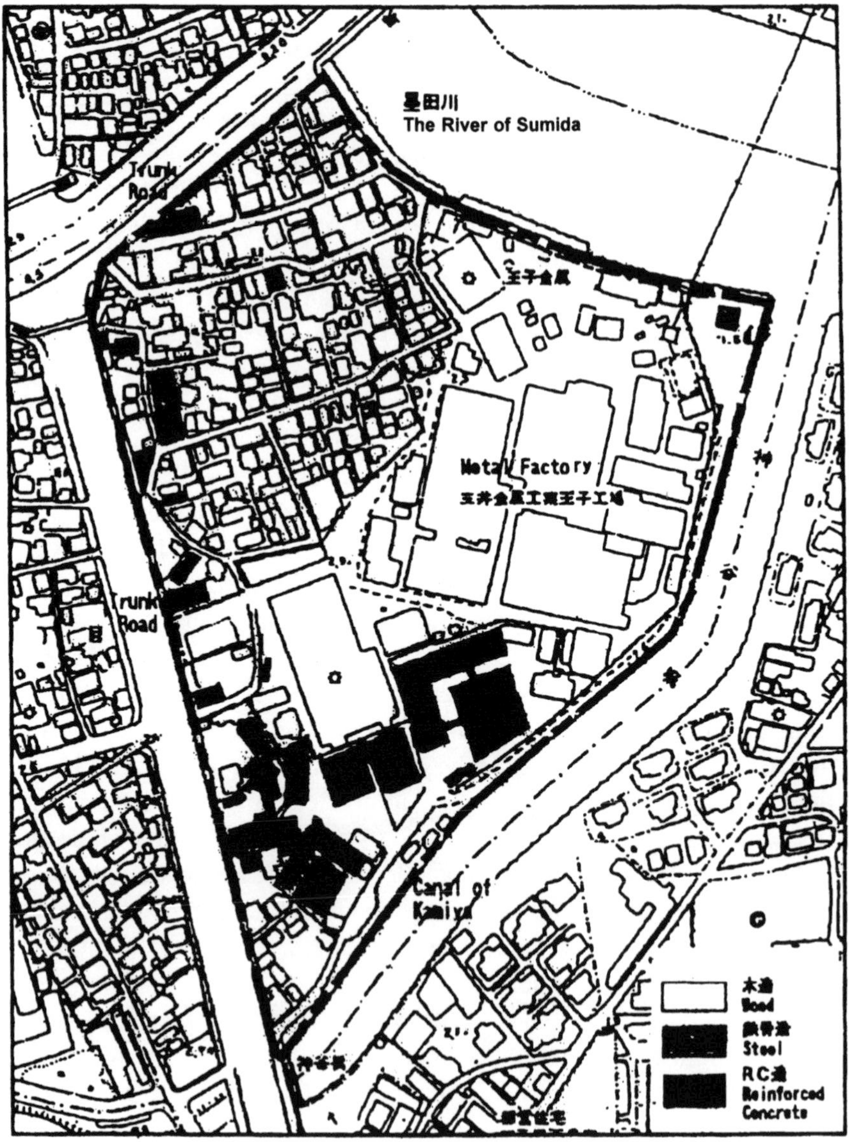

Figure 15.2   The condition of structure

constructed and the factory which once stood along the trunk road had been relocated to the industrial relocation site (Figure 15.3).

The space between the Corporation housing site and the existing residential area had been provided as the relocation sites and the Model housing for the residents who had to move. The existing access road was widened, new access roads and new small parks were created, the form of sites changed, old houses moved out, decent houses retained, and some tenants moved to Model housing.

The total cost of this project was 2.724 billion yen, including the cost of demolishing the old houses (412 million yen), roads and parks construction costs (2.288 million yen) and the Model housing cost (124 million yen). This figure does not include the cost of Corporation housing. The central government subsidised 50 per cent of the cost. Before the project, there were 167 decent houses and 179 old houses; the old houses were demolished and 275 new houses supplied. There are seven factories in two factory apartments.

Some Comments on the Japanese Urban Renewal System

The main idea of the Japanese urban renewal system is the rebuilding of wooden houses and the construction of roads and parks. For this purpose, the Japanese renewal system kept the existing residents and the rights-holders in the same district. In other words, the characteristics of the procedures of the Japanese renewal project are the conversion of the rights of the land and the buildings before and after the project (Figure 15.4). But the system can be realised only in the case where the regulated floor area ratio is sufficiently greater than the existing ratio, because the implementation cost would generally be paid by selling the surplus floor to new urban functions. The newcomers and old residents must cooperate and try to create a new image in the renewal area. The cases explained above satisfied these conditions. The fact is, however, that a large proportion of our built-up area is full of the so-called pencil buildings or high density small wooden buildings. For these areas, the difference between the regulated and the existing floor area ratios (i.e. the surplus floor area ratio) is not large enough. In the areas that have surplus FAR and are located closer to the core of the city, new floors can be obtained in a good location at low costs by the conversion system, provided that the land price is relatively high. But if we have to carry out the renewal project with a no surplus FAR area, we must pay all the costs ourselves: the clearance cost of the existing unfitted buildings and the cost of constructing new roads and parks are the major ones.

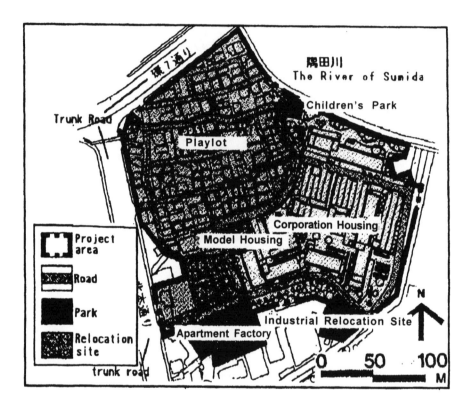

Figure 15.3 The district plan for renewal

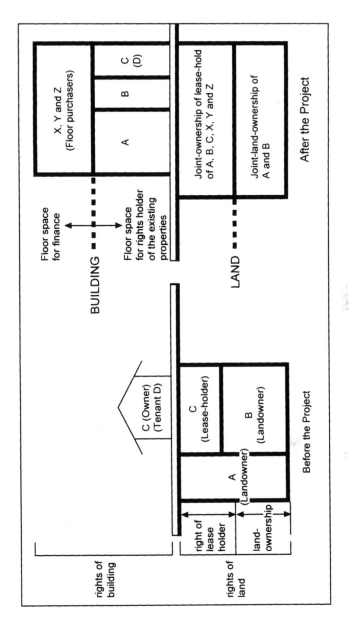

Figure 15.4 The rights conversion system in the urban renewal project

At present, Japan is a mature industrial society and the population growth has halted. The Japanese have to challenge the creation of new businesses and employment in the post-industrial society. The private firms face severe competition, so they have to reduce the cost. All organisations have to restructure their own systems. It is very important to note that the land prices in Japan still remain at a high level. The high land price forces new business to open in the suburbs where land cost is low. The core of the city has relatively lost the attraction. This tendency will cause the reform of urban pattern, which will need new transportation systems and a new life style. The urban form, which is characterised by the prosperous downtown, would only be effective in the industrial society. In the post-industrial society, we might have to find a new settlement form. In the course of this exploration, we must develop a new renewal system that will be able to change the existing built-up area in harmony with reformed settlement areas.

The aim of planning in Japan has been, and still is, to reform the urban physical conditions, but now citizens expect that city planning will create and keep employment as well as the decent living conditions. It follows the fact that the convention facilities, the incubator and the pleasant facilities, or places like amusement parks, are welcomed in every city because these are considered to be in accordance with the urban policy of the post-industrial society. Therefore, the system of urban renewal under the current planning system is forced to be modified. Is the financial support from the state (the subsidy and the loan) enough? Should direct investments be added to the menu of the state support for urban renewal project? Is a more flexible rights conversion system needed? Should there be a new organisation for the existing rights holders? These are issues we are discussing now. The public authority will have to manage the projects with more flexibility and power, just as the way the private company manages the financing, personnel, organising and planning.

Acknowledgments

Many people assisted the author during the preparation of this paper. I would like to express particular thanks to Mr Y. Motosumiyoshi of The Toshi Kikaku Koubou, who prepared the figures of the Kamiya project, and Mr H. Fujii of the Toshi Keikaku Doujin, who prepared the figures of the inner area study of Tokyo.

Note

1   *The Land Re-Adjustment System*: the purpose of the LRS is to achieve the basic urban environment and to promote effective land use through reorganisation of housing lots (sites) in coordination with the construction of public facilities.

There are several special features of the 'land readjustment' (rights conversion system of land, '*kanchi*' in Japanese) method. First, the project should be carried out without affecting the rights of existing landowners and lease holders. The second feature relates to the procedure through which the land for public use should be obtained. Before the project, a site has space 'sb' and land price 'pb', and after these will be 'sa' and 'pa'. Generally, 'sb'>'sa' and 'pb'<'pa' because the land space for public use should be generated or enlarged and the land price would rise through the project. The property value of land will change from 'sb'*'pb' to 'sa'*'pa'. If 'sb'*'pb' equal to or less than 'sa'*'pa', the site owner will not make a complaint. The growth rate of land price depends in the site conditions, such as the accessibility to public facilities, the width of the road and the environmental conditions, determined by the design of the site arrangement. A part of the difference in lot space between before and after the project should be used as public land and the rest of the difference would be sold to a third party in order to finance the project. In general, the reduction rate of lot space will be between about 20 per cent and 40 per cent, but sometimes it can go beyond this. In summary, the LRS produces land for public use and the finance for the project by reducing privately-owned land. The reshaping of lots and the construction of urban facilities such as roads and park will improve the environment of the area as a residential neighbourhood, thus enhancing the land value from 'pb' to pa'.

The Rights Conversion System in the Urban Renewal Project: as shown in Figure 15.4, the existing rights, such as land ownership, rights of land lessee, building ownership, rights of house lease and mortgages, should be converted into rights concerning the newly constructed building floor and their land through the renewal project. This is the essence of the Rights Conversion System. The property value should be kept equal between before and after the project and the surplus in the newly constructed building floor and the right of land concerning the floor would be sold to a third party to pay for the project. The principles and procedures of the rights conversion are similar to those of the Land Re-Adjustment System. In this system, the developer would not purchase the existing properties and the existing residents and owners do not have to move out from the project area. The existing community would not be destroyed and the new comers who purchase the surplus floor would obtain an incentive for the revitalisation of the area or the community.

References

Fukami, T. (1990), 'Mobility and the Attraction of Commercial Space, the Wheel Extended', *A Toyota Quarterly Review*, No. 74, pp. 3-9.

Fukami, T. (1991), 'The Shopping Street Revitalisation and the Urban Renewal', *The History of the Urban Renewal in Japan*, Tokyo: The Urban Renewal Association of Japan, pp. 139–45.

Ministry of Construction (1985), *City Planning in Japan*, Tokyo: Japan International Co-operation Agency and The City Bureau.

The City of Toyama (1993), *The Revitalisation Program for the CBD of Toyama*, Toyama: The City of Toyama.

The Department of City Planning (1990), *The Committee Report on the Comprehensive Renewal Project for the Mixed Use Area*, Tokyo: The Department of City Planning, Tokyo Prefecture.

The Department of City Planning (1991), *The Committee Report on the Comprehensive Renewal Project Model for the Residential-Industrial Mixed Area*, Tokyo: The Department of City Planning, Tokyo Metropolitan Government.

The Urban Renewal Association of Japan (1983), *The Renewal Plan for Kamiya*, Tokyo: The Urban Renewal Association of Japan.

16 Preserving Traditional Architecture: Noble Residential Area Development in Java

IKAPUTRA, KUNIHIRO NARUMI AND TAKAHIRO HISA

The Spatial Context of the Traditional Environment

Most of the previous studies of Javanese palaces have been concerned primarily with the social and political history, often discussing the way of life experienced within the context of the palaces. Any discussion of architecture has focused almost exclusively on the few high-style buildings of the palace. Moreover, preservation efforts have been biased toward 'important' buildings that are subjectively selected (Rapoport, 1991), such that preservation has acted to create museum pieces – no more than a reconstruction of the physical architecture of the past.

Javanese palaces were often selected as objects of preservation because they were traditionally the centres of culture and also because of the greatness and historic importance of their monumental architecture. However, it is important to understand that one cannot study the palaces as monumental/ historic buildings without also considering the context of their surrounding environments (the traditional urban centre), which originally functioned much as the present day's city centre: the focal point of urban life. It would also be incorrect to consider the future of the traditional urban centre outside the context of contemporary needs.

An understanding of the links between social phenomena and the palace environment (including the greatness of its architecture) is urgently needed. With the increasing development pressure, the palace environment is placed in a critical situation. The palace environment lies precisely in the domain where it could either be pulled back into the 'narrow-mindedness' of preservation as an historic monument, or undergo modern development: giving way to forces of economic rationalisation, relocating its residents to the periphery of the city, and taking on the appearance of an international city, or even demolishing the existing historic city structure and its identity. However, the dynamic life of these areas presents an opportunity to balance the opposing tendencies described above without destroying the present context.

In preserving this context, the residential neighbourhood should be understood as being composed of a set of traditional conditions which have survived until the present because they have continued to be relevant to contemporary needs (Ikaputra, 1993). The term 'contextual adaptability' will be used to refer to an ability to fit something into a specific context. The 'contextual adaptability' of something suggests how much a new condition can be appropriately placed in an existing context, or, as applied to a place, how well the existing context can accommodate the new condition. Thus, it implies the ability of old values – traditional culture and its built environment – to continue to exist alongside with modern values such as contemporary lifestyles, business interests and commercialisation.

The Set of Elements Present in the Traditional Environment in Yogyakarta

The 'contextual adaptability' of Yogyakarta, the most significant historic city in Java, Indonesia, is constantly being tested as economic pressures threatening the continued existence of residential spaces in the centre of the city. The city of Yogyakarta was formed originally by the establishment of a palace called a *kraton* which was surrounded by fortifications known as *baluwerti*. Yogyakarta has an imaginary axis in which the city pattern can be distinguished from all previous Javanese palaces (Figure 16.1).

The King of Yogyakarta, whose title was *Sultan Hamengku Buwono*, used to live in the palace which nowadays stands as a monument in Javanese culture. It serves as a cultural centre and a tourist attraction. In contrast, aristocratic families continue to live in the nobles' residences – the so-called *dalems* which are built around the palace. Most of these aristocratic families share their *dalems* with their *abdi dalems* (servants). *Abdi dalems* work as servants for the noble in exchange for a place to live, to receive the good fortune of *berkah* (blessing), and for the benefits of *magang* (study of the etiquette and practices of high Javanese culture by working for and serving the aristocratic family). A noble and his family require the services of *abdi dalems* not only for their daily needs, but also as a status symbol. For this purpose, the noble lends land

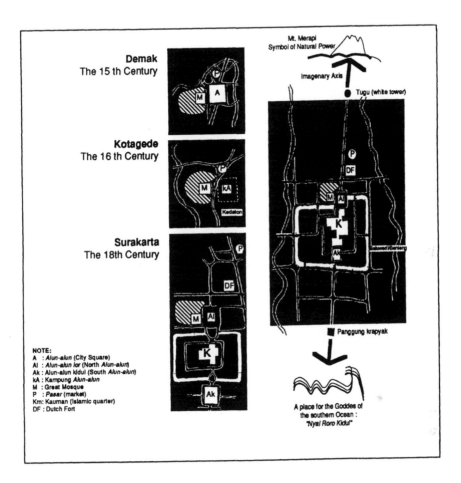

Figure 16.1 Comparison of Javanese palace cities

within his *dalem's* walls to the *abdi dalems*. The right of *abdi dalems* to occupy this land is known as *magersari*. Interestingly, the descendants of both the noble and his servants eventually expanded the numbers of occupants living within the *dalem's* walls which ultimately grew into a clearly defined community. The set of elements which together make up the traditional *magersari* community, the primary settlement pattern in the palace environment of Yogyakarta, can be defined as:

a)   the *Dalem,* which is representative of Javanese traditional architecture. Its architecture can easily be distinguished from that of the less formal structures of the surrounding environment. The *dalem* or the noble's residence is usually a complex of buildings, which is surrounded by a high wall. To enter the complex, one passes through a gate or sometimes, two gates. A gate with a roof is known as a *kori*. One without a roof is usually called *candi bentar* (a split-gate). The location of the gate depends on the direction of approaching the *dalem* from the main street. It could be on the north, south, east, or west. However, the orientation of the *dalem* is always to the north or south (Figure 16.2);

b)   the people of the *magersari* community belong to two groups: the noble family whose members belong to the *priyayi*[1] class and the *abdi dalems*[2] *families* who live and work there. Many of them still attend to royal tasks or the 'feudal' lifestyle, indicating the importance of social symbols over materialistic needs;

c)   the *magersari* system which is a symbiotic relationship of mutual dependence between *abdi dalems* and the noble family.

This set of elements which make up the traditional *magersari* community appears to be very fragile in the face of the dynamic downtown development. *Magersari* communities face the risk of being absorbed into the central business district. In fact, the historic central axis of the city has been developed into an urban commercial strip called *Malioboro* street and serves as a showpiece for visitors. In order to retain the residential areas of the city centre, which are threatened by increasing land values, the *magersari* community needs to be strong enough to resist dispossession and, at the same time, evolve in consideration of contemporary conditions.

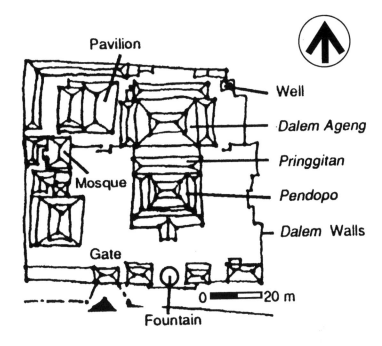

Figure 16.2   The architecture of *dalem*

The Study Objective

To examine the above statements, fieldwork had been conducted involving nearly 50 *dalems*. We traced the growth of these *dalems* based on maps from 1941 and 1987, and on their latest condition in 1991. Observations of these nobles' residences help us to understand what the *dalem* has actually been experiencing in terms of growth and in face of contemporary urban developments.

This paper will explore an approach to preserve the *dalem's* environment as a part of efforts to guide the development of the area. This approach will relate potential contributions of our past traditions to viable solutions that satisfy contemporary socioeconomic needs. This can be done by examining the growth, trends and changes involved in this traditional settlement type. Moreover, it can also be achieved by recognising current development as a continuation of living tradition and by acknowledging the possibility of utilising urban land in a more beneficial way within the traditional context. We can direct the development in a way which preserves the spatial context of the *dalems* as the primary structure of the traditional urban landscape.

Dalems as the Primary Structure of Urban Growth

Dalems were located throughout old Yogyakarta both at the inside and the outside of the *baluwerti* (the palace fortifications). Originally, each *dalem* was built in the middle of a large empty lot which nowadays is completely filled with the houses of a *kampung*. The *dalem* is like an 'island' floating in a larger settlement. Located at the centre of the lot, the *dalem* required an access drive which nowadays cuts through the houses of the *kampung* toward the gate of the *dalem*. The *dalem* still dominates its surroundings because of its access drive, the gate, and the size of the main house which made up of two large structures: the *pendopo* and the *dalem ageng*. Also, its large yard in which Banyan trees or other big trees still exist can be seen from the main road and the surrounding area.

From the 1941 base map and the 1987 aerial photograph of Yogyakarta, we can easily recognise the physical characteristics of the *dalems*. From these materials, a field study covering 49 *dalems* was conducted which focused on the issues of their functions and ownership. The sample size of 49 *dalems* represents an estimated 70–80 per cent of all *dalems* in the city (60–70 *dalems*). Of the *dalems* studied, 13 (26.53 per cent) were located inside the *baluwerti* and 36 (72.47 per cent) were outside of it. All of the *dalems* studied were part of the pattern of the city ordered by the imaginary axis of the *kraton* running from the north, through the *panggung krapyak* (a square stage-like building), to the south passing through the *kraton* (palace), *alun-alun* (square), the *tugu* (white tower), and it ends at *Mount Merapi* which is to the north of Yogyakarta (Figure 16.1).

The first ruler of Yogyakarta built the palace by first cutting the *beringan* forest. Then he arranged freely the palace's spaces according to the principles exhibited by previous Javanese palaces. Historically, the 'island-like' *dalems* were the first structures built in each area of the city from which all subsequent development grew. The freedom to plan on open land gave the first *Sultan* and the later rulers the flexibility in managing easily the lands of his nobles and aristocratic families. The nobles were each given a *dalem* and the right to use the land surrounding it.

Although the noble was still below the *Sultan* in the political hierarchy of the *kraton*, he still enjoyed autonomy. Here, it is important to understand that Yogyakarta was built after eighteenth century's Javanese civil war. As a defensive measurement, the noble surrounded his *dalem* with his loyal followers as well as the *abdi-dalem* who would protect him in times of crisis. The men of each *dalem* were grouped under one command which was directly led both politically and communally by the noble.

As conflicts with the Dutch and British colonisers and within the royal families themselves became less frequent, and even further, since Indonesia proclaimed its independence, the political power of the *kraton* in Yogyakarta decreased. As a result, the nobles' power has also been reduced. Although some *priyayis* were given the land they occupied as a gift from the *Sultan* (the ownership of the land became private), most noble families eventually returned to the *Sultan* the *dalems* and the land where their followers had lived. The noble's followers then, became royal servants and their attention were focused upon the *kraton* and its ceremonial tasks. The *magersari* land right system was still utilised, but under royal control.

Formerly, land inside the *dalem's* walls were occupied solely by the noble, his family and their *abdi dalem* (servants). However, seeking good fortune and the honour of *magang* service attracted many more Javanese, especially villagers, to work for the noble family. In addition, more and more Javanese villagers were beginning to migrate to the city for economic reasons. They filled in even the smallest spaces of the city. They were often allowed to use rooms of the existing building and the vacant land inside the walls of the noble's residence; they also became part of the *magersari* community. In this way, nobles' residences continued to be the primary structure from which urban neighbourhoods grew. Consequently, development of urban residential space is necessary.

The Historical Development of the *Dalem*

Historically, the city changed according to the development of the nobles' residences. Changes in urban culture and land values have dramatically altered the position of these residences. Land values have risen substantially with the pressures of urban development. These forces have changed the position of the traditional urban settlement. Some important questions are: Do these residences still play the same historic role in managing urban land use? How has land use shifted in response to contemporary needs?

The Transition from a Priyayi Household to a Community

Nobles had a higher social status than the ordinary people. Their political and socioeconomic positions were reflected in their better houses. They lived in *dalems* that were often twice as big as the houses of the ordinary *priyayi*, such as *priyayi* belonging to the educated class or rich farmers (Koentjaraningrat, 1984). Their *dalems* included a spacious verandah, ten or more bedrooms and many additional rooms for servants. There was usually a *pendopo,* a wide open plan building, in front of a verandah. The *pendopo* was used in entertaining guests with performances of traditional dance, formal dinners, meetings, and other social functions. Every activity of the daily life was accommodated within the *dalem's* walls which bounded and defined each noble residence. In this way, the *dalem* provided a living space for the noble family and their servants where they carried out the functions in accordance with the *priyayis'* role within the kingdom.

The noble's prestigious status in urban society did not only depend on the two poles of interests (the number of servants each noble commanded) but it was also indicated by the size of the nobles or the *priyayi's* extended family. It was common for a noble to foster seven to 10 children and often up to 20 if he had several wives. In addition, the noble household was augmented by taking on stepchildren from less well off distant relatives from the region. These children were actually not legally adopted as they still belonged to their own parents. In Javanese, this relationship is called *ngenger* (Kartodirdjo et al., 1987; Koentjaraningrat, 1984). *Ngenger* provided the manpower needed to maintain the household and to carry out the traditional ceremonies of the high Javanese culture. The social status of the *priyayi* was highly dependent upon the size of their households. A large family also helped to support a noble's social or political position when children became *priyayi* later on.

Having many children also resulted in the need for more servants (*abdi dalems*) to deal with all aspects of the 'family's' living activities. Similar to *ngenger, abdi dalems* used to come to the *dalem* because working for a noble family conveyed a higher social status. Therefore, the noble together with his wife/wives, children, stepchildren and *abdi dalems* did not only create a 'large-scale' extended family, but also a community, especially when each newly born 'member' of the family inherits the right to keep on living within the *dalem's* wall. Over the time, the *dalem* shifted from being the residence of a *priyayi's* family to that of many families, and eventually it developed into a community. In this way, *dalems* became the dominant form of traditional urban settlement.

The Changing Reasons for Living in Dalems

Magersari is a form of social assistance based on a relationship of mutual dependence between *priyayi* and *wong cilik* ('little-people' or the masses).

The *magersari* system supported the way of life of aristocratic families who lived in harmony with their servants' families for centuries. However, this study reveals that today, most *magersari* residents live in the *dalem* and benefit from the *magersari* landright system while working outside the *dalem* without serving as an *abdi dalem*.

We conducted a field survey which examined the reasons why *magersari* families came to live in the cramped spaces found in most nobles' residences. Twenty-six *dalems* using the *magersari* system were surveyed for this purpose. From these *dalems, 118 magersari* families were selected at random. These families were interviewed and were also required to complete a questionnaire. The data we gathered showed that these families moved to the *dalems* in the period between 1932 to 1991. This study has found that there are two groups of people who move to live in *dalems.* (Ikaputra et al., 1992). The first group consisted of those that had some connections with someone who already lived in the noble's household, either with the noble's family or a *magersari* family. The second group was made up of those who were not previously connected to the people living in the noble's residence.

An interesting phenomenon could be seen in other reasons for moving to the *dalems*. Figure 16.3 shows that until 1960, 80 per cent of those surveyed had some prior connection with inhabitants who were already living in the noble's residence. Even if we omit the inhabitants who were born in the house (20 per cent), the intention to work for the noble family is still high. This group of inhabitants sought work at the nobles', or they might be asked by the nobles to work for them. Consequently, these inhabitants have to live in the *dalem.*

During the period from 1961 to 1975, the number of people who decided to live in the nobles' residences, due to their proximity to school/work place, increased more than eight times (from 2 per cent to 16.28 per cent). Those people who came with the intention to work for a noble decreased by almost half (11.63 per cent) of that before 1960. And from 1976 to 1991, the major reason why people moved to *dalems* was the need to be close to their school or work place (19.15 per cent) or merely because of the need for housing (14.59 per cent).

The data showed a change in people's tendency not only to devote their lives to the aristocratic family, but also the increasing difficulty to find living quarters in the city. It indicates that there is a dialogue between the tradition of 'pride' in being an *abdi dalem*, which caused people to migrate to the city, and the changing image of Yogyakarta. They came to the city to receive *berkah dalem* (blessing from the *sultan*), to find a job, or go to school. They need a

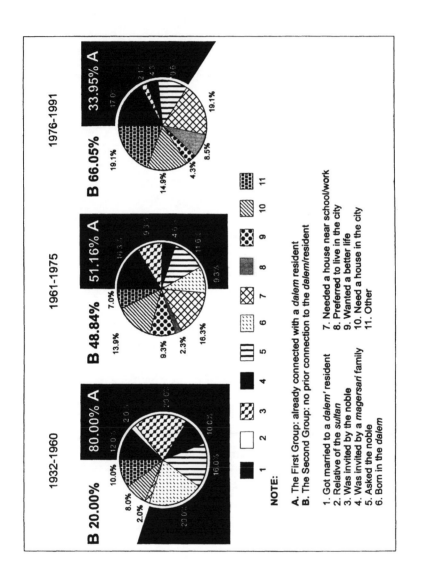

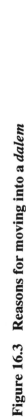

Figure 16.3   Reasons for moving into a *dalem*

place to live in the city. In this situation, *dalems* and the *magersari* system offer the possibility of providing social assistance for *wong cilik*, the masses.

The Dalems' Changing Function

Most of the 49 *dalems* we observed were nobles' residences which were built for aristocratic *priyayis* (the *Sultans* family or at least royal dignitaries). The field study began with the hypothesis that *dalems* have been facing at least three possibilities for development. One of the possibilities is the continuation of facilitating living spaces for the nobles and *magersari* families. Another possibility for development involved in the change of the house or commercial facilities function. Finally, both public and commercial facilities, together with residential uses could also be developed.

We identified four categories of functions of *dalems* from the field survey. As can be seen from Figure 16.4, two categories have in common that they still maintain the *dalem* as a noble's residence (36 *dalems* or about 73.47 per cent) with or without the *magersari* system; the *dalems* in the other two categories function solely as public (eight *dalems* or 16.33 per cent) and commercial facilities (four *dalems* or 8.16 per cent). One *dalem* has been abandoned and only its ruins remained. Although many nobles have evicted *magersari* families from their *dalems*, most *dalems* still function as the living space for noble's and *magersari* families. More than 50 per cent of the *dalems* studied (26 houses) still function as the urban residences of the aristocratic and *magersari* families.

Economic rationalisation which increases the land values of the urban area puts pressure on *dalems* to be used more beneficially. Therefore, about fifty per cent of the *dalems* with *magersari,* in practice, have also gained in exploiting the economic value of its location by creating a third function (two *dalems* have even a fourth function): commercial facilities (hotel, performing traditional dance for tourists) and renting the *pendopo* or a part of the land for private use such as for a school or a rental garage. Adoption of a third or perhaps even a fourth function served actually as preservation of the *dalem's* existence and the *magersari* system. In addition, it will give a better development for the value of the city.

This case demonstrates the validity of the *third* hypothesis above: the *magersari* system successfully facilitates both public/commercial and residential needs at the same time. This third hypothesis is more significant than the preceding two hypotheses.

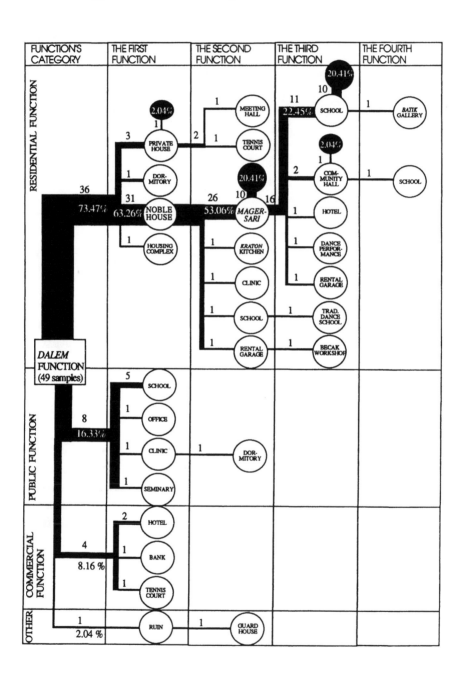

Figure 16.4 *Dalem's* **changing functions**

The 'Contextual Adaptability' of the *Magersari* Community

The palace environment with its set of elements which make up the *magersari* community (*dalems*, the *priyayi* and *abdi dalem* classes, and the *magersari* system) lies in the critical domain where, as stated in the first section of this paper, could either be pulled back into the 'narrow-mindedness' of preservation as an historic monument to 'feudalism', or undergo modern development: giving way to the forces of economic rationalisation, dislocating its residents to the periphery of the city, taking on the appearance of the international city, or even demolishing the historic city structure and its identity.

This classic conflict in historic cities yields opposing viewpoints regarding development. Should the historic city centre be a showpiece for visitors or a commercial district serving the daily lives of residents? Should it be historically preserved or recreated as an appealing new city centre? Should it become a competitive, vivacious downtown or a uniform, calm residential area? (Narumi, 1988) In answering these questions, architectural heritage should be understood with relevance to its function, identity and environment. Every step of growth or development of the palace environment must be discussed in terms of these factors. Appropriate functions will tend to be contemporary functions that are likely to change in the future. The dominant identity will be reminiscent of the 'past' enriching the contemporary city. The environment will continue to be that which regulates relations among people, between things, and between people and things.

But, we should first consider the problems faced by *dalems* as the components which structure the city centre. With certainty, the old downtown area or, in this case, the palace environment with its *dalems,* has always had a tendency to grow along with economic activity and increasing land values. These activities always create conflicts of interest. The growth of economic activities can lead to rapid physical change which is often in conflict with the goal of preserving architectural heritage.

Other problems are related to population and land use. Limited land and an increasing urban population, either as a result of migration, or of land urbanisation, create a shortage of urban residential space. This condition is worsened by the fact that many people come to the city for work but do not bring capital with them. Cheap rental land must be found to accommodate these people. Often, this population occupies land illegally. Usually, this progression results in the phenomenon of slum-like settlements. These problems become manifested in the physical architecture, and alter the identity of the urban environment. Yet, we have not discussed the effect on relationships within this environment.

The conflict of relationships in the palace environment is between the 'setting for feudal activities' which is more or less static, and the 'setting for contemporary activities' which is more changeable and more free in choice, movement and direction. And what is interesting about *dalems,* which still use the *magersari* system, is the evidence that 'feudal values' have successfully regulated the development of the city and finally resulted in the creation of more 'democratic' residential communities. This demonstrates that the *magersari* community is capable of being developed to function in the context of the contemporary city.

The 'contextual adaptability' of the *dalems* in the palace environment regarding the possibilities for both preservation and development will be discussed in the following section.

The Dalems' *Role within the Residential City*

In discussing the *dalems'* role in structuring the residential areas of Yogyakarta, we cannot disregard the reasons why people come to live in the *dalems.* The results of this study indicate that these reasons have changed. People used to come to live in a *dalem,* to work inside the *dalem's* walls as a *magang* (an intern learning the aristocratic way of life) and to receive *berkah* (blessing). Nowadays, people come to live in a noble's residence, to work outside the *dalem's* walls or to attend school. However, this change has not altered the *magersari* social function in which the noble provides social assistance to the people we have called *wong cilik* (the masses). This is because this kind of social assistance still serves as a clear status symbol for the *priyayi.*

The 'contextual adaptability' of the *magersari* system now depends on how its role is updated when the conditions under which people come to the *dalem.* Also, it depends on how social assistance changes are received. If the *magersari* system can be updated and continue to offer social assistance, then the residential areas of the city can be preserved. Preserving the residential areas of the city will make it more difficult for commercialisation to disrupt traditional urban neighbourhoods. The 'contextual adaptability' of *dalems* will result in the preservation of the most remarkable Javanese architecture; moreover, it will also maintain the existing open space to which *dalems* contribute more than any other type of urban settlement found in Yogyakarta.

The preservation of residential neighbourhoods in city centres offers a solution to the problem of downtowns which are busy during the day, but are changing to 'ghost towns' at night (when commercialisation displaces residential neighbourhoods).

The Palace Environment and Commercialisation

Most *dalems* are located in the palace environment at the centre of the city where land values have never decreased. Due to increasing land values, the biggest threat to preservation of the palace environment is commercialisation which would displace the residential function out from the old downtown. If commercialised, the land will be used more intensively, and preclude the possibility of maintaining both the *dalem's* architectural heritage and its role in contributing to the open space of the old downtown.

However, this has not happened yet in the case of the palace environment in Yogyakarta. In fact, this study found that more than seventy-percent of *dalems* still structure the urban residential neighbourhoods through its social assistance role. Moreover, the study also showed that about 60 per cent of *magersari* families (of the 118 sampled) do not pay anything for occupying their land, while the rest pay very little (Table 16.1). Thus, the noble does not benefit commercially from permitting residences on his land. From this, we can see that a mutually beneficial relationship between the noble and the *dalem's* community utilising the *magersari* system, has counterbalanced the changes that would otherwise occur given the purely economic value of the land.

Table 16.1 *Magersari*'s rental fees

No.	Type of payments	No. of households	Percentage of the total sample (116)
1	Free	62	53.45%
2	Sugar, rice, other foodstuff, etc.	9	7.76%
3	Less than Rp. 1,000/month	14	12.07%
4	Rp. 1,001–Rp. 5,000/month	13	11.21%
5	Rp. 5,0001–Rp. 10,000/month	11	9.48%
6	Rp. 10,000–Rp. 20,000/month	5	4.31%
7	More than Rp. 20,000/month	2	1.72%

Note: Rp. 1,000 = ¥67 or $1.8 or same with 1kg rice in 1992.

Then, what can the noble do to cultivate the commercial potential of the *dalem's* land value if his social assistance yields him practically no income?

The *priyayi* family's 'real' income comes from their work as civil servants of the government, as private officers, or at most, from running their own businesses within the *dalems* walls. Nobles who live in *dalems* owned by the *kraton* are allowed to make money by renting the *pendopo* (one of the biggest and most representative buildings in a noble's residence) or the front yard surrounding the *pendopo*. They benefit from the *dalems* value and its location as much as possible from this supplemental income. As a result, many nobles modify the *pendopo* and the *dalem's* yard for their family's private purposes but *magersari* families are almost never evicted. The new function in the *dalem* makes its environment more liveable, more efficient and supports a 'mixed-use' counteracting the tendency afflicting commercial downtowns throughout the world to become a 'ghost town' at night. This study has identified an attitude taken on by the nobles in which new functions have been 'transplanted' to increase the 'contextual adaptability' of *dalems*, so that they fit within the context of the contemporary city.

Physical Preservation and Community Needs

The *dalem's* change from serving primarily as the house of an aristocratic family to become the 'village' of a *magersari* community was followed by a change in the activities within the *dalem's* walls. This change compels us to identify those elements that can still be used from the existing *dalem* and what should be done to make the existing *dalem* suitable for current community needs. The land and the existing rooms are utilised to serve either the *dalem's* community or the nobles' private concerns. The increased utilisation of land and structures leads to either the erection of new buildings, the reconstruction of existing buildings, or the acquisition of existing buildings adjacent to the *dalem* which results in the *dalem's* physical growth.

In many cases, the physical growth of *dalems* threatens to change the architecture. However, preservation of the *dalem's* architecture should be considered part of its contemporary function in addition to the accommodation of a 'transplanted' function. Thus, not only is the *dalem's* architecture preserved as in the past but the *dalem* is also modified.

In other words, the attitude towards the physical preservation of *dalems* should not be based only on the architectural greatness of the 'past', but should also consider the degree to which harmony can be found, minimising the gap between preservation and development. The preservation of traditional architecture can be achieved through an effort to maintain the 'daily image' of the *dalem's* architecture, such as when a *pendopo* is used as a school, when a *dalem's* gate houses a *warung* (food stall), or when the *dalem's* yard is used as a rental garage, etc. In this way, the architecture can be modified but not to the point where either its traditional identity or its 'presentness' in the context of contemporary urban life is lost.

Yet, the problem that remains in utilising traditional buildings is the possibility of bringing damages to the existing fabric. Construction practices need to be found which are suitable for joining old and new construction without causing damages to the traditional structure. Furthermore, although the noble can control any construction on the land where *magersari* families live, he cannot afford to assist them in house construction or maintenance. As a result, many *magersari* houses still lack healthy conditions.

Government programmes such as the *Kampung* Improvement Project (KIP)[3] and social foundations should turn their attentions to *dalems* in which *magersari* communities live. Yet, it cannot be assumed to be identical to the KIPs. A '*Dalem* Improvement Project' should be understood within the following context:

a) the *dalem* is a specific community under the authority of a noble functioning as a 'landlord';

b) the *magersari* system preserves the close relationship between the noble and *magersari* families and permits him to grant the land rights which permit the *wong cilik* to live in the city;

c) the architecture of the *dalem* is part of an important cultural heritage as well as part of the possible solution to urban problems.

The Obstacles to Preservation

Dalems can be the property of the *kraton*, the government or private citizens. Formerly, all *dalem* properties belonged to the *Sultan*. The noble only had the right to use the land and build a *dalem* on it. The *dalem* and the use of the *kraton's* land would be transferred to other nobles depending on the decision of the reigning *Sultan*. However, the land could become privately owned if the *Sultan* gave the land as a gift to the noble as a political statement or as compensation. Some *dalems* have become government property or been sold to individuals outside of aristocracy.

Here, we are going to discuss the relationship between ownership and the function of *dalems*. The data relating to ownership have been collected through interviews with either the noble, the *magersari* families or the people who work at the *dalem*.

Table 16.2 shows that *dalems* owned by the *kraton* are clearly continuing to serve both as noble residences and *magersari* communities (90.48 per cent or 19 *dalems*). Of these, thirteen *dalems* (61.91 per cent) are benefiting from the increase in land value by taking on 'transplanted' functions. 66.67 per cent of *dalems* owned by the government have been changed into public facilities. The physical appearance of *dalems* belonging to government institutions (or foundations) will usually be preserved while they take on a new function (usually as public facilities). Although it may be the aim of the owner to revitalise the old building with a new function, the community is often dissolved. The social unity of the *dalem* is either lost or moved inevitably toward its abandonment to the social problems of the surrounding environment.

Table 16.2 Ownership and the *dalem*'s function (%)

Ownership	CF	PF	H/OHS + TF	NH +TF	H/OHS NH	NH+M	NH+M +TF	O
The property of the *kraton*	0	0	0	9.53%	0	0	28.57%	61.91%	0
Government	0	66.57%	0	0	22.22%	0	0	0	11.11%
Private	21.05%	10.53%	15.79%	10.53%	5.26%	0	21.05%	15.79%	0

Notes

1   Note for the function:
    NH: noble house;
    M: *magersari*;
    H: house;
    OHS: other type of housing;
    TF: *transplanted function*;
    PF: public facilities;
    CF: commercial facilities;
    O: others.
2   The number of samples is 49, of which 21 are owned by kraton, nine by government and 21 by private individuals.

Meanwhile, private ownership can lead to various possibilities for changes in the function. Of the privately owned *dalems*, which still facilitate nobles' and *magersari* inhabitants' residential needs, 15.79 per cent of the houses 'transplanted' function correctly, while 21.05 per cent do not. They experienced about the same degree of changes as those given over entirely to commercial functions (21.05 per cent). From Table 16.2, we can also see the other changes in the functions of private *dalems* to the following groups: public facilities (10.53 per cent), noble houses without *magersari* but with a 'transplanted' function (10.53 per cent), private houses with a 'transplanted' function (15.79 per cent), and to residences only (5.26 per cent).

What we can conclude so far regarding ownership is that once a *dalem* becomes a private property, it is more susceptible to development. Private ownership, either by individuals (nobles, *priyayis*, or rich people) or by institutions/companies, effects the development of the traditional environment. If the *dalem* shifts to a commercial function, it will influence people's perception of the value of the land they occupy, and they may begin to compete more aggressively to benefit from the commercialisation which is introduced by the *dalem's* change in function. As a result, it can be predicted that the land will become valued more for the money that is to be made, and less for its role in supporting a strong social relationship.

Dalems which are privately owned by a noble are also in a critical situation facing the possibility of changing ownership. Although some *dalems* owned by private nobles such as *Dalem Suryoputran* and *Dalem Suryocondro* still facilitate *magersari* families, some nobles have demanded *magersari* families to move out, while some *dalems* have even been sold to wealthy individuals or companies.

Three cases are used to describe the change of ownership of *dalems* (privately owned by nobles). In these three cases, ownership changing to private hands caused the land to become easily a commodity which will threaten the *magersari* with eviction from the *dalem*.

When ownership would be shifted from a noble to a company, the *dalem* suffered the loss of its role as urban residential space, and at the same time, some old buildings were demolished as they were perhaps deemed unsuitable for the new function. At the end, what most people ask for when faced with this situation is that we preserve the architecture of the *dalem*. If this demand is not met, we will lose the traditional setting, its social benefits, the possibility of architectural preservation along with the economic opportunity for the palace environment to evolve into the appropriate context for future development.

The Proposed Direction of Future Development

The set of conditions which make up the *magersari* community that many people understand as being 'traditional', is not only relevant to the past. This set of conditions is the result of cultural accumulation which 'has been handed down' up to the present time (Tuan, 1988; Oliver, 1988). This process occurs in two ways: through retention ('keeping') and through assimilation ('transferring') (Queysanne, 1988). Similarly, the set of conditions in the *dalem* environment includes elements which are both resistant to and accepting of contemporary needs. But one must understand: What is being resisted or accepted? And how does the traditional urban setting best accommodate it? The 'contextual adaptability' of this set of conditions, has to be understood in order to support and guide every planning process and policy effecting the *dalem* environment whether it eventually undergoes change or not.

Any development of the traditional urban residential areas of Yogyakarta, especially around the palace, should be based on a thorough investigation of this set of traditional conditions i.e. the *dalems*, the people living there and the *magersari* system. The resistance to some changes and the acceptance of others indicates an ability of urban areas to adapt to contemporary needs within their traditional contexts. To preserve the context can mean to preserve the set of traditional conditions while allowing the possibility of developing this set of conditions itself.

Below are some recommendations for the preservation of the set of traditional conditions which make up the *magersari* community:

Preserving the Role of Dalems

As the primary agent and locus of growth, the *dalem* plays an important role in structuring the residential areas of Yogyakarta. The piecemeal growth process permits the urban environment to adapt to changes. *Dalems* also contribute valuable open space to the city as its density is low in comparison with the densities of other urban environments (such as *kampungs*, Chinatown, etc.). Moreover, the *dalem* is the most remarkable residential building of the Javanese architecture and should be preserved within a contemporary context instead of attempting to freeze it in the past.

Reducing the Threat of Change in Ownership

Keeping the *dalem* as a property of the *kraton* is one of the most effective ways to maintain the *dalem's* role in structuring residential neighbourhoods. The *magersari* system is dependent upon the ownership of the *dalems*. A change in ownership usually results in a change in land use, a change of activities and the possible removal of the *magersari* system from the *dalems*. When a *dalem* no longer utilises *magersari*, it is likely to suffer the loss of its social function and become an object, which is more susceptible to development in any direction: commercial, museum, amusement centre, etc. This can result in the rapid destruction of the city, not only physically/ architecturally, but also structurally, functionally and spatially. This is due to the *dalems* position as the primary unit from which the city of Yogyakarta is constructed, especially its traditional urban centre: the palace environment.

Maintaining Control over the Environment

The clear physical boundaries of the *dalem* walls permit greater control over the growth of the *dalem* community. The noble, as the head of the household, can effectively control 'his big family' because he knows everyone who is allowed to live in the *dalem*. Although the land within its walls is the property of the *kraton*, the noble has autonomy to handle the apportionment of the *dalem's* land. This is made possible by the *magersari* system which governs the relationship between the noble and his community. This relationship facilitates good communication within the community enabling social and physical problems to be revealed and resolved directly. Maintaining the community within the *dalem* walls in turn helps to ease the pressures of growth throughout the historic district.

Benefiting from Increased Land Value

The preservation of residential areas on land which has a high potential for profit-making, can be more easily justified if accompanied by efforts to utilise the land more beneficially. A 'transplanted' function can be developed as guided by the principle of 'selective commercialisation' i.e., a commercial function is selected according to its ability to harmonise with the *magersari* community.

Modifying the Dalem

Architectural preservation is often a matter of attempting to freeze some architectural monument in a particular moment in the history. In the case of Yogyakarta, we should preserve the context of the *dalem's* architecture while recognising its ability to accommodate contemporary needs. The self-regulated growth of the *dalems* and their ability to accept 'transplanted' functions indicate that *dalems* can be modified without destroying the original character of the *dalem's* architecture. Modifying the *dalem* while preserving its architectural context will create a new chapter in a continuous history in which the *dalem* will be able to tell future generations.

Dealing with the Decrease in Environmental Quality

The decrease in the quality of many *magersari* residents' life needs to be addressed by a *Dalem* Improvement Project (DIP). Beyond improving living standards as in the Kampung Improvement Projects (KIP), the DIP should also preserve the set of conditions that make up the *magersari* community and the palace environment.

Concluding Remarks

Preservation in Yogyakarta's palace environment, especially its *dalems,* is threatened not only by economic growth and new construction, but also by the failure to comprehend the concept of traditional land rights: the *magersari. Magersari,* as the primary system of land management in *dalems,* has played a crucial role in ordering urban space in which a dialectic tension and balance have been maintained between development and tradition. The successful development of communities in *dalems* demonstrates the capability of the *magersari* system to preserve both architecture and social conditions in the contemporary context.

This 'contextual adaptability' to adapt, while continuing to support residential functions, presents the opportunity for commercial activities to be 'transplanted' to the *dalem.* Living activity ordered by the *magersari* land rights system helps to preserve traditional values in *dalems* as well as to control the rapid change that can be brought about by commercialisation. At the same time, the 'transplanted' function will promote the development of the traditional urban environment which would otherwise stagnate or decay physically resulting in valuable urban land that is under-utilised.

The results of this study demonstrate that tradition is not necessarily a static phenomenon. The 'contextual adaptability' of the conditions of Javanese tradition, particularly in Yogyakarta as described above, proves to us that maintaining traditional urban areas, especially aristocratic lands, as living space for residential neighbourhoods, will preserve tradition while serving contemporary needs.

Acknowledgments

We would like to thank Robert Cowherd from Karaton Surakarta Project Office for his valuable comments and English corrections.

Notes

1 In addition to blood-relatives, common people can become *priyayi* through marriage or through personal achievement by being appointed to a position in the *kraton* by the *sultan*. This often brings with it the right to live in a *dalem*.

2 *Abdi dalem* means literally 'king's servant' (*abdi* in Javanese means servant and one of the meanings of *dalem* is king), so that *abdi dalem* can be understood as a person who works for the king at the palace or works simply as a royal servant. When an *abdi dalem* works for a noble's family s/he is actually similar to the present day household maid in terms of duties performed. However, an *abdi dalem* has an additional role within the nobles family. *Abdi dalems* do not benefit economically but raise their social status by doing *magang priyayi* (acquiring familiarity with the aristocratic way of life by working and serving the *priyayi's* family directly). They play a role in building the noble's grand lifestyle. Their existence raises the symbolic status of the noble and his family.

3 KIP *(Kampung Improvement Project)* is a project designed to improve the quality of physical-environment *(bina Lingkungan)*, of human resources *(bina manusia)* and of economic opportunity *(bina usaha)* of the urban residents of cities throughout Indonesia. However, the government's own assessment of the programme is that while the physical environment of kampungs involved have been improved, the other two goals (human resource and job development) have not been met. See Yudohusodo, 1991, pp. 311–18.

References

Ikaputra, K.N. (1993), 'A Study on the Contextuallity of the Palace Environment', a thesis submitted for Masters degree, Osaka University, Japan.

Ikaputra, K.N., Hisa, T. and. Parimin, A.P. (1992), 'Understanding *Magersari* – A Land Right Concept, in Forming City Spatial For Living: A Big-Scale Extended Family in the Noble House', The Summary of The Planning and Architecture Seminar, Kinki Region, Osaka, May 1992.

Kartodirdjo, S., Sudewo, A. and Hatmosuprobo, S. (1987), *The Priyayis Culture*, Yogyakarta: Gadjah Mada University Press (*in Indonesian*).

Koentjaraningrat (1984), *Javanese Culture*, Jakarta: PN Balai Pustaka (*in Indonesian*).

Narumi, K. (1988), 'Senses of Beauty of Japanese for Townscape', in K. Narumi (ed.), *Urban Arrangement from Townscape Design*, Kyoto: Gakugeishuppansha Co. (*in Japanese*).

Oliver, P. (1988), 'Dwelling and Tradition', *Symposium Preliminaries on Traditional Dwellings and Settlements in Comparative Perspective*, Berkeley: University of California.

Queysanne, B. (1988), 'Architecture and Tradition', *Symposium Preliminaries on Traditional Dwellings and Settlements in Comparative Perspective*, Berkeley: University of California.

Rapoport, A. (1991), *History and Precedent in Environmental Design*, New York: Plenum Press.

Tuan, Y.-F. (1988), 'Traditional: What Does it Mean?', *Symposium Preliminaries on Traditional Dwellings and Settlements in Comparative Perspective*, Berkeley: University of California.

Yudohusodo, S. et al. (1991), *House for All People*, Jakarta: INKOPPOL, Unit Percetakan Bharakerta (*in Indonesian*).

17 New Divisions of Public/ Private Responsibilities in the Management of Urban Development and Their Implications for Planning Education in India

JAMAL H. ANSARI

Introduction

India's urban population was 217.2 million at the time of the 1991 census. Demographic growth during the 1990s is anticipated to push this figure up by another 125–130 million to the range of 340–350 million (Government of India, 1988). Increasing demographic pressure on urban areas is characterised not only by its large base but also the tendency of concentration in large urban centres. While it took eight decades for India's metropolitan cities (each with more than one million population) to multiply to 12 in 1981 from merely one in 1901, as many as 11 metropolitan cities were added during the decade of 1981–91 alone, taking the total up to 23. Their number is expected to increase to 40 by the year 2001 (Government of India, 1992). Four of the metropolitan cities have been classified as mega-cities (each with more than 5.0 million population) in 1991 and their number is expected to increase to seven by 2001 (Government of India, 1988). Metropolitan cities now account for about one-third of the total urban population, and the contribution of cities with over 100,000 population including the metropolises is 65 per cent.

Urban Growth Trends and Related Issues

The tendency towards urban concentration had begun to build up as a result of partition of the country at the time of Independence in 1947. Consequently, a mass exodus of population from across the newly created borders started, and refugees mostly resettled in existing large cities. The trend of urban population concentration was first noted by planners and policy-makers during the late 1950s when the Third National Five Year Plan was being formulated. The Plan also addressed the problem of growing regional disparities as a result of increasing concentration of industries and other economic activities in a few big cities. It was observed that cities like Calcutta and Bombay had grown too big and some of the other metropolitan cities were close to reaching that stage. The Plan thus advocated a policy of planned decentralisation of industries away from metropolitan cities. But, the pressure of economic market forces on government's decision-making process continued throughout the Third Plan period to favour industrial investment in the existing industrial concentrations of Bombay, Calcutta and Ahmedabad.

Nevertheless, during the later plan periods, central government, in conjunction with state governments, pushed forward projects and schemes to restrict the growth of large urban centres. The Pilot Research Project on Growth Centres and the Scheme for Integrated Development of Small and Medium Towns (IDSMT) were two of the more prominent efforts in this direction. Furthermore metropolitan regional planning was introduced in Delhi and subsequently supplemented in several states, mainly with the aim to deflect population migration streams away from respective metropolitan cities. Simultaneously, governments at central and state levels adopted policies of not letting large and medium size industries to locate in metropolitan cities. But the experience until now suggests that these efforts have largely failed.

Basic Problems

Scarcity of developed urban land, an accumulating backlog of housing, large deficits in urban infrastructure and services, degradation of the physical environment and an alarming number of urban poor are only some of the problems faced by cities in India. According to recent estimates, nearly 27 per cent of the total urban population have no access to safe water supplies and nearly 70 per cent are reported to be without basic sanitation. On the account of road deficits, average motorised travel speed in several metropolitan cities has declined to less than 18 km/hour. Slums and squatter settlements' population account for 17.5 per cent of the country's total urban population (Chelliah and Mathur, 1993). Environmental degradation in large metropolitan cities like Delhi, Calcutta and Bombay has reached such levels that it is becoming difficult to breathe. Respiratory diseases caused by inhaling polluted air are taking a heavy toll; infant mortality caused by diarrhoea due to unclean drinking water is on the rise.

Institutional Framework

The persistence of urban problems reflects badly, not only on the effectiveness of the planning processes, but also on the institutional framework and the administrative structures through which the problems are sought to be alleviated.

An institutional machinery has gradually developed in the country whereby central government lays down policies and priorities, and oversees the flow of central funds to states. State governments' directorates of town planning prepare town development plans. Development authorities, and a host of other agencies are responsible for implementing development plans. Municipalities ultimately take over developed areas to look after their day-to-day maintenance.

National economic policies in the 1990s have resulted in a system of governmental props in the form of subsidies for backward regions, infrastructural support for small and medium towns, and long-term low interest loans (without any recovery guarantees) for the poor. Particularly impressive is the list of programmes administered with the help of grants from central government to help the urban poor. The programmes are ultimately projected in the form of low cost housing and infrastructure, sites and services, slum upgrading, urban basic services and community development programmes. One of the more ambitious actions in the series is the Urban Land (Ceiling and Regulation) Act, implemented in 1976. It was enacted by the national parliament to exercise social control over the scarce resources of urban land, mop up excess land and pool it for providing housing and other basic facilities to the poorer sections of the society. But the Act has proved to be a principal obstacle in the operation of the land market. Large areas of urban land are entangled in legal battles and, as a result, can neither be acquired nor brought into the market. Partly it has resulted in soaring inflation in land prices and a further reduction in accessibility to develop urban land for the poor. Furthermore, it is argued that the city system acts as a redistributive mechanism whereby real income is distributed from the poor to the rich in the form of rentals (Kumar, 1994).

Similarly, the Rent Control Acts were enforced in various cities to prevent exploitation of the poor. But these too have proved to be big obstacles to the operation of the housing market. These Acts have not only taken away incentives to invest in housing at a time of rising demand but have also been significantly responsible for the problem of vacant houses at the time of increasing rentals. It is now clear that a series of public interventions which were made with a view to help the urban poor were actually misplaced. They produced results to the detriment of the poor, introduced distortions in the market for housing and land, shot up prices, abetted clandestine activities and tied up the urban development process into cobwebs of rules and regulations, and consequent delays. The observations contained in the National Housing Policy (Government of India, 1992, p. 7), succinctly describe the situation:

> ... the effective operation of the land market has been affected by the existing legal and regulatory framework, lack of infrastructure, and the slow pace of release of serviced land by public agencies. This, along with other economic factors, has led to unwarranted increase in land price and housing costs and widespread speculation and profiteering especially in larger cities. The exclusion of the majority of the poor from the formal market, and their inability to build or acquire legal shelter has led to the proliferation of squatter settlements and unauthorised colonies.

Weaknesses of Local Urban Institutions

Local urban institutions in India have also been found wanting in meeting the challenges posed by ailing cities, particularly the larger ones. Urban government is the key local institution responsible for provision of essential services in urban areas, such as water supply, sewerage, drainage, solid waste collection and disposal, street lighting, primary health care, primary education and other services. But severe problems exist in the delivery of these services, mainly due to the extremely weak financial state of local governments. While population has increased substantially, resources have not multiplied in a commensurate manner. Consequently, a general deterioration in the standard of civic amenities has been observed. Moreover, there is the problem of inequity in the distribution of the available services. The poorer sections of the population, which comprise nearly half of the total urban population, hardly benefit from any of these services. A huge gap exists between what is needed to be spent to bring services to a minimum acceptable level and what is actually available in the form of budget.

The weak revenue base of urban institutions results mainly from the constitutional and legal framework within which the urban government system in India has to operate. Urban local bodies derive their revenue mainly from house taxes and octroi. Whereas house taxes are difficult to assess and administer, octroi is being abolished in most states. The revenue base of local bodies is thus slated to shrink further making them be more dependent on state government's largesse (conspicuous by its absence). Local governments have been to date outside the statutory devolution of funds. It is not very often that the state government comes to the rescue of local government as the former are generally themselves short of funds and on the look out for usurping one or the other of the latter's revenue avenues. In addition, in view of the past record of inefficient debt servicing by local governments and the paucity of public and private savings, state governments rigidly control the extent of market borrowings in respect to the local governments.

In April 1993, central government made an amendment to the Constitution, popularly known as the 73rd Amendment. It provides for a statutory basis for distribution of funds amongst various tiers of government. This provision may help to improve the revenue base of local government. But local government is also saddled with inadequate management expertise. They are unable to improve the quality of their personnel since appropriately skilled officials and staff are not attracted to work for local government due to low salary scales and prestige. Paucity of managerial resources leads to inefficient fiscal management. Physical development plans are rarely able to provide a basis for projecting capital and current account expenditure that is needed for the implementation of plans. Project programming, thus, does not take place. The task of urban management is further impeded due to lack of an information system for policy making. The end result of all these inadequacies is that inefficient use of available resources only further aggravates the resource problems.

Another major source of administrative problem emanates from the large number of authorities. A host of public agencies – central, state and local – function at city level with overlapping jurisdictions and conflicts of interest. In addition to municipalities, development authorities and improvement trusts, a host of parastatal and quasi-government agencies are also involved. There are also a large number of government departments dealing with specific sectors of urban development. Considerable resources are wasted and sometimes projects get delayed due to overlapping functions of these agencies, which often work at cross-purposes since the necessary mechanism and jurisdiction for coordinating the works of these agencies do not exist.

The Urban Planning Process

India's Five Year Economic Plans, state policies and other programmes give priority to the urban development sector, including housing. Through these plans, the central government also sets aside grants that are to be distributed as central assistance for schemes, programme and projects formulated by the states to pursue designated priority activities. One activity that has been actively promoted over the years under this centre-state patronage is the preparation of master plans for planned development of urban areas. State Town Planning Directorates take credit for having prepared more than 1,000 master plans for towns and cities of various sizes. However, there are another 3,500 towns in the country for which planning efforts of this kind are yet to be initiated. But the fact is that the existing plans remain, by and large, unimplemented. Out of the plans under implementation, not even 5 per cent of the proposals see the light of the day.

There is an urgent need for planners to sit, pause and think why plans so meticulously prepared ended up ultimately as paper plans. Although planners promoted the concept of master plan and got rid of piecemeal planning, the master plan itself is a diagrammatic presentation of the proposed arrangement of land uses for a specified time perspective accompanied by proposals for transportation and municipal services network, etc. Although the social and economic dimensions of the proposals are not forgotten while preparing the plan, the emphasis is mainly on the physical factors. The plan is supported by a report which explains the proposals, standards and rules and regulations.

Several weaknesses of master plan technique have been noted. Firstly, it is found to be a rigid document, a once-for-all statement about the future shape of a town or city. This characteristic of the master plan makes it unamenable to change, if a change is found necessary as a corrective measure during implementation. Spatial standards incorporated in master plans are too rigidly framed and often out-of-tune with the socioeconomic realities of the community for which they are prepared.

Moreover, the methodology for preparing master plan is time-consuming, and the plans, though prepared with considerable care, are often outdated by the time they are enforced. In spite of the time and effort spent in their preparation, the plans are not supported by details regarding fiscal and institutional measures or policies needed for their implementation. No wonder then, most of the master plans have become items to decorate the shelves of planning offices rather than panaceas for urban evils that they were meant to be. The main concern of the town planning directorates all over the country remains the preparation of more master plans, even though the planners fully realise that master plans are quite inadequate for the present urban challenges. Surely, there is a need to rethink about alternatives to master plans.

Absence of an Effective Urban Land Management System

Various institutions and instruments of land management have evolved over the years, each reflecting a distinct policy context and priorities of the period in which they were made operational. However, most of these instruments are not applied in a concerted manner and many have lost their meaning. Aberrations in the choice and application of instruments and institutions of urban land management have led to more imperfections in land market. At present, numerous difficulties are involved in making legal access to land, particularly for the poor. Land registration procedures are time consuming, unduly cumbersome, and costly. Obtaining permission to develop is equally cumbersome and very often will tie up land development for several years. Besides, there are many other factors inhibiting the land market. For instance, the tardy adjudication process has contributed to large chunks of land being entangled in disputes and kept away from the market (Farvacque and McAuslan, 1992). A compulsory land acquisition procedure is practised by many development authorities, but development is slow. Thus large areas are being withheld from the market. Many times, the acquisition procedures get stalled due to compensation related disputes or complications arising out of the Urban Land Ceiling (and Regulation) Act. Furthermore, land is disposed off at exorbitant rates, leading to general inflationary trends in the land market.

Advantages exist in selling land on perpetual lease. These are in the form of continued public control over the land after its disposal, regular income over the period of lease and the opportunity to mop up unearned income at the time of secondary lease transactions. The advantages have, however, been negated by unhealthy land transaction practices. Owners of leasehold land take recourse to 'power of attorney' arrangement whereby the owner gives the 'right to use' tenure to the buyer of leased land. The user virtually becomes the owner and also has the option to transact the land if necessary to a subsequent buyer. In this manner, the owner manages to avoid paying that proportion of unearned gains on the leased land which the state intended to mop up. Furthermore, the urban land records system gets completely disjointed. No official record exists and it becomes difficult to trace who owns which piece of property. This complicates the land transaction process and further distorts the land market.

Legal and Regulatory Framework

Land use regulations suffer from being mere vestiges of the erstwhile colonial era when public intervention was motivated by the desire to exercise authority and control rather than streamlining development process. Multiplicity of legislations governing urban development, institutional fragmentation, and multiplicity of authorities initiated during the colonial era to suit the vested interests of the rulers, have been perpetuated during the post-independence era.

An inability to cope with the plethora of rules and regulations, time consuming processes in getting necessary clearances, and unrealistic standards in the background of sharply rising unfulfilled demand for residential and business premises, are the compelling reasons for the existence of a large scale illegal informal sector in the form of unauthorised residential colonies, industrial premises and business areas. Even in the authorised and planned areas, building bye-laws are violated with impunity and encroachments are continuing without any let-up or hindrance. The problem is not so much in the lack of enforcement but the lack of rationalisation of the building codes, bye-laws and zoning regulations. Rather than wasting energy and resources in bulldozing the 'unauthorised' structures, it is high time that planners and policy-makers took a hard close look at the existing legislative and regulatory framework with a view to evolving a practical urban land management policy and building control regulations.

New Economic Policy and Urban Development

During the mid-1980s, urban problems compelled policy-makers in the country to reorganise their approaches for arresting the continuing decline in the quality of life in towns and cities, particularly metropolitan cities. By then it had become quite apparent that all the efforts during the 1960s and 1970s to restrict the growth of large cities and encourage small and medium towns had not had any perceptible effect on the pattern of urbanisation. In fact, the economic policy then followed was in many ways instrumental in concentration of economic activities in a few large size urban centres. The emphasis on import substitution and domestic market, and the plethora of rules compelled the entrepreneurs to concentrate near large cities, most of which also happened to be of administrative headquarters importance. The practice of heavily subsidising the infrastructure and services in these cities was no less a contributory factor for the emergence of mega-cities at the cost of smaller urban centres.

By now the planners sought to sustain population in rural areas by diversifying the rural economy. However, exploding rural population made it very difficult to keep even the existing work force effectively engaged in the rural sector, and there seemed to be no alternative but to tap the relatively numerous opportunities for employment in the urban sector.

Economists argue for economies of scale, agglomeration economies and productivity in the urban sector. Several new trends in the macroeconomic indicators have emerged by this time. Firstly, the share of urban sector activities in the GDP of the country had begun to rise sharply. In 1950–51, the contribution of the urban sector to India's GDP was estimated at only 29 per cent; it increased to 47 per cent in 1980–81 and is likely to go up to 60 per cent by the turn of the century (Government of India, 1992). Secondly, a strong middle class comprising about 200 million people has emerged as a vast potential market for consumer products. These significant changes in the economic scene in the country open up possibilities for trying out new models for development.

Present approaches to help the weaker sections of the population, by either building low cost houses, sites and services programmes, infrastructure improvement, or various programmes for alleviating urban poverty, have not shown significant citywide effects. These measures, instead of alleviating the situation for the weaker sections, actually aggravate it.

The Government of India has thus introduced major economic policy reforms with a view to accelerate the rate of economic growth. As a consequence, industrial licensing, except for 18 industries sensitive from the point of environment or security, has been abolished in all cities with less than one million population. Even for cities of more than a million population, industrial location policy has been made more flexible leaving scope for the location of new industries if the existing ones are decaying due to technical obsolescence. The general view of the policy-makers appears to be to let industries and business prosper where these can be most productive. Urban productivity, in view of its increasing share in GDP, has to be encouraged and obstacles hindering its growth are to be removed. Furthermore, government is insisting that while pricing public utilities, the concept of cost recovery should be adopted and all subsidies and other forms of government support should be eliminated. In line with changes in policies, administrative procedures and regulations have been made simpler in order to reduce the government's detailed planning and management of the economy. Such changes have important implications for different sectors of the economy, especially the urban sector and cities where most of the new investments are expected to be made.

New national economic policies are likely to result in a change of the existing pattern of urban growth and activities over space. Delicensing of the industrial sector and appropriate pricing of infrastructure and services may induce the dispersal of economic activities leading to a more balanced spatial distribution of population. However, such changes in settlement patterns need to be anticipated and provided with investment support.

The National Commission on Urbanisation (NCU) concludes that investment must be concentrated in selected growth centres in backward areas only. The existing approach is defective because investments and other support systems extended to such centres have invariably failed to pull them out of their backwardness and the investment cannot be revolved to replicate the efforts. The Commission thus proposes threshold level investment in GEMS (Generating Economic Momentum) – towns and cities which have shown potential for growth (Government of India, 1988).

Nevertheless, implementation of the new economic policy, which is designed to accelerate the pace of economic growth, is bound to place a heavy demand on urban economic infrastructure, public utilities, land and housing. Such demands will be in addition to the requirements to service India's swift urban transition. Merely on a prorated basis, to keep pace with the demand generated by increasing population, the urban infrastructure stock and capacity must increase by about 45–50 per cent during the next 10 years. However, if we take the existing deficit in urban infrastructure, other services and housing into account, then the existing stock will need to increase by about 55–60 per cent (Chelliah and Mathur, 1993). This demand will go up further in view of the need to supply additional land, housing and infrastructure to service new economic activities which will congregate towards urban areas bestowed with comparative locational advantages in respect of these activities.

Meeting these demands will be crucial for the successful implementation of new economic policies. The urban economy and macroeconomic parameters are inextricably linked. To meet the demand for urban infrastructure, services, land and housing, major reforms will be necessary in the existing financial, institutional, and regulatory mechanisms. As discussed above, the existing mechanisms which came into being under a different set of economic and institutional conditions are inflexible and, if not downright obstructive, are at least not helpful in ameliorating the shortfalls in developed land, housing and infrastructure. International aid agencies, such as the World Bank, also feel the necessity of moving towards a broader view of urban issues, one that looks beyond housing and residential infrastructure, and emphasises the productivity of the urban economy and the need to alleviate existing constraints (World Bank, 1991).

New Division of Public/Private Responsibilities in Managing Urban Development

Planning and development agencies, mainly in the public sector, are finding it difficult to cope with the continuing decline in the quality of life of urban dwellers. These agencies are faced with lack of public finance to support urban development programmes and projects. The Eighth Five Year Plan (1990–95) has made a total allocation of Rs 52,770 million for urban development, Rs 57,573 million for urban water supply and sanitation and Rs 49,230 million for housing (including rural housing), which together account for less than 4 per cent of the total plan outlay.

But it must be noted that these plan allocations have, over the years, stayed in percentage terms at approximately the same level, even though a huge gap has existed in each plan period between the investment requirement and actual allocations. For instance, against the total proposed outlay of Rs 49,230 million for housing (including rural housing) in the Eighth Five Year Plan, the Plan itself estimates that the investment required to achieve the goal of housing for all will be Rs 975,530 million.

> If the present conditions are any guide, the overall allocations comprising of plan allocations, subsidies, and allocations that accrue to the sector via directed credit or statutory requirements could, in fact, decline, and present a new challenge to the urban sector (Chelliah and Mathur, 1993).

In this situation, government has no alternative but to go for a higher level of private sector involvement in the development and maintenance of urban infrastructure, housing and social and cultural services. More thought is being given by policy-makers to pinpoint public tasks that can be transferred in a responsible manner to the private sector (Ansari, 1990a).

As far as housing provision is concerned, the share of the public sector is only 15 per cent of the total housing stock. The bulk of new housing in urban areas is produced by the private sector and individual households themselves. Recognising this, federal government in its national housing policy advocates an enabler's role for itself, rather than directly involving itself as a producer or provider. To promote the facilitative role of the government, in the National Housing Policy, it is further suggested that reforms in urban land policy, legislation and procedures related to rent control, land transactions, land taxation, land revenue codes, preparation of master plans and development control rules, are necessary. It is also recommended in the Policy that housing finance institutions should work more for the mobilisation of household savings and development of secondary mortgage markets through greater integration with financial markets rather than depending on increased flow of directed credit to raise resources for housing finance.

Furthermore, there is a need to understand the operation of local land and housing markets, if local and state level agencies intend to play an enabler's role. Local governments today do not even have adequate information on the number of developed plots of land and housing units constructed within their jurisdiction. If the objective is to generate a conducive environment for greater participation of the private sector, it is necessary to ensure that market transactions become more transparent than they are at present, and serve all segments of the society. For this purpose, information about markets (production, consumption, inventory, prices, etc.) should be readily available to both the consumers and suppliers. Builders, real estate developers and agents should be registered and listed according to their performance ratings. A good information base also helps in generating buoyancy in land and housing markets by attracting more investors and making it possible for the inter-mediaries to perform in a more professional manner. Thus building up a suitable information bank would be one of the most necessary steps by the public sector to promote private sector investment in land and housing (Mehta, 1993).

The public sector could also work vigorously towards ensuring an increased supply of serviced plots of land in the market through appropriate legislative, regulatory and procedural reforms. Under the new division of public/private responsibilities in land development and housing, the official development agencies should concentrate on vigilance and provision of bulk services. They should lay down broad parameters for development of land in the form of norms and simple rules and regulations and formulate mechanisms which would allow private developers of repute to engage in large scale acquisition, development and disposal of land, and construction of houses (Ansari, 1990b). This has started to happen already in some parts of the country, and the approach can be further promoted by floating joint sector schemes, permitting private developers to assemble properties for urban renewal of highly congested pockets of towns and cities, and introducing single window clearance system. As regards the role of the private sector in the provision of services, society at large will have to determine to what extent it is prepared to accept profit in connection with the provision of urban services and education. One may argue that in a situation of scarce public resources that exists in India, the issue is not one of private versus public sector, but of directing resources where they are most needed. Perhaps, the most reasonable approach would be to direct public programmes towards areas that are likely to be adversely affected by market principles, and let the private sector concentrate on the needs of the rest of the population and areas.

The World Bank has suggested a wide range of tasks to be performed by government agencies to bring about reforms in the urban sector and create suitable environment for strong private sector participation. Citywide regulatory reforms will have to be pushed through to increase the efficiency of land and housing markets in urban areas. Regulatory reforms should play a key role in expanding supply and lowering costs of housing, finance, infrastructure, developable land and increasing business opportunities. The Bank suggests two techniques which should be utilised in developing a reform strategy. Firstly, market assessments should be undertaken. They should identify patterns of supply, demand, and constraints to supply. Secondly, regulatory audits should be carried out to assess the costs, benefits, and distributional consequences of specific regulations at the city level. The next objective would be to formulate regulatory reform programmes; devoting particular attention to the impact of different choices of national, state or local instruments to be used, either singly or in combination. These include in the case of land:

1) fiscal and monetary instruments such as tax incentives, capital gains taxes, land transfer taxes, property taxes, betterment levies, and targeted subsidies;

2) legal instruments, such as land-use controls, land tenure, accepted practices and procedures for contractual law, land registration, land allocation, flexible mortgage instruments, and procedures for public acquisition;

3) institutional measures to facilitate transactions in land markets such as land valuation systems and information systems to monitor land ownership patterns, prices and transfers.

Regulatory reforms should lead to a better environment, which will attract private sector entrepreneurs to participate even in infrastructure development where they are generally reluctant to invest. This includes electric power generation and distribution, water supply, waste collection and disposal, and the management of related activities, such as maintenance, administration, metering, monitoring or billing and collection. Under this new environment, the role of government would be limited to monitoring and supervising private sector operations (World Bank, 1991).

Status of Planning Education in India

Modern town planning education in India started to develop after independence in 1947. But the orientations and contents of the education programmes adopted during this period were very much shaped by the planning ideologies of the 1920s and '30s. Patrick Geddes, the renowned British town planner, visited India during 1915–20 to advise on town planning works in various parts of the country. His ideas about diagnostic surveys, conservative surgery and integrated planning helped set the guiding parameters of the practice of planning as it came to be established during the later years. Planning and development of New Delhi, the seat of the colonial government, also took place around this time. The classic urban design concepts, as well as some of the key garden city principles which were used in its planning, must have weighed heavily with the professionals and academics who pioneered town planning education and practice in the post-Independence era. Moreover, Dr. Koenisberger and Sir Walter George, two highly acclaimed British town planners and architects, who set-up planning practice during the colonial period, and who stayed on even after independence, and the early group of Indian town planners, were no less instrumental in establishing both the teaching and the practice of town planning on models that are of British origin.

The Institute of Town Planners, India (ITPI) was formed in the year 1959 on the pattern of the Royal Town Planning Institute of the United Kingdom. It set definite directions and priorities for adopting codes of practice and formulating programmes of planning education. The Institute's objectives as enunciated in its constitution are:

1) to advance the study of town planning, civic design and kindred subjects, and of the sciences and arts as applied to those subjects;

2) to promote planned, economic, scientific and artistic development of towns, cities and rural areas;

3) to foster teaching of subjects related to town and country;

4) to promote the general interests of those engaged in town and country planning;

5) to inculcate and propagate the physical planning aspects of development.

By pursuing these objectives, the Institute has clearly committed itself to the promotion of physical aspects of planning and development with a peripheral interest in related subjects. Immediately after its establishment, the Institute was helped by the Government of India in establishing the School of Town and Country Planning in 1955. The main aim of this School was to produce a crop of indigenous planners to cater to physical planning needs arising out of the implementation of the Five Year Plans for the economic development of the country. In the early 1960s, the Department of Architecture of Delhi University's Polytechnic Institute had merged with the School of Town and Country Planing which was then renamed as the School of Planning and Architecture. This event had a significant contribution in outlining the main thrust of planning education, which has since developed with a close affinity towards architectural and civic design components.

The School has since grown into one of the foremost institutions of planning in Asia. The School's curricular structure has been used as a model by other planning schools which came up in other parts of the country (see Table 17.1).

In addition, the Institute of Town Planners, India, has also constituted a Town Planning Examination Board which conducts regular examinations leading to the award of a Diploma which is now recognised by the Government of India as equivalent to a postgraduate degree.

Most of the postgraduate degree courses could be completed within the stipulated period of one and a half years divided into three semesters. The total intake of all planning schools is 250 at present.

Specialisation

Earlier, only generalised programmes in town and country planning were offered. But these have now been diversified into various programmes leading

Table 17.1   Planning education in India, 1993

Sl. no.	Name of institution/ department	Year of establishment	Programmes offered	Permitted intake	Enrolment (1993)
	Master's level course				
1	School of Planning and Architecture, New Delhi	1955	UP, RP, TP, HSG, EP, LA, AC, UD	105	93
2	Department of Architecture and Regional Planning, IIT, Kharagpur, WB	1956	CP, RP	16	4
3	School of Architecture and Planning, Anna University, Madras	1964	TCP	20	15
4	Department of Architecture and Town Planning, Bengal Engineering College, Howrah	n.a.	TRP	8	n.a.
5	Faculty of Architecture and Planning, GNDU, Amritsar, Pb	1972	CRP	23	9
6	Centre for Environmental Planning and Technology, Ahmedabad	1972	URP, EP, HSG	38	35
7	Town Planning Department, Government College of Engineering	n.a.	TCP	26	n.a.
8	Department of Architecture and Planning, University of Roorkee, Roorkee	1973	URUP	5	n.a.
9	Centre for Town and Regional Planning, RECT, Surat, (NR)	1987	TPR	10	8
10	Department of Civil Engineering, REC, Nagpur	1989	UP	10	7
11	School of Planning and Architecture, Technical University, Hyderbad	1990	URP	16	n.a.
12	Institute of Town Planners, India, New Delhi.	1955	AITP	n.a.	n.a.
	Bachelor's level courses				
1	School of Planning and Architecture, New Delhi	1989	P	20	20
2	Faculty of Architecture and Planning, GNDU, Amritsar, Punjab	1991	P	30	12

Notes

UP	–	Urban Planning	RP	–	Regional Planning
HSG	–	Housing	TP	–	Transport Planning
EP	–	Environmental Planning	LA	–	Landscape Architecture
UD	–	Urban Design	AC	–	Architectural Conservation
URP	–	Urban and Regional Planning	CP	–	City Planning
CRP	–	City and Regional Planning	URUP	–	Urban and Rural
CRP	–	City and Regional Planning	TRP	–	Town and Regional Planning
P	–	Planning	n.a.	–	not available
X	–	Examining Body	NR	–	not yet recognised by ITPI

to specialisations in substantive areas of study. The School of Planning and Architecture in New Delhi runs five specialisation courses at postgraduate level: Urban Planning, Regional Planning, Housing, Transport Planning and Environmental Planning. Besides, the School also offers independent though related master's level programmes in Urban Design, Architectural Conservation and Landscape Architecture. The School has also plans to start new master's level programmes in District Planning and Rural Development, Urban Development Management, Housing Design and Management, Real Estate Management, and Transport Systems Engineering and Management. Clearly, the trend is to institute management courses related to various specialised disciplines of planning. See Table 17.1 for more details.

Undergraduate Planning Education

The School of Planning and Architecture at New Delhi instituted the first undergraduate programme in planning in 1989. Two batches have by now graduated from this programme. Only one other institution in the country is presently running a bachelor's programme though a few more may be started in the immediate future. The programme extends over four years.

Doctoral Programmes

Almost all of the planning schools now register candidates for doctoral programmes. Currently, very few planners hold a PhD degree, but the number of doctoral programme candidates is on the rise and about 50 candidates are presently enrolled at various planning schools for doctoral degrees.

Entry Qualifications

Admission to the postgraduate programmes is generally open to bachelor's degree holders in architecture or civil engineering, and master's degree holders in geography, economics or sociology. The School of Planning and Architecture at New Delhi is the only institution which admits students even with backgrounds in statistics and operations research to its transport planning programme. The Madras School also takes master's degree holders in social work or law. As a recent development, students with bachelor's degree in planning have also begun to get admission not only in the Master's level programmes in planning but also in landscape architecture, urban design and architectural conservation. At the undergraduate level, the entry is after completing education up to twelfth standard with physics, chemistry and mathematics.

Model for Curriculum Development

Planning curricula are generally structured around theory subjects, projects and a terminal assignment in the form of a thesis. Theory subjects can be generally divided into five categories related to topics listed below, namely:

a)   Urban, regional, and rural planning theories, processes and systems including history of settlements and evolution of planning thought;

b)   quantitative methodologies and analytical tools, including statistics;

c)   substantive areas dealing in components of settlements such as housing, transportation, network systems and services;

d)   substantive areas by aspects such as heritage, urban conservation and design, environmental and socioeconomic bases, ecology and landscape, planning legislation and administration, information system and remote sensing;

e)   substantive areas pertaining to problems, e.g., disaster area management, metropolitan area planning, housing for the masses, rural development, urban renewal, informal sector, backward area development and other problem related topics.

Recently, planning schools have realised the importance of subjects such as project formulation and evaluation, fiscal programming and budgeting, energy conservation, and operations research. These subjects are selectively offered at various planning schools. Generally, the curricula are designed to include only compulsory subjects and the choice of subjects offered in a particular programme depends very much on the specific orientations and available resources of individual schools. However, with time, curricula are being made more flexible and students have a limited option to choose elective subjects and/or pick programmes offering specialisations in one or other substantive areas. Programmes leading to specialisation are structured around a common first semester of core subjects which are compulsory and then second and third semester subjects are oriented to specific areas of specialisation.

Project oriented courses aim at applying theories and techniques for problem solving and design. The other important aim of these courses is to impart skills to the students in presentation and communication techniques. Thesis work is an essential component of all planning programmes and aims at the teaching of research methodology.

The curricula of undergraduate planning programmes are built around five components: the groups of subjects belonging to the related disciplines of architecture, civil engineering, environmental sciences, planning and social sciences. A major portion of the curriculum is oriented to imparting skills in design and communications.

Need for Reforms in Planning Education and Practices

While the number of planning graduates every year is only about 225, they always cannot find suitable jobs within a reasonable time after graduating. This is a sad phenomenon given India's size. It not only reflects the poor status of planning profession in the country, but also the quality of planning education. A mismatch seems to have developed between the kind of graduates that are being produced by planning schools and the nature of demand generated by the fast changing socioeconomic parameters. The increasing diversity of issues and the complexities involved in dealing with them in view of the need to do more with less financial resources have thrown up a challenge to planners to initiate suitable changes in planning practices. This has also led to an increasing awareness for quality planning education.

Limitations of the Existing Models of Planning Education

Heavy orientation towards physical aspects of planning in the planning curricula and the kinds of model adopted from the West in the 1950s are not relevant any more. Whereas the schools in the West have moved onto a multi-disciplinary mould, planning curricula in Indian planning schools continue to follow traditions predominated by architecture and civic design. As a result, a large part of the students' time is spent on acquiring drafting and presentation drawing skills in studio courses. This leaves little scope for intellectual development and innovative thinking.

Due to the heavy emphasis on physical design solutions without much consideration to financial, fiscal, administrative and political dimensions of the problems, planning graduates invite the stigma of being equipped with less relevant planning skills. This leads to the isolation of physical planners from the main stream of planning and development process in the country. Planning in India at the national and sub-national levels is geared to sectoral economic planning where physical planners have very little to contribute. At the settlement level, the development works are implemented mostly by the concerned sectoral departments, by development authorities or special purpose agencies wherever they exist. These agencies generally prefer to involve architects and engineers rather than planners, since the former are more useful for the kind of work the agencies carry out. Nowadays, the physical planners' main achievement is in the form of preparing hundreds of master plans for towns and cities all over the country. But almost all of these plans remain substantially unimplemented, which further diminishes the creditability of physical planners in the eyes of the decision-makers as well as the people.

Thus, it is most important today that planning education should be reoriented primarily with the aim of breaking down the isolation of the town planner from the mainstream of the institutions and organisations engaged in the formulation and implementation of policies, programmes and projects for urban development. Today, town planners need to be educated to become perceptive about the textural changes in the society, such as the changes in socioeconomic and political milieu, and the changes in technology. Moreover, they should develop an understanding of the elements that are amenable to intervention through public or private initiative, to help the society move towards the 'agreed' qualitative goals. The planner's understanding of the planning, implementing and monitoring processes, instruments, tools and techniques should dovetail into this framework with continuing realisation of the need for flexibility, rationale and applicability of proposals in achieving urban development objectives.

The broad based undergraduate planning programmes, in combination with possibilities of developing in-depth understanding of specific components and aspects of urban development at the postgraduate level, presents varied possibilities for reorienting planning education towards a format which would help town planners play a much more useful role in society.

Towards Making Planning as a People's Movement

The formal regular programmes, whether at undergraduate or postgraduate level, may have to be supplemented with continuing education programmes in the form of workshops and short-term courses for the in-service professionals in planning to tune them with the changed context and prepare them for the kind of roles they may be required to play vis-a-vis planning process and development (Saini, 1990).

Planning schools should simultaneously work with the people at large with a view to changing the latter's perception, knowledge and attitude towards planning ethos and inculcate in them a sense of participation for the better development of the society, as well as the promotion of people's self-interest. The mounting challenges in urban areas can be met by adopting processes and techniques for better planning and management of development, and more so through wide-ranging reforms in the society, institutions, and political processes. It is, therefore, essential that planning schools should not confine themselves to in-house teaching but also undertake to disseminate new knowledge amongst decision-makers and people at large through meetings, seminars, extension works, publications, media and other channels for soliciting people's participation and making planning a people's movement.

Towards Sustainable and Equitable Development Approaches

The Indian society and economy are on the threshold of a big and positive change. Most importantly, the feeling of despondency and despair which had pervaded the body politic of the Indian society has given way to hope and determination to innovate and improve for a better future. The government has begun to realise that they must play a catalytic role to support the initiatives. The number of non-governmental organisations (NGOs) is growing and is ready to deliver goods. Community based activities are on the rise. These changes have opened up opportunities for alternative approaches for development which emphasise the goals of equity, sustainability and people's participation in addition to the traditional pursuits of growth, efficiency and productivity (United Nations, 1993). There is now a need to evolve in perfect equity that would lead to sustainable development strategies, which would foster partnerships built on trust with community organisations, NGOs and the private sector. At the same time, planning curricula also need to be restructured not only to cover the growth, efficiency and productivity aspects of the urban processes, but also to give equal emphasis to the considerations of equity, sustainability and public participation. This would mean a substantive change in the courses offered. New courses may tentatively include topics such as linkages between urban processes and macroeconomic parameters, and vice versa, the role of economic liberalisation and deregulation on the buoyancy and efficiency of markets for land, housing and infrastructure and their combined impact on supply positions, potency and otherwise of various instruments presently being used for urban land management and possible alternatives. Moreover, courses covering the areas of project formulation and evaluation techniques, investment priorities and lending instruments should be implemented. Also, topics such as the roles of community-based organisations (CBOs), NGOs and the private sector in the management of urban development are important as well. Finally, subjects related to the processes of planning, implementing, monitoring, and evaluating projects that use community involvement approaches, implications of community involvement in design and planning, and physical design solutions for sustainable urban development, are also necessary.

The above is, however, only a demonstrative list of topics stated to highlight the need for reform in the planning education for a more useful multi-disciplinary role of town planner in urban planning and development processes. Details may be worked out to suit the specific orientations of a particular planning school. But, to start with, it appears that entry qualifications for various planning education programmes may have to be relaxed in preparation for a more broad-based and multi-disciplinary approach to education in planning.

References

Ansari, J.H. (1990a), 'Public/Private Responsibilities in Management of Urban Development', *ITPI Journal*, Vol. 8, No. 3, pp. 77–8.

Ansari, J.H. (1990b), 'New Divisions of Public/Private Responsibilities in Housing', *ITPI Journal*, Vol. 8, No. 3, pp. 36–40.

Chelliah, R.J. and Mathur, O.P. (1993), 'Reforms in the Urban Sector: Some Perspectives', unpublished paper presented at the National Symposium on *Managing our Metropolis: New Directions for 21st Century*, 29–31 March, School of Planning and Architecture, New Delhi (mimeo).

Farvacque, C. and McAuslan, P. (1992), *Reforming Urban Land Policies and Institutions in Developing Countries*, Washington DC: The World Bank.

Government of India (1988), *Report of the National Commission on Urbanisation*, New Delhi: Ministry of Urban Development.

Government of India (1992), *National Housing Policy*, New Delhi: Ministry of Urban Development.

Kumar, A. (1994), 'A Metamorphosis of Shelter', *The Economic Times*, 23 June, p. 8, New Delhi.

Mehta, D. (1993), 'New Policies and Urban Housing', *Indian Institute of Urban Affairs*, New Delhi (mimeo).

Saini, N.S. (1990), 'Emerging Dimensions in Planning Education in India', *Space*, Vols 5, 3 and 4, pp. 78–85.

United Nations (1993), *State of Urbanisation in Asia and the Pacific*, New York: Economic and Social Commission for Asia and the Pacific.

World Bank (1991), *Urban Policy and Economic Development – An Agenda for the 1990s*, Washington DC: The World Bank.

18 A Continuing Commitment to Better Planned Environment: The Experience of the University of the Philippines

ASTEYA M. SANTIAGO

The Academic's Response to the Mandate for a Better Planned Environment

Background

The year 1990 was a milestone for the School of Urban and Regional Planning, one of the academic units in the Graduate School of the University of the Philippines (UP), the country's premier University. It marked its twenty-fifth anniversary since it was by the legislative fiat in 1965. *Republic Act 4341* passed by the Philippine Congress authorised the University of the Philippines to create the first planning school in the country, entrusted with capability building, and, developed a cadre of planning professionals to assist in solving the then worsening development problems of the country. As of August 1994, there were close to 450 graduates. By the beginning of the twenty-first century, it is expected that 2,000 planning trainees will make their commitment to a better planned environment.

While buoyed by the confidence arising from the accumulated planning knowledge and experience of its faculty and support staff, the School's administrators and the faculty envision an increased and more varied demands on their academic, research and training resources. Since the School's creation in 1965, an undergraduate degree programme in environmental planning has been launched in the Miriam College, a private university in Metro Manila. Much earlier, two other private schools (the University of St Thomas in Manila and the University of St Louis in the country's summer capital, the city of Baguio) also introduced courses intended to eventually lead to a degree in

344

urban and regional planning. For various reasons, these efforts were not sustained. After a few years of existence, these programmes, except for that in the Miriam College, had been discontinued. Later, however, other academic institutions had introduced a number of specialised degree courses in the environmental field (Table 18.1).

In the meantime, it became clear urban development problems of most third world countries, more particularly those affecting the physical environment, had geometrically increased as the twenty-first century approaches. In the case of the Philippines, these problems had arisen within the backdrop of a volatile sociopolitical economic scenario. Paradoxically, it was this same fast moving scenario which downplayed and made the School's role superfluous. The major issue was that there were not enough graduated professionals to cope with the country's demand. Thus, recruitment to various planning positions, even the critical ones, was not confined to those who had formal training in the profession

Early Beginnings

The School had a modest beginning in 1965, characterised by its initial reliance on the national government for financial and other forms of support. A Memorandum Order (*Memorandum Circular 156*, 1968) from the Office of the President of the Republic exhorted various agencies of the government to send their selected scholars to the School. This was done during office hours and at the expense of the agencies. Consequently, a regular quota of students in its graduate degree programme had been guaranteed (Carino and Santiago, 1993, pp. 103).

Much earlier, the School's administrators launched a fellowship programme through which they set up a core of their teaching staff. United Nations Development Fund (UNDP) fellowships were made available to the faculty members whom they recruited, who then obtained graduate degrees in urban and regional planning from the Colombo Plan countries such as Australia and Canada. At this early stage, the School's decision-makers already mounted a comprehensive research programme with the funding assistance of the UNDP. They regarded activities like faculty recruitment and undertaking of research as some of the prerequisite steps in carrying out their legislative mandate to assist the officials of the local governments and the private organisations in achieving a better physical environment for their constituents (Institute of Planning, 1967–74).

Baptised in 1965 as the Institute of Planning, the School underwent during

Table 18.1 Listing of academic institutions offering specialised degree courses in the environmental field

Academic institutions	Degree courses specialising in the environmental field
Camarines Sur State Agricultural College, Camarines Sur	BS Agricultural Engineering (Farm Structure and Environmental) Master in Resource Management
Holy Trinity College, Palawan	BS Sanitary Engineering
Mapua Institute of Technology, Manila	BS Environmental and Sanitary Engineering
Miriam College, Quezon City	BS Environmental Planning
National University, Manila	BS Sanitary Engineering
Pamantasan ng Lungsod ng Maynila	Master in Civil Engineering (Water Resources)
Philippine Normal College, Manila	Master in Health Science Master in Public Health
Philippine Union College, Cavite	BS Health Education BSE (School Health Education)
University of Baguio, Baguio City	BS Sanitary Engineering
University of Northern Philippines, Ilocos Sur	BS Sanitary Engineering
University of the Philippines Diliman, Quezon City	MS Biology (Environmental Biology) MS Environmental Engineering MS Energy Engineering PhD Energy Engineering Master in Urban and Regional Planning PhD in Urban and Regional Planning
University of the Philippines Los Banos, Laguna	BS Biology (Ecology) BS Human Ecology (Human Settlements Planning) MS Environmental Studies
University of the Philippines Manila	BS Public Health MS Public Health Master of Public Health Doctor of Public Health Master of Occupational Health

Source: Environmental Education Network of the Philippines, 1989; Villavicencio, 1984. Table reprinted from Environmental Management Bureau (EMB), 1991, pp. 4–21.

the first few years of its existence several changes in its name. This was part of an identity search *vis-à-vis* the other academic institutions in the University. It also served as the response to the changing requirements of its academic and non-academic constituencies. The renaming process also coincided with the School's quest for an appropriate place in the academic community. When the School joined the University's academic family, there were other degree granting units nationwide. These institutions offered similar or analogous academic programmes. Therefore, it was imperative to issue the needed clarifications. Thus, it was legislatively acknowledged that the course offerings of the University of the Philippines' School of Economics and the College of Architecture were already nationally recognised programmes before the need for an academic institution to produce a core of professionals to plan, develop and enhance the country's natural and man-made environment.

Not unlike the other developing countries, in the Philippines, economic planning as a government function took precedence over physical planning, although the School of Economics was established in the University of the Philippines only in the same year (1965) as the School of Urban and Regional Planning. On the other hand, by the time the School was organised, the architecture programme which is one of the traditional preparatory courses toward a second degree in physical planning, had been offered as early as in 1956 (Office of the University Registrar, 1991, pp. 55 and 83). Furthermore, architecture was already a well recognised professional course in the country and had attracted a respectable share of students nationwide.

It was not surprising, therefore, that there were those in the University of the Philippines who initially believed that a degree programme in environmental planning was unnecessary and dispensable,, while others regarded it as a course offering which could be subsumed under the existing academic programmes of the University. Its introduction as a separate degree programme, as some observed, could be considered as luxurious in view of the many competing demands for the University's manpower, infrastructure, financial resources and expertise.

These perceptions slowed down, to a certain extent, the recruitment process undertaken by the School to build up a core of highly qualified faculty members and staff. In fact, some invited professionals had expressed reluctance to join this new academic unit in the University because either they were not familiar with the career prospects in environmental planning or, they did not think that the career opportunities in this field were promising. Added to this was the relatively unknown factor of the kind of theoretical and practical preparation needed by both the faculty and the students in participating in the newly mounted academic programme to enhance the country's natural and man-made environment.

The Macro Environment in the 1960s

A report prepared by the UNDP experts, who were invited by the Philippine government in 1958 to conduct a survey regarding the housing and planning situation in the country (Abrams and Koenigsberger, 1959), carried a strong endorsement for capability building among the country's professionals. Therefore, the creation of an academic institution for this purpose had taken place. This report catalogued a plethora of urban development problems affecting land use, housing and transportation which, although equally serious, had to compete for immediate remedial attention and for funding support. The report revealed a dismal environmental picture of communities characterised by mushrooming of squatters. Numerous residential subdivision lots, while isolated from the rest of the community by the absence of roads and adequate facilities, were gobbled up in the market at soaring prices. Towns and villages strung in an unsightly fashion along highways imposing great economic and social costs to the community; and the congested roads were clogged with an assortment of transportation vehicles (Faithfull, 1969, p. 3).

Not unlike many other commissioned reports, the UNDP report was not widely disseminated and ended up in the hands of a select group of decision makers in the country, the University of the Philippines faculty and researchers. However, there was nothing in the report which those who were not privy to it, more particularly the ordinary people in the street, did not already know. With the rest of the teeming population, they had all along, been witnesses to, if not hapless victims of the deteriorating environment of the cities and urban centres in the country.

Phases in the School's Involvement in the Planned Environment

1965–72: The Continuous Quest for Identity and Relevance

The genesis of the School from an Institute of Planning, to the Institute of Environmental Planning, before emerging as the present School of Urban and Regional Planning, chronicles almost on a parallel basis: the twin quest of its faculty and staff. One is to secure for the School a recognised place of academic relevance within the University; the second is for its planning degree programme in order to attain a secure position of practical usefulness to the national community. After all, the School was dependent to a large extent on the regular budgetary appropriations from the Philippine government and had to convincingly demonstrate that it was deserved to have this support.

The choice of its original name, the Institute of Planning, was prompted by the need to provide an umbrella term which allowed ample flexibility in covering, in a comprehensive manner, the multifaceted nature of its various programmes in physical development. Its appropriation of this all encompassing name was made possible as there was no other existing academic institution that offered courses either in physical or land use planning, or environmental planning. When the School started, the environmental planning field was totally unexplored and its organisational name and programme coverage became an exclusive concern to the school's administrators.

Its renaming in 1974 as an Institute of Environmental Planning signalled a change in both the macro and the micro environment in which the School operated. The change was brought about by the administration's realisation that the course was, curiously, being confused with the country's and the University's other existing concerns. One was family planning which, at that time, had started to generate a strong interest among policy makers because of problems associated with the country's high population growth. The other was the ongoing academic programmes of the School of Economics in the same University. The University of the Philippines officials ruled that there was a need to dichotomise clearly between physical planning and the other concerns (Institute of Environmental Planning, 1975–76).

Clarifying its Concerns Vis-à-Vis the Other University Units

The milestone decision to change the name of the Institute of Planning to the Institute of Environmental Planning was triggered by the representations made by a then member of the Board of Regents, the top policy making body of the University. This official was also a faculty member of the School of Economics and concurrently the country's Secretary of Economic Planning and head of the National Economic and Planning Authority or NEDA, the central economic planning body in the country. His intention was to ensure that the subject turf of the School of Economics would not be encroached upon, actually or seemingly, by another institution in the same University adopting an all encompassing name.

This friendly 'skirmish' became the first concrete challenge to the School officials to delineate their actual field of concern. In this process, they had to define more clearly their role *vis-à-vis* their counterparts in other institutions and agencies, within and outside the University. The School's decision-makers welcomed this opportunity to firmly draw the lines between the field of the professional concern of its graduates with those of economists and economic planners who were being trained by the School of Economics. In fact, they were relieved that at the academic level, the issue of the functional delineation between economic and physical planning was already resolved, even if they were not satisfied about the use of the term 'environmental'. In fact, they had entertained serious reservations on whether this was the most appropriate nomenclature of their academic offerings.

Involvement in National Physical and Economic Planning

At the level of the Philippine bureaucracy, however, the overlapping concerns of economic and physical planning remained a continuing issue in the seventies, downplayed by the concerned decision makers with varying levels of success. The officials of NEDA and the then Ministry of Human Settlements (MHS) sometimes found themselves at loggerheads on this issue. Simple as it appeared, the issue festered the relationship of these two agencies because concomitant to it was the determination of the nature and scope of the planning functions of their respective staff. While the School was involved in the joint training and research programmes of NEDA and MHS, the conflict between the latter's officials did not affect any of the School's corroborative activities with them. Neither did it diminish the 'marketability' of its graduates who continued to find employment in these two central planning bodies.

That the lack of a firm delineation of functions between NEDA and MHS somehow had adverse consequences on the country's programme for planned development. Those at the forefront of plan formulation and implementation in these two agencies often found themselves involved in clarifying their respective functions. For instance, the two agents' respective technical staff were required to prepare the delineation of the nature and effects of plans. While sometimes, these problems were more apparent than real and were a by-product of conflicts in personalities. Nevertheless, it resulted in some of their staff's counterproductive activities.

Amidst these uncertainties, the School's faculty and researchers, by themselves or through their graduates and the trainees who were active participants in the various programmes of these national institutions, continued to assist national and local government officials in solving their development problems, and in planning their towns and cities. The 'location' of the School within the University had placed its faculty and staff in a 'neutral' position because they had been vested with an aura of objectivity that generally surrounded the academicians all over the country. Through the various problematic situations, the School's faculty and researchers had acted out the role of 'disinterested third parties' who had avoided being drawn into any kind of bureaucratic 'micropolitics' which they knew could sidetrack them from their main task. They had realised that this was the ideal position to take if they were to maintain their credibility before their academic and non-academic constituents. So the focus was on the appropriate application of the School's resources in services which extended to various government agencies involved in land use planning, urban development and environmental management.

Research and Extension Services with Government Agencies

It was in the early 1970s that the School's staff initiated activities complement to its primary function of teaching, such as research and consultancy services which were undertaken with some government and private agencies. Among others, these resulted in the production of the three major research outputs of the School known as the National Physical Framework Plan (NPFP), the Mindanao Regional Development Study (MRDS) and the Manila Bay Metro Region Plan (MBMRP) which received funding support from the UNDP. More than twenty years later, these studies remain relevant reference materials which address contemporary planning and development situations.

The National Physical Framework Plan, used as a guideline, identified the political, social, economic and physical determinants of the growth of the country, from both the national and regional points of view. The Mindanao Regional Development Study resulted in an integrated development plan for this major region of the country. It was supported by an analysis of the physical and socioeconomic factors which influenced its urban growth. The Manila Bay Metro Regional Plan, on the other hand, was a strategic development plan for the Manila Bay Region which, among others, provided a suitable form of metropolitan planning and development organisation and a workable growth plan for the mentioned region. The study also involved the preparation of plans, and feasibility and pre-investment studies for particular projects in the problematic areas of the region. The study had also provided appropriate strategies for the implementation of such plans and projects. These documents continue to be used as reference materials of students and researchers at the University, the bureaucracy, and the policy makers in the legislative and executive branches of the government.

Training constituted an informal component of the capability building mandate of the School. This programme has brought the School's faculty and staff into useful contacts with other institutions with similar interests and concerns. From these, reciprocal benefits accrued to the said parties (Carino and Santiago, 1991, p. 115). This had also ensured a wider reach and ambit of influence of the School through its trainees who have come from various regions of the country. Upon their return, many of these trainees have been able to apply their acquired knowledge and skills in improving the environment.

1972–85: The Role of the School Under the Martial Law Regime

Seven years after the School's establishment, President Marcos declared martial law in 1972, bringing with it radical governmental and political changes in the country. These affected to a large degree the directions of the School's various programmes and activities, including its faculty and staff development programme. The following major political changes influenced the School's programmes: firstly, the consolidation of law making and execution powers in the President of the Republic which greatly facilitated policy making, more particularly in the field of physical planning; secondly, the emergence of a dominant role for Imelda Marcos, the First Lady of the land, who was appointed to the concurrent positions of the Governor of the Metropolitan Manila Commission (MMC) and the head of the Ministry of Human Settlements (MHS); and thirdly, the designation of the President of the University of the Philippines as the concurrent chairman of the predecessor agency of the Ministry of Human Settlements, namely, the Task Force on Human Settlements and the Human Settlements Commission.

Of the above three factors, the one which triggered the assumption by the School of additional roles which equally benefited its non-academic constituents was the keen concern by the First Lady on the planning, design and development of various towns and rural settlements in the country, more particularly Metropolitan Manila. She translated this interest into concrete plans and programmes after she was sworn into office as the Governor of the metropolis, and later, as the Minister of Human Settlements. Both organisations played critical roles in the planning, improvement, and development of various types of urban and rural human settlements.

The above organisational changes at the national level took place in the early 1970s when the country was just starting to build up its pool of professional planners through graduates of the School and various universities abroad. Attention became focused on the School as the main, if not the only, source of expertise that these two major organisations, the MMC and the Ministry, were directly needed. The appointment of the University of the Philippines' President as the concurrent chairman of the Task Force on Human Settlements (TFHS) helped ease the problem of the dearth of planners which confronted the officials of the TFHS (Executive Order 419, 1973). From the Task Force which was primarily a research institute, spun off the Human Settlements Commission or HSC to which the U.P. President was the appointed chairman (Presidential Decree 933, 1974). The HSC was later transformed into a full blown ministry called the Ministry of Human Settlements or the MHS (Presidential Decree 1396, 1978).

The University of the Philippines' president authorised the 'detail' to the various research and project undertakings by the HSC to some of the senior faculty members of the School. Eventually, when the HSC evolved into the MHS, the same secondment arrangement of the School's faculty to the MHS was effected, thus helping ease the Ministry's problem of professional planners. For instance, at least three of the senior faculty members of the School became project heads of the MHS while the others became part time consultants (Institute of Environmental Planning, 1974–83). The School's graduates were likewise harnessed into these organisations, helping beef up their personnel requirements.

The direct involvement of the School's faculty and staff in the joint research and consultancy services of the government agencies, and their designation to decision making roles paved the way to an unlimited range of opportunities for them to serve the community in different capacities. The assignments also proved beneficial to the faculty who found the first hand experience in these government offices to be valuable inputs to their classroom lectures. Their exercise of line functions, either as project assistants or project directors, provided the much needed practice of the profession, where the bureaucracy was virtually transformed into a practical laboratory, with the public which it served, acting as real life clients. Planning activities were 'acted out' in real life situations and not just lectured in classrooms and seminar halls. This mutually beneficial arrangement became a precedent which was replicated later in the School's relationship with other government agencies.

The active participation in the implementation of programmes of MHS and of its predecessor agencies became more imperative in the case of the School's faculty members. This was because all of them were harnessed into teaching immediately after their acquisition of their masters degrees abroad. This was in contrast to its student population, a high percentage of whom were professionals already engaged in planning or planning-related work when they gained admission to the School. This uncommon combination of faculty and students would have posed a vexing problem, if not an awkward situation, if not because of the opportunity for the faculty and staff to be placed on secondment in these agencies.

The involvement had, to an immeasurable degree, bolstered the faculty's confidence in their teaching capabilities and helped enhance their credibility as teachers and researchers. For they had, in the process, become active practitioners of their profession. This involvement of the School's academic and research staff also manifested the School's continuing effort towards maintaining its relevance to its non-academic constituents. This was a commitment that needed to be sustained, especially when the programmes continued to be supported mainly by government appropriations.

Contemporary Developments and Future Concerns

Relevant Post-Martial Law Developments (1986 to the present)

The post martial law period (1986 to the present) has strengthened the School's position of relevance on matters affecting various aspects of planning. For, by this time, technical expertise and/or extensive practical experience accrued by bureaucrats or academicians had gained wide public acceptance.

Other factors also contributed to the key roles that the School had assumed. During the pre-martial law period, members of the Philippine Congress and the technical personnel of the MHS, and various housing and environmental agencies under it, were the repository of information and experience on government actions taken on the various aspects of urban development such as land use, housing, transportation and environmental management. Upon the abolition of the legislative body during martial law and its disbanding, access to information and resource persons originally available in these institutions had become extremely difficult, if not impossible. This had placed the School in a unique position in providing the needed institutional memory. The School's faculty and staff helped fill the vacuum and bridged the gap created between the time these organisations went out of existence till the period when their replacements were phased in (School of Urban and Regional Planning, 1983–92).

Involvement in Policy Formulation

The School's faculty involved in the policy formulation activities of the legislative branch of the government in order to make themselves more relevant to their non-academic constituents. The post-martial law activities of the Congress of the Philippines focused on the passage of legislation affecting the priority concerns of the government such as urban land reform, pollution control, decentralisation and local autonomy, and metropolitan planning. In the deliberation of all these proposed bills, faculty members of the School, together with those from other academic units in the University, had lent a helping hand. The School's faculty were armed with the stock of knowledge and accumulated experience which dated back in the 1960s when they were involved in the drafting of the first comprehensive planning legislation of the country in the late 1960s (Institute of Planning, 1968).

The participation of the faculty and researchers of the School included joining workshops, seminars, and caucuses which were also attended by officials from the government and non-government agencies, the private sector and lobby groups which subjected these proposed legislation to extensive discussion and debate. Aside from submitting position papers and actually attending deliberations held among the legislators in aid of legislation, the School had helped to produce alternative versions of bills under legislative consideration.

The above-mentioned activities had kept the School's faculty abreast of recent developments and have provided valuable inputs to their teaching and research activities. This acquired greater significance in view of the fact that participation in policy making in these executive and legislative ad hoc or regular bodies has been quite limited and was usually by invitation only. The beneficiaries of the School's involvement had not only been the administrators and the legislators themselves, but also its own faculty and students who experienced first hand the 'dynamics' and 'politics' of planning and of improving the physical environment of urban and rural settlements. They became privy to the formulation of planning policies and solution of controversial issues which they generally only heard or read about.

Involvement in Policy Execution

During and after the martial law, the School's faculty were recruited to serve as consultants or resource persons in various departments and offices of the government. They participated actively in the inter-agency councils and committees created to deliberate on and resolve various issues in housing and urban development, land use, transportation and environmental protection and management. Among these major inter agency bodies was the National Land Use Committee (NLUC) constituted under the National Economic Development Authority.

The membership of the NLUC included the representatives of almost all the departments constituting the President's cabinet, and its policy coverage encompassed the entire range of the country's land use concerns. Another inter-agency body was the Task Force on Land Use, spearheaded by the Department of Agriculture which sought to reconcile or resolve land use conflicts, especially those adversely affecting the country's prime agricultural land and the network of integrated protected areas in the country. The third inter-agency undertaking was the Town Planning Assistance Programme, spearheaded by the Department of Local Government which sought to synchronise all town planning efforts at the local level undertaken jointly by the national and local government officials.

In the above three major intergovernmental committees, the School's representative served as the only regular member coming from the academic world. The School was also involved in the policy and programme execution activities undertaken by the Presidential Commission on Monuments and Cultural Objects and the National Disaster Preparedness Committee of NEDA. Complementary to the assistance in the School's extension on policy and programme formulation were training programmes conducted by the School itself or jointly with these government bodies.

Under the above circumstances, and to the extent that the School's faculty and staff directly participated in the planning and implementation of activities with decision-makers of the executive and legislative branches of government, its library had been the recipient of the outputs of their involvement. These included the usually limited editions of technical reports and documents of the government sectoral agencies, the up-to-date and advance copies of legislation, and summary of the proceedings and congressional deliberations attendant to their enactment. Generally of limited edition and circulation, these have proven to be valuable research materials used by the students of the School and other research units of the University of the Philippines.

Involvement in Local Government Planning

In the past two-and-a-half decades since the School's establishment, most of the School's faculty and research staff had been active and direct participants and observers in the formulation and implementation of various programmes that were intended to promote and enhance the planned development of the country's urban and rural communities. For instance, one of the major activities initiated by the Human Settlements Commission in the seventies was the National Town Planning, Housing and Zoning (NTCPHZ) programme which launched a campaign directing local government units (LGUS) nationwide to plan for and zone their respective areas. The leaders and technical staff of a number of these LGUS had sought the assistance from the School's faculty and researchers in preparing their development plans, thus providing the latter with valuable and practical experience. This experience later served them in good stead when the revived national legislative body and reorganised institutions for planning, such as the Housing and Urban Development Coordinating Council (HUDCC), the Housing and Land Use Regulatory Board, and the Metropolitan Manila Authority, among others, took over where their predecessor agencies left off. The officials of these agencies had likewise availed themselves of the Schools' faculty and research resources.

Parallel Developments in the University of the Philippines

The dynamic changes in the School's 'external' environment were matched by progressive policy changes in its micro environment. Of the three complementary functions of the University of the Philippines, namely teaching, research and extension services including training, the last activity had invariably, because of budgetary constraints, been accorded with the least attention and resources. The realisation especially about the state-supported universities should take on missions, which should approximate, more or less their almost unlimited vision, had challenged the Universities' decision makers to extend their reach much further.

Thus, the University of the Philippines officials had conceded that the University's relevance had to be measured not only in terms of the kind of formal education that it offers, but also in the scope and the impact of its outreach programme. There was an impelling need to share more widely the institution's manpower resources, technical expertise and accumulated experience with the larger constituency, way beyond the confines of its various campuses.

The above mentioned University policy changes had paved the way for the School's more extensive participation in consultancy and extension services in both the national and local governments, whose leaders had become involved in more meaningful interventions in settlement development and environmental management.

Harnessing the School's Graduates

The graduates of the School had actively participated in translating these interventions into specific programmes and projects. Many of them, by this time, had found themselves in the forefront of decision making in a number of agencies such as the planning and housing agencies (the National Economic and Development Authority, the Housing and Urban Development Coordinating Council, the National Housing Authority and the other housing finance institutions); the infrastructure departments (the Department of Public Works and Highways, and the Department of Transportation and Communications); and the environmental resources management and regulatory bodies such as the Department of the Environmental and Natural Resources, and more particularly, its Environmental Management Bureau, the Department of Agriculture and the National Land Use Committee (NLUC) under the NEDA.

The School's graduates had found themselves working side by side with their mentors in a venue other than the classroom. These extension services had resulted in many unexpected benefits. Firstly, the wide range of practical services that they rendered had become a part of a 'continuing education' programme for both the School's faculty and its graduates. Secondly, these had brought the faculty and the staff face to face with the ultimate beneficiaries of the School's educational programme administered for more than two decades, namely the decision makers and the public. Thirdly, these services had helped the students translate and apply their book learning into pragmatic day to day activities and had also provided the faculty members with the unique personal experience which they input into their classroom teaching.

Collective Efforts with Other Academic Units

Closer to home, the School has likewise been involved in extension services rendered jointly with other academic and research units within the University. These units are particularly those involved in similar extension services. For instance, early in the history of the University of the Philippines, its top decision makers seriously considered establishing a Public Affairs Complex. The Complex was intended to be the physical site for a number of academic and research units of the University of the Philippines which were also involved in policy making and /or project development in the service of the general public. Among these units, aside from the School, were the College of Public Administration, the College of Social Work and Community Development, the Institute of Small Scale Industries, and the School of Labour and Industrial Relations.

The rationale for the Public Affairs concept was the need to bring together in one site the decision makers with more or less similar concerns affecting the bigger community. Its specific objectives were the following:

a) to consolidate, synchronise and maximise the efforts and resources of these decision-makers;

b) to make it more convenient for the 'public' to transact their businesses with the officials and staff of these units to visit their offices and facilities in one central location;

c) in the process, to save and reduce effectively personal and bureaucratic costs.

While the Public Affairs Complex project was well conceived and favourably endorsed by the University officials, inadequate resources and higher priorities prevented it from actually being taken off. However, this did not preclude the officials of the concerned units from getting together to undertake joint research, training and extension programmes, or even to just consult, 'interact' and share their experiences. The concept of a common location, shared facilities and manpower resources appealed to them, and even when the concept was shelved, the idea had already gelled and had inspired collective efforts among the officials of the concerned units in the University of the Philippines.

Unable to put up the physical infrastructure needed for this purpose, the University of the Philippines' top officials had encouraged and actually facilitated this regular face to face interaction through other means. They created several committees consisting of the representatives of these units as regular members, together with others who were similarly interested, and provided the venue for the sharing of expertise and resources among them in the promotion of commonly identified programmes and projects. Among the committees created were the Management Education Committee (MEC), the Campus Planning Committee, the Disaster Management Group and the Environmental Education Network of the Philippines (EENP).

Multi-Disciplinary/Inter-Campus Activities

A quick review of these multi-disciplinary or inter-campus collaborative efforts reveals the evolved role of the School and its contribution in furthering the various concerns of the University and the general public. Among these interests was environmental management. The Management Education Committee is a clustering of all the degree and the non-degree granting units in the five campuses of the University of the Philippines all over the country which offer management or administration courses. This Committee has become the venue for the exchange of experience in management education, research and training. Its officials have sought to ensure that the overlapping or duplication of efforts in these common areas of concern are avoided; there is a joint effort in teaching, research and training which resulted in sharing of facilities. The Committee has sponsored major research projects with the government as a client, such as disaster management and the monitoring of the implementation of diverse projects pertaining to agrarian reform and public health. In these projects, researchers of the member units were placed in charge of the sectors pertaining their respective areas of expertise.

The membership of the Campus Planning Committee, organised in the University of the Philippines' flagship campus in Quezon City, on the other hand, consists of the School, the College of Architecture, the College of Engineering and the Business Office of the University. It is in charge of over-seeing the implementation of the University of the Philippines Diliman Campus plan, in the process in which its members review and act on the applications by individuals and organisations for all types of development within the campus. It is its members' expectation that this pioneering experience would be replicated in the other campuses of the University. Eventually, it has been projected that its other campuses nationwide would likewise improve their respective immediate and 'impact' physical environment with the assistance extended either by the Committee or their individual members. The School's planners, for instance, has already assisted in a modest manner in the campus planning of another University in Central Luzon and requests for similar assistance are under consideration. Hopefully, where time and resources permit, analogous technical support could be given to other colleges and universities that are also concerned with environmental enhancement.

The Environmental Education Network in the Philippines or the EENP is a bigger group of academic units in the University of the Philippines whose membership had been extended to other colleges and universities in the various regions of the country. Established in 1989, it had only eight members. Four years later, the membership swelled to almost three times its original number. The EENP has enabled the School to actively initiate and support activities concerning the environmental aspects of development, without duplicating the efforts of other units of the University and other academic institutions in the country. Through these inter-academic committees, the School's faculty and research staff have the opportunity to undertake joint projects with their counterparts in academic and research units, both within and outside the University. These have gone a long way in strengthening its network of co-workers in servicing its academic and non-academic constituents.

Conclusion

In the past 27 years, the School had brought up close to 500 professional planners and had trained almost 2,000 government officials and private practitioners who had been harnessed by the country in its development activities. These numbers may not be impressive. In fact, they are obviously inadequate to respond to the requirements of 60 million people and over a thousand towns and cities with urban development problems multiplying rather than diminishing throughout the years.

The newly installed government of the Republic of the Philippines, headed by President Fidel V. Ramos with its emphasis on rational land resources management and sustainable development, has encouraged the School to take a critical look at its future commitments in order to respond to these new thrusts of the government. Thus, the School's officials project the days of their own piecemeal contribution to the comprehensive efforts of the country toward a better natural and man-made environment for the Filipinos will soon be over. In its place, it becomes inevitable for them to prepare their own medium- and long-term plans, and to synchronise them with those plans of the government and the private sector.

The School's faculty and staff will likewise have to sustain its pro-active approach in coordinating and synchronising its activities within and outside the academic community; they also have to strengthen their roles in providing the needed institutional memory in the field of urban and regional planning. They propose in continuing to consolidate and maximise the efforts of the School's graduates through their collaborative involvement efforts. All these will ensure that the School, in collaboration with its graduates, will be able to render a more meaningful contribution to the environmental planning programme for the country.

The above-mentioned plans, however, will need more than a determined will and a ready commitment of manpower and logistical resources. Equally important is the need for the School to generate its own financial resources to augment its budgetary outlay and the University's funding support. This would require the formulation and implementation of a fund generation scheme which is fully supportive to its academic, research and extension functions. This financial scheme could be formulated and undertaken jointly with other units of the University and/or other similar academic institutions in the country. With the establishment of the Asian Planning School's Association or APSA, the prospects of undertaking this programme with other academic institutions within the ASEAN region looms favourably and holds a lot of promises.

References

Abrams, C. and Koenigsberger, O. (1959), *Report on Housing in the Philippine Islands*, Manila, Philippines: prepared for the Government of the Philippine Islands, United Nations Technical Assistance Administration (mimeo).

Carino B.V. and Santiago, A.M. (1993), 'Institution Building: the SURP Experience', *Institution Building for the Study and Practice of Public Administration*, Lecture Series in Honour of Professor Carlos P. Ramos, College of Public Administration, University of the Philippines, Diliman, Quezon City, October, pp. 99–120.

Environmental Education Network of the Philippines (EENP) (1989), *The Role of Educational Institutions and Non-Government Organisations in the Sustainable Management of the Environment and Natural Resources in the Philippines*, Los Banos, Philippines: Institute of Environmental Science and Management (IESAM), University of the Philippines.

Environmental Management Bureau (EMB) (1991), *Philippine Environment and Development*, Quezon City, Philippines: Department of Environment and Natural Resources.

Executive Order No. 419 (1973), Manila, Philippines: Office of the President of the Philippines, 19 September.

Faithfull, W.G. (1969), 'United Nations Assistance in Environmental Planning in the Philippines', *Philippine Planning Journal*, Vol. I, No. 1, pp. 2–9.

Institute of Environmental Planning (1974–83), *Annual Reports of Institute of Environmental Planning (1974–1975 to 1982–1983)*, Quezon City, Philippines: School of Urban and Regional Planning.

Institute of Planning (1972–74), *Annual Reports of Institute of Planning (1967–1968 to 1973–1974)*, Quezon City, Philippines: School of Urban and Regional Planning.

Institute of Planning (1972), *Papers and Proceedings in the Preparation of the National Physical Planning Legislation of 1968, Occasional Paper No. 5*, Quezon City: Institute of Planning.

Memorandum Circular 156 (1968), Manila, Philippines: Office of the President of the Philippines, 9 February.

Office of the University Registrar (1991), *U.P. Diliman General Catalogue (1990–1991)*, Quezon City: University of the Philippines, Diliman.

Presidential Decree 933 (1974), Manila, Philippines: Office of the President of the Philippines.

Presidential Decree 1396 (1978), Manila, Philippines: Office of the President of the Philippines.

Republic Act 4341, October, 1965, Manila, Congress of the Philippines, Republic of the Philippines.

School of Urban and Regional Planning (1983–92), *Annual Reports of School of Urban and Regional Planning (1983–1984 to 1991–1992)*, Quezon City, Philippines: School of Urban and Regional Planning.

Villavicencio, V. (1984), 'The Role Tertiary Institutions Should Play to Provide the Necessary and Relevant Manpower to be involved in Solving Environmental Problems with Particular Reference in the Philippines', paper delivered in Indonesia.

19 Whither Asian Planning Education?

CHUNG-TONG WU

Introduction

Asian economies on the whole are booming. Though the rate and extent of growth are uneven, the impact of this growth on Asian urban development is relatively similar. Rapid economic growth has a number of urban consequences: increasing rural to urban migration, mounting pressures to develop the urban fringe and rising demands for infrastructure. The skylines of major Asian cities and even the not-so-major cities are all being transformed at an unprecedented rate. It is an exciting and exasperating time for planners and planning educators in Asia. Economic growth is opening vast opportunities for the planning practitioner and perhaps it will increase the awareness of the need for high-quality planning education. Planning education in Asia is poised for a new era of expansion. This paper seeks to examine the challenge facing Asian planning educators and to raise questions about the type of planning education that might be equal to the tasks ahead.

Countries in Asia are diverse. Differences in terms of territorial area, population size, history of urbanisation, level of economic development, political system and culture can be vast. The following discussion recognises these diversities and, where appropriate, will make specific references. The term Asia is used in preference to the term 'Third World' because that term conjures up images inappropriate to many of the east- and southeast-Asian nations that this paper covers.

The next section of the paper briefly reviews the evolution of planning education in North America as background to a discussion of developing an Asian vision for planning education. The choice of North America is deliberate, since it has the largest number of planning schools in both undergraduate and postgraduate programmes. The combined number of faculty members in the planning schools is also the largest, and they along with other researchers produce the largest volume of the planning literature.[1] Debates about the

364

planners' roles and planning education are summarised as a backdrop to a discussion of the same in the Asian context.

The third section contrasts the context in which the Asian planners work against that of their counterparts in the West.[2] The final section, intended to stimulate discussion, includes suggestions on the broad directions for clarifying a vision of Asian planning education.

Overview of Planning Education in North America

The following overview of planning education in the West is admittedly broad-brush and based largely on the evolution of planning education in North America over the last 30 to 40 years. The basic approach adopted in this paper is that a discussion of planning education cannot be isolated from the social and economic context, planning practice or research. A number of detailed discussions are available (Perloff and Klett, 1974; Feldman, 1994; Hemmens, 1987; Alterman, 1992; and Niebanck, 1987). This review will be restricted to a few major themes: the economic context in which planning practice and planning education have evolved, the perceived major planning issues, the type of planners that the education system has tended to produce, the focus of planning style and practice, academic roots and specialisations available in university planning programmes, the latter's research foci and the state of planning education. A summary is presented in Table 19.1.

Economic Context

The national economic context is fundamental to planning practice and planning education. A growing economy generates pressures for urban change that will in turn impinge on planning practice and planning education, in terms of the major issues to be addressed and of special problems which skilful planners are expected to solve. The size, scale and scope of urban changes are also intimately tied to the degree to which the public sector is willing to engage in planning. Depending on the political climate, both economic prosperity and economic austerity can stimulate a great deal of planning activities, though the nature and the extent of these activities could be quite different.

During the economic boom of the 1960s and early 1970s, North American governments focused their attention on expanding public-sector involvement in providing infrastructure, dealing with urban decay and initiating programmes to stimulate depressed regions. The goal was that of sharing the nation's

Table 19.1 Overview of the evolution of planning education

	1950s	1960s	1970s	1980s	1990s
Economic context	Economic recovery and expansion	Economic growth and prosperity	Oil shocks and the beginning of economic restructuring	Economic restructuring	Economic austerity (for the West)
Major planning issues	Land use; zoning; city growth	Urban decay; urban renewal	Urban poverty; inequality	Unemployment; growing inequality; infrastructure decay	Environment; mega-cities; limited public finances
Planning style	Practitioner	Analyst; social critic	Advocate	Critic	Aware practitioner 'designer'
Planning practice	Unified; well-defined scope – master plans, regions	Staff analyst – politically neutral; planning as 'wicked problem'	Planning is multifaceted; broad scope; planning for the poor; critic (academic planners as outsiders)	Planner as packagers and negotiators of private and public joint ventures	Refocusing on the built environment
Specialisations in planning education	Land use Housing	Transportation	Social policy planning; environmental planning	Local economic development; property development	Environmental planning; physical and land use planning; public finance; international planning
Academic roots	Regional planning	Policy science Geography	Social sciences	Finance; management	Environmental sciences
Research foci	Architecture; physical design	Applied social science	Rational model: large scale models	Critiques of practice	Research based on practice; research on practice
Planning education	Regional planning first established in University of Chicago	Beginning of expansion of enrolments and programmes	The heyday, due to expansion of government programmes	Lack of government funding; slight decline in enrolment	New expansion? New integration

prosperity with the less fortunate. Public-sector spending meant increased public-sector involvement in planning and led to the expansion of both planning practice and education.

Starting with the late 1970s, North America and much of the world experienced a long period of economic restructuring and adjustment. Economic restructuring, coupled with a market-oriented political philosophy, ignored the need for a national urban policy and this impacted on public sector planning in two ways. Firstly, public-sector programmes were reduced or abandoned, ostensibly because of economic austerity but largely because such programmes were regarded as incompatible with the prevailing political philosophy (Levy, 1992). Secondly, the public sector retreated from proposals or proposed actions which might be seen as 'obstructing' economic growth – the pursuit of which has too often become an all-consuming passion in an era of economic recession. Consequently, planning agencies were faced with budget and staff cuts, and planning education no longer attracts the best and the brightest, who see careers in other fields as more attractive and financially more rewarding. Faced with budgetary constraints, some universities are making choices about the restructuring of planning programmes (Galloway, 1992).

Major Planning Issues

There was a time when the planner prepared a master plan, identifying land-uses and zoning regulations, and could expect little or no public dissent. A kind of physical determinism prevailed. Planners believed that an orderly physical environment would lead to a better and more efficient society. While the preparation of a master plan or some similar plans, land-use plans and development controls are still on a planner's agenda, the planner's role is no longer so simple. Recognition of the limited impact of land-use planning was accompanied by a gradual realisation of the huge social and economic problems in major urban centres. Starting with problems dealing with urban decay in the 1950s and 1960s, planners are now engaged in issues ranging from unequal access to public facilities, urban poverty, housing provision, homelessness, local economic development and regional planning to issues of sustainable environmental development (Levy, 1992). The explosion of issues and the complexity of dealing with them are compounded by the problems of having to do more with fewer resources.

Planning Style and Planning Practice

In the uncomplicated days when planners were concerned almost exclusively with land-use planning, development control and little else, the planning profession had a clear and seemingly unified set of ideas about the city, the role of planners and the validity of master plans (Levy, 1992; Hemmens, 1987). Life for the planner was probably never that simple, but compared to the pluralistic society and the multiplicity of political, ethnic and community interests and practical demands on the planners in the 1990s, it surely must have seemed almost idyllic. Planners have moved from the role of a staff analyst, providing information for the political decision-makers, to one intimately involved in advocating the positions of the poor and the disadvantaged in society. During a time when economic development and employment generation ruled the day, the planner was reduced to a 'judge of plans and the negotiator of planning compromise' (Levy, 1992, p. 82) rather than the initiator of plans. Many planning educators seemingly retreated to criticising the failures of planning practice, as government cut back funding programmes. Research focusing on the failures of the system partly contributed to the perceived split between the practitioners and the educators.

Academic Roots and Specialisations in Planning Education

A traditional planning-education programme, linked closely to its roots in architecture or in landscape architecture, often treated planning as merely large-scale design.[3] This approach suited the practice and specialisations in land-use and physical planning of the time. As the role of planners evolved and the tools that planners required to perform their duties changed, so did the disciplines from which planning had to incorporate new ideas and specialisations in its education programmes (Perloff and Klett, 1974). Policy science introduced planners to economic analysis and analytical models about decision making, and the social sciences became widely recognised as the basis for planning education as planners found themselves involved more and more with delivery and analysis for social programmes and services.[4] More recently, skills in financial analysis and economic management have helped established specialisations in local economic development and real-estate development (Levy, 1992). Into the 1990s, planners who expect to be involved in environmental planning will have to acquire skills and knowledge from the environmental sciences.

Research

Applied social science was probably a good description of much of planning research. Although some planning educators adhered to it, others engaged in research on large quantitative modelling during the quantitative revolution of the 1960s. In the 1970s and 1980s, much of the planning research focused on critique of policies. Increasingly, though, much of the research carried out by planning educators was criticised as being divorced from practice. Planners who had to deal with issues of the day found little in planning research either that was helpful to the issues they had to contend with, that could inform their practice, or that was based on the experience of practice (Levy, 1992; Hemmens, 1987). This criticism is, however, questioned by Mandelbaum (1993), who challenges the profession (which he calls the 'field') to recognise its education role.

In some ways, the evolution of research in planning reflects the background of those who engage in research. Planning researchers and educators come from a whole variety of disciplinary backgrounds and their interest and identification may not be with planning as a profession (Susskind, 1993).

Planning as an area of practice and as an area of academic endeavour is constantly in a state of flux. This is both its strength and the root of its problems (Brooks, 1988). Recent debates amongst practitioners and educators in the United States indicate that there is again a process of sober soul-searching in progress (Blakely and Sharpe, 1993), as practitioners and educators look towards the shape of practice and education in the next century.[5]

The debates about planning education in the West stem not merely from the threat to planning education by a perceived dwindling career-opportunities but also from the recognition of a lack of clear identity for planners (Levy, 1992; Susskind, 1993). As planners ventured more and more into policy analysis and away from the 'traditional' links with land-use planning, they found any claim to special expertise to be tenuous (Galloway, 1992). At the same time, as the planning educator's research focuses more on policy, the traditional areas of land-use planning/physical planning, in the case of North America, are being taken over by landscape architects. The recent debates therefore include calls for more focus on practice, on the environment and land planning, and on educating planners who are able to make proposals, to provide an urban vision and to lead rather than merely criticise or regulate (Galloway, 1992; Niebanck, 1987).

The Asian Context for Planning Education

The broad canvas of the evolution of planning practice and education in North America presented above is intended as a device to contrast the present situation facing Asian and North American planning practitioners and educators. The public sector in most Asian countries is, in general, no match for the private sector in bringing about change in the urban environment. The exception is perhaps Singapore. Unlike their counterparts in the West, Asian practitioners and educators are now faced with rapid urban change and rising demand for urban expertise due to rapid economic growth. Rapid economic growth and associated urban changes present tremendous opportunities and pitfalls. The young Asian professionals are usually thrown into the thick of the action almost immediately after graduation. The pace is hectic, the time span for decisions is extremely limited. This is a propitious time to take stock of planning-education programmes and ask whether they are indeed providing an education, which prepares our students for the challenges ahead. A summary of the following discussion is provided by Table 19.2.

Economic Growth

Much of Asia is in the grip of rapid economic growth accompanied by significant industrialisation. As economic restructuring in the West progresses, the global shift of manufacturing is benefiting those Asian countries that have the labour resources and capacity to attract such industries. Not all Asian countries are experiencing the same level of economic growth, and growth does not necessarily lead to the same urban pressures. The Newly Industrialised Economies (NIEs)[6] are facing labour shortages and rapid wage rises and are therefore looking towards the high-tech sectors for the transformation of their industries. Some of the Association of Southeast Asian Nations (ASEAN) member countries are well on the way to industrialisation, but with varying capability to supply skilled labour.[7] The enlarging middle class in both the NIEs and ASEAN are able to provide an increasing market, which adds to the economic expansion. Some of the other Asian nations are able to take advantage of the growth in their midst by supplying cheap labour and capturing the spill-over effects of their prospering neighbours. Other Asian countries, such as China, are moving from central planning to more market-oriented policies, resulting in spectacular economic growth that considerably increases pressures on urban centres.

Table 19.2 Contrasting issues associated with planning in Asia and the West

		Asia	West
1	Economic growth	– rapid and sustained growth over a number of years; – largely based on industrialisation; – some Asian nations are facing acute shortage of labour; – regional differences.	– slow economic growth; – economic restructuring; – de-industrialisation and re-industrialisation; – focus more on service sectors; – high and sustained unemployment.
2	Urbanisation and population growth	– rapid urbanisation and suburbanisation; – sustained rural to urban migration; – still a very large young population; – some Asian nations facing significant international migration of labour; – low but rising car ownership.	– steady state population; – sunbelt migration; – aging population; – use of private cars.
3	Urban development	– mega projects; – vast investments for new infrastructures.	– urban consolidation; – continuing suburbanisation.
4	Income growth	– rapid rise of the middle class; – employment expansion; – still vast number of the poor especially in the rural areas.	– squeeze of the middle class; – growing income gap between rich and poor.
5	Provision of infrastructure	– attention given to the overall poor infrastructure provisions now regarded as an obstacle to economic growth; – more and more attention to the possibilities of private sector provision of infrastructure.	– deterioration of some infrastructure but lack of public capacity to provide for replacement; – looks towards the private sector and/or user pays for funding; – increasing demands placed on developers to provide infrastructure as part of the development consent process.
6	Environmental awareness	– environmental concerns amongst the public still nascent; – large and growing environmental/pollution problems.	– very strong environmental concerns amongst the citizens; – increasing political response to incorporate environmental concerns into planning decisions.
7	Professional identity	– professional identity still emerging; – in some countries lack of official recognition of professional standing; – private planning consultancies emerging; – planning practice largely restricted to land use planning; – use of outside expertise often required by funding agencies; – legal framework usually ignored.	– professional recognition and professional bodies well entrenched; – in some countries, recognised professional qualifications a requisite to employment and promotion in the public planning agencies; – thriving private sector planning consultancies; – legal framework strictly enforced.
8	Planning education	– few separate planning programs; – a mixture of undergraduate and postgraduate planning programs or planning majors at universities; – planning education still often tied to architecture and engineering or some technical field; – need for recognition by overseas (colonial) professional bodies.	– most planning programmes at the postgraduate level though there are some large undergraduate programs; – planning programs facing funding and staff constraints; – planning education linked to research.

Urbanisation and Population Growth

Unlike Asia, most Western economies are faced with aging populations and steady-state demand for goods and services. Rapid urbanisation is not an issue, except perhaps in those areas, such as the 'sunbelt', which are favoured by internal migration.[8]

In Asia, rapid urbanisation is proceeding apace, with tremendous impacts on demand for land use changes and urban services. This is stimulated by economic growth and the uneven distribution of economic opportunities, leading to sustained rural to urban migration. The fact that much of the population is still young means that a wave of household formation with, by implication, an attendant wave of demand for housing and urban services, is yet to come. By the year 2010, it is estimated that 11 of the world's largest cities will be located in Asia (Vernon, 1992).

The urban fringe is particularly under pressure. Suburbanisation of both population and industries means that more and more land on the urban fringe has to be converted to urban uses, with the attendant demands for infrastructure investments, urban services and displacement of the rural labour force. Ginsburg, Koppel and McGee (1991) identify 'extended metropolitan areas' in various parts of Asia as the locations where this growth is most rapid. Rapid urbanisation is resulting in significant environmental impacts and the loss of arable land. Regions in Southern China are cases in point (Wu, 1993).

Urban Development

Urban development in the West has focused on making more intensive use of existing infrastructures, such as urban consolidation, gentrification and urban renewal. In Australia, 'urban conservation' is a policy seen as necessary to minimise public outlay for new infrastructures and to contain urban sprawl. In Asia, urban development is likely to be spearheaded by mega-projects that involve very large areas and significant changes to the existing stock of housing, offices and commercial space. This path is taken partly because inexpensive land is available at the urban fringe and is probably easier to obtain. With increasing car ownership due to rising incomes, these mega-projects are increasingly viable. At the same time, due to the lack of infrastructure investments in the past, many Asian countries are on a drive to catch up with the basic requirements. Large-scale public transportation projects, highway-building programmes, sewerage and other infrastructure projects such as new airports, are underway or planned in almost every country in Asia.[9] Infrastructure projects are likely to lead to further urban expansion and changes in the urban structure over the next few decades, as more and more land becomes accessible for urban development. In many aspects, the coming urban 'explosion' will far outstrip those of the past, for the next wave will be driven by higher overall incomes and greater consumer expectations.

Income Growth

National economic growth is filtering to a wider portion of the Asian population and there is now emerging a strong and expanding middle class in a number of Asian nations, excluding Japan. Rising income leads to higher vehicle ownership and demand for better housing. There is also a rising expectation for a level of public services and facilities that is more commensurate with increasing national economic growth. The pressures on the urban fabric through the demands generated by rising incomes will require public-sector responses.

In North America, the middle class is shrinking due to the loss of long term employment opportunities (Harrison and Bluestone, 1988). The gap between the poor and the rich is also widening. The demands generated by a growing population and growing income are probably no longer to be repeated in the West. On the whole, urban development in the West will tend to focus on replacing what is obsolete rather than coping with significant expansions of economy or income.

Provision of Infrastructure

According to one estimate, some US$100 billion worth of infrastructural projects are being planned and implemented in various parts of Asia.[10] These range from mega-projects such as power plants,[11] urban transit systems,[12] inter-regional super highways[13] and major airports,[14] to much smaller urban-scale projects including sewerage systems,[15] traffic management, housing projects and other public facilities. These infrastructure projects open new areas, and make others more desirable, for urban development. Real-estate development of varying sizes, from small estates to major new towns, have been and are being developed in various parts of Asia due to the new or planned infrastructure projects.[16]

While part of the new infrastructure is being funded through bilateral or multilateral assistance, much is being constructed through private-sector funding. Investing in basic infrastructure has become attractive to the private sector if concessions in terms of tolls and charges, and perhaps even rights to develop some of the land near or next to the infrastructure and facilities, become available. In some way then, as in North America, the initiation of urban development and major plans for change have shifted from the public to the private sector.

Public-private sector joint ventures are common in the West. The United States and other countries have also pioneered the concept of development-impact fees and development charges, so that the public sector can recoup the infrastructure costs necessary to enable private sector developments to go ahead. The permission to develop (a housing estate, a business park, etc.) is contingent on an agreed payment by the developer to the government, to cover the major infrastructure costs such as the extension of water-distribution systems, head works and others. In contrast, Asian countries encourage the private sector to invest in major infrastructure that will enable other urban development to occur.[17] This variety of schemes calls for different types of skills and expertise from the planners whether they are involved in the decision-making or planning process.

Environmental Awareness

It would be fair to say that there is a quite uneven awareness of environmental issues amongst the Asian countries, but it is clear that very few Asian countries have at present a system of environmental assessment and control for urban development as elaborate and legislatively enshrined as those found in North America. The general public in Asia may be increasingly aware of environmental issues, but they are neither well informed nor as organised as the citizens of the West. Even more so than the West, the environmental movement may be seen more as a movement of the well-off and the upper-middle class. In addition, the electoral and legislative processes do not usually provide the opportunities to express environmental concerns.

Today in the West, politicians and planners ignore at their peril the potential environmental issues associated with any project or proposal. Developers and politicians still often blame undue environmental concerns or regulations for long delays in development approvals contributing to cost increases. These same excuses are often used in Asia as warnings not to copy the West and possibly impede the rapid economic growth.

The Asian planner who is concerned about environmental planning has a much more difficult task than his/her counterpart in the West. The Asian planner must first create coalitions with concerned citizens and groups to ensure that environmental issues are not brushed aside in the rush towards economic development. He/she needs to demonstrate that what appear to be gains may be short term and will be far outweighed by long-term undesirable consequences. Planning education in Asia therefore has to address a much more basic set of issues than in the West, where putting environmental concerns on the agenda is seldom an issue. At the same time, the Asian context implies that certain methodologies for planning might even be inappropriate for Asia, and that assumptions about the legal framework for planning actions and the role of the citizenry could be quite inappropriate.[18]

Professional Identity

The United Kingdom has the oldest professional body for planners, which has had a strong role in setting the planning agenda in the country (Alterman, 1992). By comparison, the United States has a large number of practising planners but professionalism is not as strong. In Australia, the public sector is beginning to insist on Institute membership, or eligibility for membership, for planner, but there has been a history of entry via a variety of professions, and different levels of government tended to have different requirements.

In Asia, there is little legislative recognition of the profession of planning in the sense that 'registration' of planners be a requirement for practice.[19] In spite of the colonial history and the inherited basic planning framework of countries such as Malaysia and Singapore, professional bodies of planners do not necessarily attract the kind of prestige or public recognition that other allied professionals, for example architects and engineers, commonly enjoy.

The legislative framework for planning is also well developed in the West, with either 'Town and Country Planning' or 'Environment and Planning' legislation being the foundation for planning actions and regulations.[20] Under such a system not only is there a clear process that is known to both the proponents of development and the regulators. This process also provides the basis on which appeals for specific cases or class actions might be mounted. In the United States, individual appeals and even class actions can be mounted in the courts against public-sector planning decisions. In the state of New South Wales in Australia, the state government, local authorities or the individual citizen can take a matter of dispute to the Land and Environment Court.

These frameworks are generally not available in most Asian countries. In some respects, the public-sector planner has more power relative to his/her counterparts in the West, in that there are few opportunities for the developers or citizens to challenge planning decisions. At the same time, the lack of a clear legal framework for the planner's actions means that there is little recourse for the planner to the enforcement of the plans. Under such circumstances, the developer will tend to ignore the plan, resort to subterfuge, or go around the plan by appealing directly to the politicians. None of the above circumstances makes for good planning.

Planning Education

In North America, most planning programmes are taught at the graduate level, but there are a number of large and well-respected undergraduate programmes.[21] There are also numerous doctoral programmes. If one were to include the related fields, then the number of universities offering planning programmes in North America would be much larger. With universities facing increasing budgetary constraints, there is a real danger of planning programmes and departments being merged, restructured or even closed down in North America.[22] The worst fears of some planning educators are now being realised (Galloway, 1992).

In Asia, planning education is not as well established. Many programmes are still struggling to establish an identity separate from architecture and other allied fields such as engineering and geography. There is a mixture of undergraduate and graduate programmes, mostly of relatively short history. Without a comprehensive survey, it is difficult to assess the full situation. My impressions from visiting a number of Asian countries is that there is a real mixture of programmes, of varying sizes and with staff members from a variety of academic backgrounds.[23] Overall it is fair to say that planning education is still at the developing stage in most Asian nations.

At a time of rapid economic growth and associated urban development in Asia, it seems that unparalleled opportunities exist for the planning profession's educators to establish their necessary identity. Furthermore, there are also opportunities to expand and strengthen the demand for planning education in order to provide the well-educated and skilled personnel to cope with present and future urban changes. The question is what kind of planning education should we provide and how might it differ from what is being offered in the West? In other words, instead of continuing to send Asia's best and brightest to the West, is there a case for developing planning courses in the home countries so that students can train for the profession without going abroad? Furthermore, is there a need for planning education that is more sensitive to the links between culture and the built environment? Note that the argument here is not against sending students overseas for their planning education.

This should and must continue. Such opportunities are unfortunately limited and often made available only to the socially and politically well-connected. The question is really about developing planning-education programmes of excellence in Asia, so that a large number of students can obtain a high-quality planning education that is appropriate to the context of their country.

Towards a Vision for Asian Planning Education

In their introduction to a series of articles about the planning profession, Blakely and Sharpe (1993) summarised the choices facing the American planning profession in the following manner:

> ... the profession stands at an important crossroads. Down one path, planners merely assembly information for others to make decisions. Down the other, they become the forgers of a new national/international socio-economic and geographic-based destiny.

The choices facing Asian planning professionals and planning educators are equally clear, but the Asian planning profession has to find additional clarity about whether there is an Asian planning practice that is different from North America and therefore whether its planning education must also be different from the West's. Available discussions of planning education in non-Western countries tend to be cast in terms of the 'Third World'. For example, Sanyal (1990) argues that there is a 'one-world' approach to planning and that the West can also learn from the Third World. Others such as Qadeer (1990) and El-Shakhs (1990) reject the notion that there is a 'one-world' planning approach and are quite critical of what they see as the deficiencies of Western planning education.

There is no doubt that the West can learn from the Third World, and I submit it would be wrong to assume the Third World is necessarily either backward in terms of economic development, or poor in ideas.[24] In fact, the term Third World holds little real meaning, except perhaps as a shorthand for non-industrialised or industrialising economies. Within the so-called Third World there is great diversity in stages of economic growth and in legal frameworks for planning actions, and significant variations in terms of infrastructures for planning education. Some of the Asian economies are now the fastest-growing economies in the world, and while by the conventional measurements of per capita GNP some might be considered poor, they are perhaps not as poor as the term 'Third World' tends to imply.[25]

For this discussion, which focuses on the situation in East Asia and Southeast Asia, I would suggest the elements included in Table 19.2 are much more important in terms of understanding the differences between Asia and the West in respect of the significant issues and of the diverse institutional settings in which planning has to take place. Five aspects are particularly germane to a discussion of planning education in Asia.

The Role of the State in Asia

The Asian State is often interventionist in economic and industrial development. Compared to most nations, the state in Asia is much more active in a leadership role in a variety of sectors, and urban development is no exception. Private and public sector joint-ventures and other forms of cooperation are common in Asia in a whole variety of settings, from housing and industrial development to the provision of public infrastructure. Planning education must recognise this interventionist role of the state and the importance of private-public interactions in planning practice. While private-sector planning consultancy is on the rise in some Asian countries, most planners find employment in the public sector. Depending on the specific country, public-sector planners tend to have varying degrees of influence over urban development in their cities. We need to educate our students to be conversant with working both in and with the private sector, but still able to achieve public interest objectives.

Planning Time Frame and Methodology

Public-sector programmes and private-sector projects tend to be implemented in a much shorter time-frame than in the West. While the West talks about 'fast-tracking', such a mode of operation seems to be the norm in Asia. The transformation of the urban landscape in large Asian cities testifies to the rapid pace of both public and private development. This state of affairs reflects partly the rapid rise in land values in many Asian cities but often also the lack of impediments to development actions. Planning practice in Asia, particularly in the economies that are now experiencing rapid growth, operates at a pace and on a scale few Western planners would be accustomed to. Methodologies for planning must come up with answers within the short time span required. Careful and scholarly methodology that carefully considers all alternatives but takes a long time to produce results may satisfy our academic needs but unlikely to be well-received in the 'boom' town mentality of rapid economic growth. By all means teach our students the careful analytical techniques, but teach them so well that they know how to adapt these to the 'quick and dirty' situation and be able to identify the critical data and performance indicators within the short time available.

Planning Institutions

Urban development in Asia, for example, in China, is often carried out in a setting where planning controls are few and even these seldom implemented. In this kind of setting, windfall gains to the private developer can be huge and effective public planning is difficult. Professional practice in Asia in the public sector is therefore much more difficult and often must be carried out with less authority than the Western planner would experience. In this setting, the effective professional is not one whose authority and powers of persuasion rely solely on rules and regulations. The Asian setting in fact is much more difficult and demanding. It calls for professionals who can command the moral authority to persuade both the politicians, who want swift results, and the anxious developer, who wants a quick return, of a planning solution that will serve their interests and yet not completely abandon the public interest. It calls for the education of planners who are able to discuss what is in the public interest in a language understandable to the developer, able to negotiate, and able to make proposals that will satisfy the demands of the private sector as well as maximising gains for the public.

Planning Practice and Physical Planning

Day-to-day planning issues in Asia are still very much related to land development and infrastructure investment. While many planning graduates do obtain employment in a variety of public and private sector organisations, it would be fair to say the great bulk of the work identified as planning in Asia is very much related to plan preparation and development control. Ability to prepare plans and to explicate in a rational manner development controls, zoning regulations and urban design are highly sought after skills. Planning schools in Asia that did not offer physical planning or did not insist on exposing students to physical planning would be remiss.

Environmental Planning

In a context where the planning profession is not well recognised or defined,

Asian planning educators have the added responsibility of helping to define and advance the profession. This must be done jointly with the practitioners. The planning profession is too new and too fragile in Asia and can ill afford a schism between the practitioners and educators. One of the areas in which both could work to promote the cause of planning is to ensure sound environmental considerations are included in planning decisions. The educators could do this best by ensuring that environmental planning is included as a key component of the curriculum. Together, planning practitioners and educators have to educate the public about environmental issues.

Continuing Professional Development

In Asia, many of the professionals working in the field of planning are allied professionals and have graduated some time ago; some have returned from overseas. Planning education in Asia has not confronted the need for continuing professional development, to assist those who are already in the field to further the development of their skills and address the issues of the new development context. Asian professionals often find themselves overwhelmed by the rapidly changing pace of development. The planning schools in Asia can provide programmes that will assist them to develop new perspectives and update their skills. In this way, continuing professional development can also provide the necessary mechanisms for defining a professional identify where this is still weak, and to assist professionals in developing their awareness of current issues.

Conclusions

At this juncture, planning practice in Asia is more than ever both difficult and challenging. Most planners who work in Asia will have to spend their time navigating the unknown, negotiating the next new town and designing the next mega-projects. The boom in Asia, driven by international capital and promoted by willing governments, is faster and more furious than what has been experienced by the West for a long time. Planning education in Asia has to fully recognise the rapidly changing context of planning action and the extreme demands that the pace of development places on the new graduates and those who are already in the profession.

Planners working in Asia must be able to deal with both the public and private sectors, to come up with answers quickly, to balance the interests of both developers and the public, to be capable of good physical planning and to understand the requirements of good environmental management. The specifications are ambitious because the tasks are formidable. Designing programmes that will inculcate in planners even a modicum of the required knowledge and skill is extremely difficult. Unless we meet this challenge, planning in Asia will never be more than a sideshow, while the main game is captured by those who are concerned about neither the public interest nor the environment.

It would be both arrogant and naive to claim the above has defined a vision for Asian planning education. Such a vision must involve a great deal more discussion and probably will not emerge for a number of years. There may well be a number of 'visions', to accommodate the diverse Asian experience. What I have tried to do is to contrast the planning context and the planning issues faced by the West with those faced by people in Asia, and to present for examination some possible avenues for the development of planning education in the Asian context.[26] We need to ask whether a conventional Western education in planning is still appropriate for the new Asian reality. Are the planning issues and planning practice in Asia so fundamentally different that they call for approaches different to those tried in the West? Should Asia continue to send its best and brightest to the West for their planning education, or can we develop an Asian vision and a planning education that will prepare them to share their ideas and experience with their Western counterparts as equals? Do we see sufficiently changed circumstances in Asia to argue for a uniquely Asian approach or approaches to planning education, at the dawn of the twenty-first century?

Acknowledgments

Comments and suggestions from Brian Hudson and John Minnery are gratefully acknowledged. This paper was first prepared while the author was a member of the Department of Urban and Regional Planning, University of Sydney. The author alone is responsible for any shortcomings.

Notes

1   See Alterman (1992) for a comparison of North American and European views of planning education and practice.

2 In this paper, the term 'West' is used interchangeably with the term 'North America'.

3 The 'City Beautiful Movement,' exemplified by the Chicago Plan and later the plan for Canberra, is an example of this type of approach.

4 The first planning programme based on the social sciences was started at the University of Chicago in 1951 but failed a few years later due to the lack of support. See Perloff's autobiographic note in Burns and Friedmann, 1985.

5 This perceived divergence between research and relevance to solving real world problems is also a subject of much discussion in the field of regional science (Bailly and Coffey, 1994).

6 Hong Kong, Taiwan, Singapore and South Korea.

7 Brunei, Indonesia, Philippines, Malaysia, Singapore and Thailand.

8 In North America, Southern California, parts of Arizona and Florida bear the brunt of the internal migration. In Australia, southeast Queensland is the fastest growing urban region.

9 For example, major new airports are planned for Hong Kong, Malaysia, and various cities in China. Major public transit projects are planned for Bangkok, Kuala Lumpur, Taipei, Shanghai, Guangdong and several other cities in China.

10 Asia excluding Japan.

11 For example, new power stations being constructed in the Philippines and in southern China.

12 Such as the ones being constructed in Taipei and Shanghai and planned for Bangkok.

13 Such as the ones under construction in the Pearl River Delta area of China linking it to Hong Kong and Macao.

14 Hong Kong, Pusan and Kuala Lumpur are all in the process of building new international airports.

15 For example, the World Bank is financing major sewerage systems in Shanghai.

16 For example, in Malaysia, private sector proposals for several new towns are being stimulated by the planned construction of a second causeway between Singapore and Malaysia.

17 One prime example of this is the variety of schemes in China to entice foreign investors to invest in a whole variety of infrastructures such as power plants, bullet trains, highways and bridges.

18 This is for example particularly true in the field of social impact assessment in which much of the methodology assumes a citizenry ready, willing and capable of participating in the planning process and that there is some legislative framework to which the social impact assessment could related to. See a discussion by Ip (1990) regarding the application of social impact assessment in China.

19 Although Hong Kong has a long established planning institute, registration of planners has only begun in Hong Kong.

20 In the United States, in addition to the local and state government legislations such as master plans and zoning plans, which are legal documents, there is the Environmental Planning Act which is a national legislation which requires compliance with specific environmental considerations. In Australia, the state of New South Wales, for example, has a Environmental and Planning Act that specifies the powers and duties of planners and the requirements for the proponents of any development. It also has a Land and Environment Court to adjudicate disputes.

21 According to the 1994 edition of the Guide to Graduate Education in Urban & Regional Planning (Association of Collegiate Schools of Planning, 1994), there were 88 members,

12 Affiliates and 22 Corresponding members. Of these, 59 of the full member programmes accredited by the Planning Accreditation Board were listed in the Guide. In the 1991 edition of the Guide to Undergraduate Education in Urban & Regional Planning and Related Fields (Association of Collegiate Schools of Planning, 1991), 37 North American universities are listed as offering undergraduate planning degrees. Of these only a handful have enrolments of over 100. In Australia, all except one of the eleven recognised planning programmes offer both undergraduate and graduate degrees.

22 For example, in 1994 the University of California at Los Angles closed the Perloff Graduate School of Architecture and Urban Planning and merged the urban planning programme with two other departments into a new School of Public Policy. The Regional Science programme of the University of Pennsylvania was also closed in 1990.

23 Planning education programmes in Asia is varied. There are countries like Singapore which does not have a urban planning programme in any of its universities and there are large countries such as China where there are about 19 programmes serving a country of over a billion population

24 Sanyal (1990) also argues that the Third World is not short of ideas but my disagreement is about framing the question purely in terms of the so called 'Third World' which is assumed to be economically backward. I believe framing the discussion in such terms ignores the variety of issues in non-western economies and, in particular, confuses the important differences between the west and the east.

25 For example, a recent proposal to use equivalent purchasing power would place China as the third largest economy in the world.

26 I have included Australia in the Asian context. The southeast Queensland region faces many of the same issues faced by Asian urban governments. Southeast Queensland has the fastest population growth in Australia. The demand for infrastructure, residential, commercial and industrial development is outstripping the capacity of local and even state governments and creating serious environmental challenges.

References

Alterman, R. (1992), 'A transatlantic view of planning education and professional practice', *Journal of Planning Education and Research*, Vol. 12, No. 1, pp. 39–54.

Association of Collegiate Schools of Planning (1991), *Guide to Undergraduate Education in Urban and Regional Planning and Related Fields*, ACSP.

Association of Collegiate Schools of Planning (1994), *Guide to Graduate Education in Urban and Regional Planning*, ACSP.

Bailly, A.S. and Coffey, W.J. (1994), 'Regional Science in crisis: A plea for a more open and relevant approach', *Papers in Regional Science*, 73 (1), pp. 3–14.

Blakely, E.J. and Sharpe, S.M. (1993), 'Planners, heal thyselves: planning education, educators and practitioners in the next century', *Journal of the American Planning Association*, Vol. 59, No. 2, pp. 139–140.

Brooks, M.P. (1988), 'Four critical junctures in the history of the urban planning profession: an exercise in hindsight', *Journal of the American Planning Association*, Vol. 54, No. 2, pp. 241–48.

El-Shakhs, S. (1990), 'The language of planners: a central issue in internationalising planning education' in B. Sanyal (ed.), *Breaking the Boundaries: A One-World Approach to Planning Education*, New York: Plenum Press.

Feldman, M.M.A. (1994), 'Perloff revisited: Reassessing planning education in post-modern times', *Journal of Planning Education and Research*, Vol. 13, No. 2, pp. 89–103.

Galloway, T.D. (1992), 'Threatened schools, imperilled practice: a case for collaboration', *Journal of the American Planning Association*, Vol. 58, No. 2, pp. 229–34.

Ginsburg, N., Koppel, B. and McGee, T.G. (1991), *The Extended Metropolis: Settlement Transition in Asia*, Honolulu: University of Hawaii Press.

Harrison, B. and Bluestone, B. (1988), *The Great U-Turn: corporate restructuring and the polarisation of America*, New York: Basic Books.

Hemmens, G.C. (1987), 'Thirty years of planning education', *Journal of Planning Education and Research*, Vol. 7, No. 2, pp. 85–90.

Ip, D.F.-K. (1990), 'Difficulties in implementing SIA in China: methodological considerations', *Environmental Impact and Assessment Review*, 10, pp. 113–22.

Levy, J.M. (1992), 'What has happened to planning?', *Journal of the American Planning Association*, Vol. 58, No. 1, pp. 81–4.

Mandelbaum, S.J. (1993), 'The field as educator', *Journal of the American Planning Association*, Vol. 58, No. 2, pp. 140–2.

Niebanck, P. (1987), 'Preparing leadership for the twenty-first century: report of the Santa Cruz Conference on Planning Education', *Journal of Planning Education and Research*, Vol. 7, No. 2, pp. 121–3.

Perloff, H.S. and Klett, F. (1974), 'The evolution of planning education' in D.R. Godschalk (ed.), *Planning in America: Learning from Turbulence*, Washington, DC: American Institute of Planners, reprinted in L.S. Burns and J. Friedmann (1985) (eds), *The Art of Planning: Selected Essays of Harvey S. Perloff*, New York: Plenum Press.

Qadeer, M. (1990), 'External perceptions and internal views: the dialectic of reciprocal learning in third world urban planning' in B. Sanyal (ed.), *Breaking the Boundaries: A One-World Approach to Planning Education*, New York: Plenum Press.

Sanyal, B. (ed.) (1990), *Breaking the Boundaries: A One-World Approach to Planning Education*, New York: Plenum Press.

Susskind, L.E. (1993), 'Preparing the next generation of planners', *Journal of the American Planning Association*, Vol. 57, No. 1, pp. 20–1.

Vernon, R. (1992), 'The cities of the next century', *Journal of the American Planning Association*, Vol. 57, No. 1, pp. 3–6.

Wu, C.-T. (1993), 'Hyper-Growth and Hyper-Urbanisation in Southern China', *Australian Development Bulletin*, (1993) Vol. 57, pp. 30–4. Australian Development Network, ANU.

Printed and bound by CPI Group (UK) Ltd, Croydon, CR0 4YY

22/10/2024

01777625-0014